METHODOLOGICAL APPROACHES TO DERIVING I
OCCUPATIONAL HEALTH STANDARDS
    Edward J. Calabrese

M000115307

NUTRITION AND ENVIRONMENTAL HEALTH—Volume I: The Vitamins
    Edward J. Calabrese

NUTRITION AND ENVIRONMENTAL HEALTH—Volume II: Minerals and Macronutrients
    Edward J. Calabrese

SULFUR IN THE ENVIRONMENT, Parts I and II
    Jerome O. Nriagu, Editor

COPPER IN THE ENVIRONMENT, Parts I and II
    Jerome O. Nriagu, Editor

ZINC IN THE ENVIRONMENT, Parts I and II
    Jerome O. Nriagu, Editor

CADMIUM IN THE ENVIRONMENT, Parts I and II
    Jerome O. Nriagu, Editor

NICKEL IN THE ENVIRONMENT
    Jerome O. Nriagu, Editor

ENERGY UTILIZATION AND ENVIRONMENTAL HEALTH
    Richard A. Wadden, Editor

FOOD, CLIMATE AND MAN
    Margaret R. Biswas and Asit K. Biswas, Editors

CHEMICAL CONCEPTS IN POLLUTANT BEHAVIOR
    Ian J. Tinsley

RESOURCE RECOVERY AND RECYCLING
    A. F. M. Barton

QUANTITATIVE TOXICOLOGY
    V. A. Filov, A. A. Golubev, E. I. Liublina, and N.A. Tolokontsev

ATMOSPHERIC MOTION AND AIR POLLUTION
    Richard A. Dobbins

INDUSTRIAL POLLUTION CONTROL—Volume I: Agro-Industries
    E. Joe Middlebrooks

BREEDING PLANTS RESISTANT TO INSECTS
    Fowden G. Maxwell and Peter Jennings, Editors

CHEMICAL PROCESSES
IN LAKES

Marty, I thought you might
appreciate this book. We are forever
in your debt for the great job that
you have done at the ERF!

Merry Christmas,
Jerry Schnoor
12/11/92

CHEMICAL PROCESSES IN LAKES

Edited by

WERNER STUMM

Swiss Federal Institute of Technology, Zurich, Switzerland

A WILEY-INTERSCIENCE PUBLICATION

JOHN WILEY & SONS

New York   Chichester   Brisbane   Toronto   Singapore

Library of Congress Cataloging in Publication Data:

Main entry under title:
Chemical processes in lakes.

   (Environmental science and technology, ISSN 0194-0287)
   "A Wiley-Interscience publication."
   Includes bibliographies and index.
   1. Limnology.   2. Water chemistry.   3. Biogeochemical
cycles.   4. Lakes.   I. Stumm, Werner, 1924–
II. Series.

QH96.C44   1985   551.48′2   84-17321
ISBN 0-471-88261-5

Printed in the United States of America

10   9   8   7   6   5   4   3

CONTRIBUTORS

Senior Authors

PETER BACCINI, Associate Head, Department of Multidisciplinary Research, Swiss Institute for Water Resources and Water Pollution Control (EAWAG), Zurich, Switzerland; Associate Professor of Analytical and Environmental Chemistry, University of Neuchâtel, Switzerland

BOŽENA ĆOSOVIĆ, Senior Research Associate, "Rudjer Bošković" Institute, Center for Marine Research, Zagreb; Chief of the Laboratory for Physicochemical Separations; Member of the Postgraduate Program of Oceanography, University of Zagreb, Yugoslavia

WILLIAM DAVISON, Principal Scientific Officer, Windermere Laboratory, Ambleside, Great Britain

RENÉ GAECHTER, Senior Research Scientist, Swiss Institute for Water Resources and Water Pollution Control (EAWAG) Zurich, Associate Head of Multidisciplinary Limnological Research, and Lecturer in Hydrobiology, Swiss Federal Institute of Technology (ETH) Zurich, Switzerland

EVILLE GORHAM, Professor, Department of Ecology and Behavioral Biology, University of Minnesota, Minneapolis, Minnesota

DIETER M. IMBODEN, Senior Research Scientist, Swiss Institute for Water Resources and Water Pollution Control (EAWAG) Zurich; Associate Head of the Department of Multidisciplinary Limnological Research, Senior Lecturer in Aquatic Physics, Swiss Federal Institute of Technology (ETH), Zurich, Switzerland

JEAN-MARIE MARTIN, Senior Scientist, Department of Geology, Ecole Normale Supérieure, Paris, France; Head of the Land/Sea Interaction Coordinated Group, CNRS

JUDITH A. MCKENZIE, Senior Research Scientist and Lecturer, Geology Department, Swiss Federal Institute of Technology (ETH), Zurich, Switzerland

FRANÇOIS M. M. MOREL, Professor of Civil Engineering, Massachusetts Institute of Technology, Cambridge, Massachusetts

JAMES J. MORGAN, Professor of Environmental Engineering Science, California Institute of Technology, Pasadena, California

CHARLES R. O'MELIA, Professor of Environmental Engineering, Department of Geography and Environmental Engineering, The Johns Hopkins University, Baltimore, Maryland

DAVID W. SCHINDLER, Project Leader, Experimental Limnology, Canadian Department of Fisheries and Oceans, Winnipeg, Canada

JERALD L. SCHNOOR, Professor of Civil and Environmental Engineering, Division of Energy Engineering, University of Iowa, Iowa City, Iowa

EDWARD R. SHOLKOVITZ, Member, Scientific Staff, Woods Hole Oceanographic Institution, Woods Hole, Massachusetts

LAURA SIGG, Head, Analytical Section, Swiss Institute for Water Resources and Water Pollution Control; Lecturer, Swiss Federal Institute of Technology (ETH), Zurich, Switzerland

HANS-HENNING STABEL, Research Associate in Limnology, Department of Biology, University of Constance, Constance, Federal Republic of Germany

JOHN M. WOOD, Professor, Department of Biochemistry, Gray Freshwater Biological Institute, University of Minnesota, Navarre, Minnesota

WERNER STUMM, Professor, Swiss Federal Institute of Technology (ETH); Director of its Institute for Water Resources and Water Pollution Control (EAWAG), Zurich, Switzerland

Junior Authors

WARIS ALI, Department of Geography, The Johns Hopkins University, Baltimore, Maryland

STEVEN J. EISENREICH, University of Minnesota, Minneapolis, Minnesota

JESSE FORD, University of Minnesota, Minneapolis, Minnesota

ROBERT J. M. HUDSON, R. M. Parsons Laboratory, Massachusetts Institute of Technology, Cambridge, Massachusetts

ENDRE LACZKO, Swiss Institute for Water Resources and Water Pollution Control (EAWAG), Zurich, Switzerland

JUERG RUCHTI, Swiss Institute for Water Resources and Water Pollution Control (EAWAG), Zurich, Switzerland

MARY V. SANTELMANN, University of Minnesota, Minneapolis, Minnesota

RENÉ P. SCHWARZENBACH, Swiss Institute for Water Resources and Water Pollution Control (EAWAG) Zurich, Switzerland

ALAN STONE, The Johns Hopkins University, Baltimore, Maryland

MICHAEL STURM, Swiss Institute for Water Resources and Water Pollution Control (EAWAG), Zurich, Switzerland

Hong Kang Wang, Gray Freshwater Biological Institute, University of Minnesota, Minneapolis, Minnesota

Ulrich Weilenmann, Department of Geography and Environmental Engineering, The Johns Hopkins University, Baltimore, Maryland

Mark Wiesner, Department of Geography and Environmental Engineering, The Johns Hopkins University, Baltimore, Maryland

SERIES PREFACE
Environmental Science and Technology

The Environmental Science and Technology Series of Monographs, Text-books, and Advances is devoted to the study of the quality of the environment and to the technology of its conservation. Environmental science therefore relates to the chemical, physical, and biological changes in the environment through contamination or modification, to the physical nature and biological behavior of air, water, soil, food, and waste as they are affected by man's agricultural, industrial, and social activities, and to the application of science and technology to the control and improvement of environmental quality.

The deterioration of environmental quality, which began when man first collected into villages and utilized fire, has existed as a serious problem under the ever-increasing impacts of exponentially increasing population and of industrializing society. Environmental contamination of air, water, soil, and food has become a threat to the continued existence of many plant and animal communities of the ecosystem and may ultimately threaten the very survival of the human race.

It seems clear that if we are to preserve for future generations some semblance of the biological order of the world of the past and hope to improve on the deteriorating standards of urban public health, environmental science and technology must quickly come to play a dominant role in designing our social and industrial structure for tomorrow. Scientifically rigorous criteria of environmental quality must be developed. Based in part on these criteria, realistic standards must be established and our technological progress must be tailored to meet them. It is obvious that civilization will continue to require increasing amounts of fuel, transportation, industrial chemicals, fertilizers, pesticides, and countless other products; and that it will continue to produce waste products of all descriptions. What is urgently needed is a total systems approach to modern civilization through which the pooled talents of scientists and engineers, in cooperation with social scientists and the medical profession, can be focused on the development of order and equilibrium in the presently disparate segments of the human environment. Most of the skills and tools that are needed are already in existence. We surely have a right to hope a technology that has created such manifold environmental problems is also capable of solving them. It is our hope that this Series in Environmental Sciences and Technology will not only serve to make this challenge more explicit to the established professionals, but that it also will help to stimulate the student toward the career opportunities in this vital area.

Robert L. Metcalf
Werner Stumm

PREFACE

The aim of this book is to give an account of current research on chemical processes in lakes. Above all, it is intended to emphasize those processes that regulate the distribution of elements and compounds, utilizing kinetic information wherever possible and often using steady-state and dynamic models. Special attention is given to the solid–solution interface and to an assessment of the dominant roles of settling particles and the sediment–water interface in regulating concentrations of heavy metals and other reactive substances. An attempt is made to show how the lacustrine ecosystem responds to human impact, especially to stresses resulting from chemical perturbation. Measures for restoring eutrophied lakes are included. Although the nonbiological side of limnology is emphasized, lakes must be viewed and studied as microcosms. Therefore, the common goal of many of the authors is to determine both how the chemical environment interacts with organisms and ecosystems and to assess how they relate to one another.

The processes discussed and the concepts presented are applicable to all natural waters (oceans, estuaries, rivers). Lakes may be used as "test tubes" to study processes occurring in other aquatic systems. These habitats are characterized by different inputs, outputs, productivities, particulate carrier phases, sedimentation rates, time scales of mixing, and redox conditions. In estuaries and rivers, mechanical and thermal energy inputs are large, chemical and biological gradients are transient, and chemical and biological processes overlap and may be dominated by heterogeneous processes occurring at the particle–water interface. In contrast, water renewal times in oceans are millions of times longer. Nonetheless, oceans cannot always be readily utilized for data acquisition. The experimental exploration of marine systems is very involved and expensive. In addition, the interpretation and applications of such information are difficult since "the ocean" has less systemic diversity than other aquatic habitats.

On the other hand, lakes (especially lakes in which steady-state approximations are still possible) are useful experimental systems to study many processes under relatively simple conditions. In order to understand the separate contributions of physical, chemical, and biological processes that are operative in all aquatic systems, one needs to compare information obtained from lakes that differ widely in morphology, time scales of mixing, productivities in inputs and outputs, sedimentation rates, and redox gradients.

Furthermore, lakes make excellent models to study the interdependence of several biochemical cycles and their interlocking feedback mechanisms. The aquatic biomass in marine and freshwater systems is characterized by a surprising compositional constancy (Redfield ratios of nutritional elements). The rates of cycling of C, N, P, S, O, and some biophile metals are within certain limits given by the uptake and release of these elements by living organisms. These rates, in turn, reflect the variation in C:N:P:S ratios in biotic tissues. Phosphorus appears to be capable of controlling major parts of the N, C, S, and Si cycles. Furthermore, the input of phosphorus also appears to enhance trace-metal and metalloid elimination from waters, since the resulting higher productivities and larger particle sedimentation rates increase the efficiency of scavenging.

Understanding how geochemical cycles in lakes are coupled by organisms may aid our understanding of the global ecosystem. Especially valuable would be information on the role of the relatively small mass of biota in regulating and interconnecting the circulation of materials through land, water, biosphere, and atmosphere, and the sensitivity of these interacting systems to disturbances caused by civilization.

The authors—natural scientists, oceanographers, limnologists, and environmental engineers—have attempted to write their chapters in such a way as to assist the readers (students, limnologists, college teachers, environmental engineers) in understanding general principles, as well as to guide research in chemical limnology. Lakes, and the chemical processes occurring within them, are looked at more from a dynamic and mechanistic point of view, rather than a descriptive one. Emphasis is on explanation and intellectual stimulation, rather than on comprehensive documentation.

Most of the authors met in September 1983 in Stans, Switzerland, for a workshop. Background papers formed the basis for the discussions. This volume is not a "proceedings of a conference," but the offspring of the workshop and its stimulating discourses and interactions.

ACKNOWLEDGMENTS

I am most grateful to many colleagues who have reviewed individual chapters and have given useful advice on the organization of the book. I am especially indebted to Diana Hornung, who has carried a large burden in helping me edit this book. Credit for the creation of this volume is, of course, primarily due to its authors.

WERNER STUMM

Zurich, Switzerland
December 1984

CONTENTS

1

SPATIAL AND TEMPORAL DISTRIBUTION OF CHEMICAL SUBSTANCES IN LAKES: MODELING CONCEPTS

Dieter M. Imboden and René P. Schwarzenbach

Institute for Water Resources and Water Pollution Control (EAWAG), Swiss Federal Institute of Technology (ETH) Zurich, Switzerland

Abstract

The understanding of the dynamic behavior of chemical species in a given aquatic ecosystem is a prerequisite for the risk and hazard assessment of chemical pollution. Modeling concepts are provided to evaluate the relative importance of reaction processes (i.e., chemical and biological reactions, water–air exchange, and interactions between the dissolved and particulate phases) versus transport phenomena (i.e., water flow and mixing), and their impact on the distribution and residence time of pollutants in lakes. A one-dimensional vertical mixing model which includes the topography of the lake is discussed by using hydrophobic organic compounds as an example.

1. INTRODUCTION

One of the major goals in environmental chemistry is to assess the residual concentration of chemicals originating from natural or anthropogenic inputs into an ecosystem (1). In lakes, the spatial and temporal distribution of substances depends on numerous factors which can be divided into two groups:

1

the characteristics of the physical environment, and the specific reactivity of the chemical substance considered.

In this article, conceptual tools are presented which are essential in order to assess the simultaneous influence of mixing processes and chemical reactions in lakes. The basic idea is to formulate a mass balance equation for a (finite or infinitesimal) test volume in the lake by taking into account all sources and sinks within the test volume (internal processes), and all mass transport processes across its boundaries (external processes). This results in an equation for the temporal change of the concentration within the test volume, where each (external or internal) process is characterized by a first-order rate constant k (or by its inverse—the residence or response time). The rate constant k describes the relative change of the concentration due to the respective process. Therefore, the comparison of the relative size of the rates allows the determination, among a large variety of possible mechanisms, of those processes which dominate the behavior of the compound in the specific volume under investigation (2).

In Section 2 typical ranges of time and length scales of various transport mechanisms in lakes are presented, such as water renewal by rivers, horizontal and vertical turbulent diffusion, advection by lake currents, and settling of particles. In order to demonstrate how the resulting transport rate constants are compared to the *in situ* reactivity of the compound, we shortly discuss the one- and two-box models, an approach which is equivalent to the completely mixed reactor model as used in chemical engineering.

In Section 3 a more realistic model is developed for the description of the concentrations of chemical compounds based on both the chemical properties of the species and the physical characterization of the lake. In contrast to the box concept, this model accounts for the specific bottom topography of lakes and for the interaction between water column and sediments—two important factors for chemical compounds with a nonnegligible particulate phase.

In Section 4 this model is illustrated by applying it to hydrophobic organic compounds, which include important environmental pollutants. The selected compounds cover a wide range of chemical processes, such as *in situ* reaction (hydrolysis, photolysis, etc.), adsorption on particles, and exchange at water–air and water–sediment interfaces. As an example, the model is employed to interpret the measured distribution of tetrachloroethylene (PER) and the hypothetical long-term behavior of hexachlorobenzene (HCB) in Lake Zurich (Switzerland).

2.   MASS TRANSPORT IN LAKES: TIME AND LENGTH SCALES

2.1.   Transport Mechanisms

Mass transport of a solute through and within a lake is driven by the following processes: rate of water renewal by rivers (inlets, outlets); transport by molecular diffusion, by water currents, and by particle movement; and mass transfer at boundaries (fluxes at air–water and sediment–water interface). Since currents in lakes are always turbulent in nature, their effect on transport is usually separated into the long-term mean advective flux and into turbulent diffusion. The advective mass flux F_a per unit time and area along a given axis is mathematically described by

$$F_a = v_a C \quad [\text{g m}^{-2}\text{d}^{-1}] \tag{1}$$

[v_a is the advection (or current) velocity along the given axis and C is the concentration of the compound], and the turbulent mass flux F_t by an equation analogous to the Fick's first law:

$$F_t = -K \frac{\partial C}{\partial x} [\text{g m}^{-2}\text{d}^{-1}] \tag{2}$$

where $\partial C/\partial x$ is the concentration gradient along the direction of transport and K is the coefficient of turbulent diffusion (eddy diffusion coefficient). K has the same dimension as the molecular diffusion coefficient but is commonly larger by several orders of magnitude. Therefore, compared to turbulent diffusion, transport by molecular diffusion is seldom important in the free water column.

The distinction between advective and turbulent mass flux is arbitrary in the sense that it strongly depends on the time and length scales of interest. As a simple example, imagine a cloud of a (natural or artificial) tracer with typical spatial dimension Δx_A [Fig. 1.1(a)]. The effect of an eddy on the cloud A would be to move it along the current direction without significant influence on the shape of the cloud. This represents an advective transport process. In contrast, the same eddy has a fairly different effect on the cloud of another tracer with size similar or larger than the scale of the eddy: The water current distorts the tracer path; this is a dispersive or diffusive process.

A similar situation is found if different time scales are employed [Fig. 1.1(b)]: Usually, the current velocity measured at a fixed location exhibits a complex temporal behavior with fluctuations of various amplitudes and frequencies. Fairly large mean velocities are found if the currents are averaged over a short time interval (Δt_A). The net advective effect usually becomes smaller (and may also change sign) if the averaging interval becomes larger (Δt_B). This means that the influence of turbulent transport is growing at the expense of advective transport. In fact, in most lakes, especially in smaller ones, all mean advection velocities are extremely small or even zero if averaged over long-enough time periods. In consideration of the relationship between length (or time) scales and diffusivity, it comes as no surprise that Okubo (3) has found a nearly linear relationship between the horizontal diffusion coefficient in the ocean, K_h, and the length scale L_h over which the diffusive mixing is effective. The linear approximation of

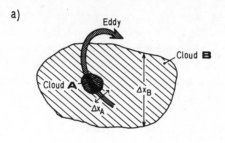

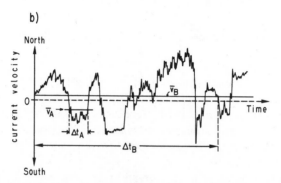

Figure 1.1. Distinction between advective and turbulent transport from (*a*) a spatial point of view: the larger the tracer patch, the larger the upper size limit of eddies which merely distort the eddy without significant advective transport; and from (*b*) a temporal point of view: if current velocities measured at a fixed point are averaged over a large time interval the resulting mean velocity is usually very small, and most of the current action becomes of a turbulent nature.

Okubo's diffusion diagram is

$$K_h = v_t L_h \qquad [\text{m}^2\text{d}^{-1}] \tag{3}$$

The turbulent diffusion velocity v_t lies between 50 and 1000 m d^{-1} (0.06–1 cm s^{-1}), where the high values describe the mixing during extreme wind conditions.

In the case of unidirectional advection, a characteristic mixing rate k_{mix}^a can be defined by

$$k_{\text{mix}}^a = v_a/L \qquad [\text{d}^{-1}] \tag{4}$$

where L is the dimension of the system in the direction of transport under consideration. In the case of turbulent transport, the characteristic mixing rate is

$$k^t_{mix} = K/L^2 = v_t/L \qquad [d^{-1}] \qquad (5)$$

where Okubo's scale-dependent diffusion concept (Eq. 3) is used.

As can be seen from the range of characteristic lake parameters listed in Table 1.1, the long-term unidirectional advection velocity is usually much larger than the turbulent velocity. It is the lake-wide circulation pattern which is responsible for the transport of matter over distances comparable to the size of the lake. Turbulent diffusion, in contrast, works more slowly but also more smoothly; it is responsible for the local compensation of concentration inhomogenities and acts also against the direction of the local mean flow.

Table 1.1. Definition and Range of Characteristic Parameters Determining Mixing Rates in Lakes

Symbol	Unit	Definition	Typical Range	Lake Z^a
τ_W	d	Water renewal time	$10-10^4$	400
L_h	m	Horizontal dimension of lake	10^2-10^5	3×10^4
L_z^{max}	m	Maximum depth	$1-500$	135
L_z^{mean}	m	Mean depth	$1-200$	50
h_{mix}	m	Depth of mixed layer (epilimnion)	$1-20$	10
L_h^{in}	m	Horizontal dimension of inlet or discharge pipes	$0.1-100$	—
		Horizontal Advection Velocity (Short Term, Unidirectional)		
v_a^E	md^{-1}	Epilimnion or mixed lake	$10^3-2 \times 10^4$	5×10^3
v_a^H	md^{-1}	Hypolimnion	10^3-10^4	2×10^3
		Horizontal Turbulent Diffusion Velocity		
v_t^E	md^{-1}	Epilimnion or mixed lake	$50-10^3$	100
v_t^H	md^{-1}	Hypolimnion	$50-10^3$	100
v_{in}	md^{-1}	Velocity of inflowing water	10^3-10^5	5×10^3
		Vertical Eddy Diffusivity		
K_z^E	m^{2}d^{-1}	Epilimnion or mixed lake	$1-10^4$	10^3
K_z^H	m^{2}d^{-1}	Hypolimnion	$0.1-10$	0.2
		Settling Velocity of Particles		
v_s^E	md^{-1}	Epilimnion	$0.1-10$	2.5
v_s^H	md^{-1}	Hypolimnion or lake during overturn	$1-50$	2.5

^{a} Simplified version of the main basin of Lake Zurich, Switzerland.

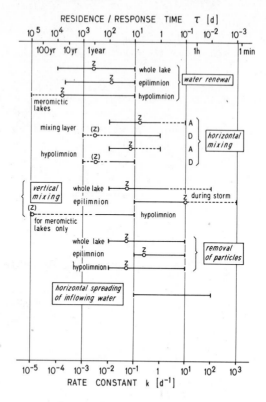

Figure 1.2.   Rate constants k and their inverse—the residence or response time τ—for various mixing processes in lakes. Z represents values for test Lake Z, a simplified version of the main basin of Lake Zurich, Switzerland. (See Table 1.2 and text for details.)

2.2.   Rate Constants

The parameters from Table 1.1 are employed to calculate various mixing and removal rate constants (Table 1.2). This information is also plotted in Figure 1.2. Note that the top axis gives the inverse value of k, that is, the response (or residence) time τ with respect to the different mechanisms. The definition of τ implies, somewhat arbitrarily, that for a linear process the response time is the time at which 63% of the process is completed (see Section 2.4, in particular, Eq. 8). In order to exemplify the idea and use of these tables, we have chosen a synthetic system, Lake Z, which is a simplified version of the main basin of Lake Zurich (Switzerland).

Based on the information given in Table 1.2 and Figure 1.2, simple estimations on the distribution of chemicals in this lake can be made prior to the tedious application of more complicated models. From Table 1.2 we can conclude for the case of Lake Z:

1. Buildup and disappearance within the whole lake of a conservative (non-reactive) substance without interaction with particles occurs with the response time of water, τ_w equal to 400 d.

2. Horizontal mixing is fast compared to water renewal (6–14 days in epilimnion and hypolimnion, respectively). In contrast, during the stag-

Table 1.2. Definition and Range of Mixing Rate Constants. Characteristic Parameters are Taken from Table 1.1

Mixing Process[a]	Rate Constant		
	Definition	Range [d⁻¹]	Lake Z [d⁻¹]
Whole Lake			
Water renewal	$1/\tau_w = k_w$	$10^{-4}-10^{-1}$	2.5×10^{-3}
Horizontal mixing	(A) v_a^E/L_h	$10^{-2}-10$	0.17
	(D) v_t^E/L_h	$10^{-3}-1$	3×10^{-3}
Vertical mixing (during circulation)	(D) $K_z^E/(L_z^{max})^2$	$10^{-2}-10^{2\,b}$	0.05
Removal of particles	(S) v_s^H/L_z^{mean}	$10^{-2}-10$	0.05
Epilimnion during Stagnation			
Water renewal	$(1/\tau_w)(V_{tot}/V_E)^c$	$2\times10^{-4}-10^{-1}$	0.012
Horizontal mixing	As whole lake		
Vertical mixing	(D) K_z^E/h_{mix}^2	$0.1-1000^b$	10
Removal of particles	(S) v_s^E/h_{mix}	$0.1-10$	0.25
Hypolimnion during Stagnation			
Exchange with epilimnion	(D) $K_z^H/(L_z^{max}h_{mix})$	$10^{-5}-10^{-1\,d}$	1.5×10^{-4}
Horizontal mixing	(A) v_a^H/L_h	$10^{-2}-1$	0.07
	(D) v_t^H/L_h	$10^{-3}-0.1$	3×10^{-3}
Vertical mixing	(D) $K_z^H/(L_z^{max})^2$	$10^{-5}-10^{-1\,d}$	1×10^{-5}
Removal of particles	(S) v_s^H/L_z^{mean}	$10^{-2}-10$	0.05
Inflowing Water (Inlet or Discharge)			
Horizontal spreading	$0.02\, v_{in}/L_{in}^{\ e}$	$0.1-100$	—

[a] (A) represents mixing by (long-term) advection; (D) mixing by turbulent diffusion; and (S) transport by settling particles.

[b] Lower values occur during periods of low input of wind energy; such conditions, however, do usually not extend over long time periods. Upper limits apply to very shallow lakes (depth ~ 1 m) during strong wind input.

[c] V_{tot}/V_E is the ratio between total and epilimnic lake volume.

[d] Exchange rates with epilimnion and internal hypolimnic mixing rates are hypothetical if they are below about 3×10^{-3} d⁻¹, and if the lake is completely mixed at least once per year. For meromictic and amictic lakes, however, extremely low hypolimnic exchange rates may occur.

[e] Dilution of inflowing water (river, discharge pipe, etc.) by entrainment; estimation with entrainment coefficients by Fischer et al. (4).

nation period vertical mixing across the thermocline and within the hypolimnion is much slower than water renewal. Thus, regardless of the spatial distribution of the sources of the conservative compound, one would expect to find significant vertical, but no horizontal, concentration gradients.

3. If the substance is predominantly adsorbed on particles and does not undergo other reactions, the response time is dominated by the residence time of particles ($\tau \sim 20$ d). The vertical distribution of the substance would be determined by the vertical transport of particles, which is much faster than the vertical mixing time.

Similar conclusions can be drawn for reactive species. Here, three different removal mechanisms (outflow, particle settling, and reaction including water–air exchange) have to be compared with the horizontal and vertical mixing rates. The horizontal distribution of chemicals with reaction rates faster than horizontal mixing rates would not be homogeneous. However, such very reactive compounds would not reach significant concentration levels within a lake (though they could create local pollution problems close to inlets or discharges); such compounds will not be further discussed here.

Note that these considerations do not yet include the sediment column which—especially for highly adsorbing species—can be the major mass reservoir of the combined lake-sediment system. As will be demonstrated in Section 4, the time scales of the sediment–water interaction may often be the most important factor for the long-term response and memory of the lake.

2.3.   Choice of Model

There exists a large spectrum of models ranging from the simplest one-box to the three-dimensional continuous model, each providing a different spatial resolution. Yet, the choice of the model structure should not be guided by the effort to obtain maximum complexity; instead, a reasonable model structure should be based on the analysis of the relative size of mixing and reaction rates as discussed above. For instance, if the total reaction rate turns out to be much smaller than a given transport rate, then the corresponding spatial structure is not essential for this compound: the system is "completely mixed" with respect to this spatial subdivision.

The transport rates given in Figure 1.2 show that for conservative substances the simplest approach, the completely mixed (1 box) model, would be suitable. A first spatial differentiation, necessary to describe a compound with a significant surface reaction (e.g., photolysis or air–water exchange), would lead to the concept of a vertical separation of the lake into epilimnion (surface) and hypolimnion.

2.4. Box Models

The typical characteristics of simple box models and their application to hydrophobic organic pollutants have been described by Schwarzenbach and Imboden (5). Application of more complicated models, such as the one discussed in Section 3, should only be attempted if the potential of the simple approach has been exploited. A short summary is given in order to demonstrate how these models can be used to estimate, for instance, the relationship between input rate and mean concentration of a given compound without a significant particulate phase.

As mentioned before, the basic units from which more complex models are built, commonly consist of a well-mixed box connected to its environment by advective and diffusive exchange processes and by particle settling. Therefore, the one-box model should be the starting point of the analysis, whether the box portrays the whole lake or just a tiny fraction of its water body.

The variation of the concentration C in a well-mixed box of constant water volume V is derived from the simple mass balance equation [Fig. 1.3(a)]

$$V\frac{dC}{dt} = J - QC - R \;[\text{gd}^{-1}] \tag{6}$$

The terms on the right-hand side mean total input, loss through the outlet (flow rate Q), and loss by reactions. It is always possible to write the total input as the product of the water input Q (which is equal to the water output since $V = $ const) and a mean concentration C_{in}: $J = QC_{in}$. In a similar way the reaction rate per volume can be written as $R/V = k_r C$, where k_r is a rate constant [d^{-1}]. Only if k_r is independent of C (first-order reaction) does Eq. 6 acquire a simple mathematical form (linear differential equation):

$$\frac{dC}{dt} = k_w C_{in} - k_w C - k_r C \quad [\text{gm}^{-3}\text{d}^{-1}] \tag{7}$$

with analytical solution

$$C(t) = C_\infty + (C_0 - C_\infty)\exp[-(k_w + k_r)t] \quad [\text{gm}^{-3}] \tag{8}$$

In Eq. 8, $k_w = Q/V$ is the water exchange rate, C_0 is the concentration at time $t = 0$, and

$$C_\infty = \frac{k_w}{k_w + k_r} C_{in} \quad [\text{gm}^{-3}] \tag{9}$$

the steady-state concentration in the box.

As explained in Chapter 17, the relative size of the removal terms on the right-hand side of Eq. 7 can be employed to assess the faith of the compound in the system: On the one hand, if the water exchange rate is much faster than the reaction rate ($k_w \gg k_r$), the compound passes through the box like a nonreactive species, and $C_\infty \sim C_{in}$ (see Eq. 9). On the other hand, if $k_r \gg k_w$, the major fraction of the compound reacts within the box; the concentration in the box is small compared to the input concentration ($C_\infty \ll C_{in}$); and loss through the outlet is negligible.

While the ratio of k_w and k_r determines the relation between C_{in} and C_∞, it is the sum of k_r and k_w which determines the response time of the system with respect to variations in the external input. As can be deduced from Eq. 8, the response time to reach 63% of a new steady state is equal to $(k_w + k_r)^{-1}$.

An additional degree of complexity is introduced into the model if, in addition to water exchange and reactivity, either internal mixing rates or transfer rates between different forms of the compound (e.g., dissolved and particulate phase) are considered. The role of the solid phase will be discussed in Section 3; here a simple example will be used to demonstrate the effects of internal mixing.

A model with at least two adjacent boxes is needed to introduce spatial variability into the description of the concentration field C [Fig. 1.3(b)]. The new feature in the two-box model is the mixing between the boxes described by the water exchange Q_{ex} [$\mathrm{m^3\ d^{-1}}$]. Since these water masses carry the compound concentration of the respective box from which they originate, the net effect on the mass in box 1 is $-Q_{ex}(C_1 - C_2)$, the negative value of the effect on box 2. The total mass balance equations for the boxes are

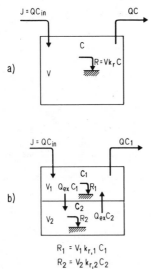

$$R_1 = V_1 k_{r,1} C_1$$
$$R_2 = V_2 k_{r,2} C_2$$

Figure 1.3. (a) One-box and (b) two-box models for lakes. (See text for explanations.)

similar to Eq. 7, completed by the extra terms for internal exchange. For simplicity it is assumed that no external input occurs into box 2:

$$\frac{dC_1}{dt} = k_w\, C_{in} - k_w\, C_1 - k_{r,1}\, C_1 - k_{ex,1}\, (C_1 - C_2) \tag{10a}$$

$$\frac{dC_2}{dt} = -\, k_{r,2}\, C_2 - k_{ex,2}\, (C_2 - C_1) \tag{10b}$$

where $k_{ex,1} = Q_{ex}/V_1$ and $k_{ex,2} = Q_{ex}/V_2$ are the water exchange rates $[d^{-1}]$ for the boxes.

The solution of Eqs. 10a and 10b is more complicated (but still possible analytically) since the expressions are coupled through the exchange term. Here, only the steady-state solution is presented; for details we refer to Schwarzenbach and Imboden (5) and any textbook on linear differential equations. The steady-state concentrations are calculated by setting the left-hand sides of Eqs. 10a and 10b equal to zero. Then from Eq. 10b the concentration ratio is

$$\frac{C_{2,\infty}}{C_{1,\infty}} = \frac{k_{ex,2}}{k_{ex,2} + k_{r,2}} \tag{11}$$

Inserting Eq. 11 into Eq. 10a yields

$$C_{1,\infty} = \frac{k_w\, (k_{ex,2} + k_{r,2})}{(k_{ex,1} + k_w + k_{r,1})\, (k_{ex,2} + k_{r,2}) - k_{ex,1}\, k_{ex,2}}\, C_{in} \tag{12}$$

Though the relationship between C_{in} and the steady-state concentration (Eq. 12) is more complicated than for the one-box model (Eq. 9), $C_{1,\infty}$ is still closer to C_{in} if the reactivities $k_{r,i}$ are small. Also, for a compound with $k_{r,1} = k_{r,2} = 0$, both box concentrations approach C_{in}. In other words, for a conservative substance internal mixing is of no significance. In fact, the degree of spatial heterogeneity, expressed by the ratio, Eq. 11, depends on the ratio between mixing and reactivity in box 2, $k_{ex,2}/k_{r,2}$, but does not depend on rate constants of box 1. This result corresponds to the statement made in Section 2.3 on the optimal choice of the spatial complexity of a model. Note that Eqs. 9 and 11 have the same structure, since both reflect the relative importance of reactivity versus mixing (water renewal or internal exchange).

In cases where the interaction between the sediments and the water becomes important, an extension to a continuous vertical transport model is needed as developed in the next section. However, as demonstrated by the simple mixing rate concept derived earlier, a further extension of the model into three dimensions would only be justified under very special circumstances.

3. ONE-DIMENSIONAL VERTICAL MODEL FOR REACTIVE SPECIES IN LAKES

3.1. Model Structure

The basic assumption of the one-dimensional vertical (1DV) model is that horizontal mixing is fast compared to reaction and boundary exchange processes. Thus, spatial distribution of a chemical is expected to exhibit no significant horizontal gradients. However, for describing the interaction between the sediment and water column, the three-dimensional bottom topography of the lake cannot be completely ignored. In the 1DV model, as developed by Imboden and Gächter (6) for the dynamics of phosphorus in lakes, bottom topography is accounted for by incorporation of the lake cross section as a function of depth, $A(z)$. One essential consequence of this model scheme is the availability of sediment surface in every—not only the deepest—horizontal layer. As will be demonstrated by the examples in Section 4, for all chemicals with sink or source at the sediment–water interface, the local ratio of sediment area to volume

$$\frac{dA}{dV} = \frac{1}{A}\frac{dA}{dz} \qquad [\mathrm{m^{-1}}] \tag{13}$$

is important for the shape of the vertical concentration profiles.

The model shown in Figure 1.4 serves as the physical base for the description of a variety of parameters, such as water temperature, phosphorus, oxygen, trace metals, or trace organic compounds. The model combines system-specific processes with the properties of the chemical under investigation. The following numbers refer to the numbers used in Fig. 1.4.

System (lake)-Specific Processes

① Inflow of water at the surface or at various (time variable) depths: Inlets act as one potential source for chemicals and particles.

② Outflow from lake (usually from the surface): Outlet acts as one sink for chemicals and particles.

③ Complete mixing of the surface water to the depth h_{mix} (epilimnion depth). h_{mix} usually varies with time.

④ Vertical mixing of water by turbulence described by a (time and depth dependent) vertical eddy diffusion coefficient: Affects dissolved and particulate species.

⑤ *In situ* production of particulate matter, usually in the surface layer (e.g., growth of phytoplankton) but possibly also below the thermocline.

⑥ Settling of particles by gravitation, either into the deeper water layer or into the sediment surface at the side of each layer: This sink term acts only on the particulate or adsorbed phase of the compound.

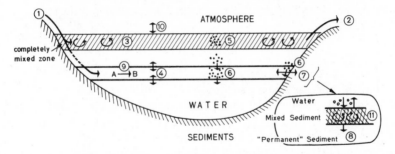

Figure 4. One-dimensional vertical (1DV) model for the description of reactive chemicals in lakes with dissolved and particulate phases. (The numbers refer to the processes explained in Table 1.3 and in the text.)

⑦   Exchange between sediment and water by diffusion or resuspension of particles.

⑧   Transfer of sediment material into the "permanent," noninteracting sediment column: Acts as a permanent sink for particulate or adsorbed chemicals.

Compound-Specific Processes

⑨   *In situ* reaction of chemicals, for example, hydrolysis, photolysis, redox processes: Since such reactions often depend on key variables like water temperature, light intensity, or pH, the reaction rates are usually time and depth dependent. A special reaction is the transfer of the chemical between the dissolved and particulate phases as described below.

⑩   Mass transfer at the lake surface.

⑪   Reaction of compound in the sediments.

The algebraic expressions describing the model are summarized in Table 1.3. Two processes, both related to the interaction between solutes and particles, are discussed in greater detail below. For the special case of hydrophobic organic compounds, the reaction processes are discussed in Section 4.

3.2. Transfer between Dissolved and Particulate Phases

Since transport and reaction processes affect dissolved and particulate species in a different manner, it is important to distinguish between the two phases. This requires an adequate description for the phase transition. Of course, for a planktic nutrient such as phosphorus the phase transition is primarily controlled by biological factors (primary production and degradation). However, for many trace chemicals like heavy metals or hydrophobic organic compounds, as a first approximation, adsorption equilibrium may be assumed and described by a linear adsorption isotherm:

$$C_s = K_p C_l \quad \text{or} \quad C_{oc} = K_{oc} C_l \tag{14}$$

C_l is the dissolved concentration [gm^{-3}]; C_s or C_{oc} is the concentration of the chemical per particle mass [g g$_s^{-1}$] or per particulate organic carbon (POC) [g g$_{oc}^{-1}$], respectively; and K_p and K_{oc} are the corresponding partition coefficients [m^3 g$_s^{-1}$ or m^3 g$_{oc}^{-1}$]. For the hydrophobic organic compounds discussed in Section 4, the equation referring to POC will be used since it is more adequate in describing the sorption equilibrium for compounds as long as the organic carbon content of particles exceeds 0.1% (7).

Provided that the sorptive processes are completely reversible and much faster than all other relevant processes, the system can be completely described by two concentration fields—for example, the total (dissolved and particulate) concentration of the chemical (C_t) and by the concentration of solids or POC: S [g$_s$ m^{-3}] or S_{oc} [g$_{oc}$ m^{-3}], respectively. The single-phase concentrations are related to C_t by

$$C_l = fC_t \; ; \; C_p = (1 - f)C_t \; ; \; C_t = C_l + C_p \tag{15}$$

$$f = [1 + K_p S]^{-1} \quad \text{or} \quad f = (1 + K_{oc} S_{oc})^{-1} \tag{16}$$

C_p [g m^{-3}] is the particulate concentration of the chemical per unit water volume, and f is the fraction of the dissolved compound relative to the total concentration. Note that for a nonadsorptive chemical all processes which are related to the particulate phase or to the sediments can be disregarded (# 5 to 8 and 11 in Fig. 1.4).

3.3. Interaction between Sediments and Water

The exchange between sediment and water is a very complicated process which, depending on the specific lake and compound properties, may either be dominated by physical parameters (growth of sediment column due to particle settling, resuspension of particles, and bioturbation of sediment column) or by compound-specific parameters (rate of chemical transformation in the sediments, release into pore water, and molecular diffusion in pore water). For many chemicals the vertical variation of the redox potential within the sediment plays a key role for the exchange flux. (See Chapter 9 by P. Baccini et al. and Chapter 2 by W. Davison).

Obviously, for situations dominated by chemical reactions a refined model would be needed which describes the sediment column as a one-dimensional vertical system (8,9), quite similar to the model used for the water column. However, due to the depth-dependent properties of the water column overlying the sediments, different one-dimensional concentration profiles would be needed for different lake depths. Such an approach would lead far beyond the scope of this chapter.

In order to concentrate on those aspects of the sediment–water interaction which are fairly independent of specific chemicals, a physical model is adapted which consists of a completely mixed (i.e., homogeneous) surface layer, characterized by the mass of particles (g_s m^{-2}) or POC (g_{oc} m^{-2}), overlying a sediment column not in interaction with the lake ("permanent" sediment). The state variable C_s (compound concentration per mass of particles or POC in the mixed sediment layer), a function of lake depth, is driven by the following three processes (the numbers refer to Fig. 1.4 and the equations in Table 1.3):

⑥   Input of settling particles carrying the compound in adsorbed form: The compound concentration C_s of the settling particles is usually different from C_s in the mixed sediment layer; it is assumed that the compound is instantaneously mixed into the surface layer until homogeneity is reached.

⑦   Exchange flux between mixed sediment layer and overlying water: This flux is assumed to be linear, expressed as the rate constant R_{ex} (d^{-1}] multiplied by the "adsorption deficit" of the particles in the mixed layer relative to the dissolved compound concentration in the water. From a mathematical viewpoint it is not necessary to specify whether the flux is purely diffusive or whether it involves resuspension of particles, equilibration between dissolved and solid phases, and immediate resettling of the particles at the same depth. It would also be possible to include transport of the resuspended particles to other lake depths.

⑧   Transfer of particles from the mixed layer into the "permanent" sediment: Since the particle mass in the mixed layer is assumed to be constant, this particle transfer is equal to the settling of fresh particles at the sediment surface. However, the particles leaving the mixed layer carry the compound concentration of this layer which is not equal to the concentration of the settling particles. If POC (instead of total solids) is used for describing the matrix of the particulate phase, transfer of particle mass from the mixed layer into the permanent sediment is significantly smaller than the input of settling POC since a large fraction of POC is decomposed in the fresh sediment. This effect is accounted for by the "preservation factor" β—the relative amount of POC reaching the permanent sediment.

3.4. Model Equations

The model equations of the IDV model, expressing mass balance in every horizontal section of the lake, are given in Table 1.3. The same numbers as introduced in Figure 1.4 are listed below every term to indicate the process.

Table 1.3. Equations for One-Dimensional Vertical Lake Model Describing a Chemical with Instantaneous Adsorption Equilibrium with Suspended Particles. Numbers under Terms Refer to Processes (See Fig. 1.4).

The (solute and particulate) compound concentration: C_t [g m^{-3}]

$$\frac{\partial C_t}{\partial t} = \frac{K_z}{A}\frac{\partial}{\partial z}\left(A\frac{\partial C_t}{\partial z}\right) - (1 - f)v_s\frac{\partial C_t}{\partial z} - fR_lC_t + \frac{F_l}{A}\frac{dA}{dz} + J_t \qquad \text{(M1)}$$

$$\qquad\qquad (4) \qquad\qquad\qquad (6) \qquad\quad (9) \qquad\quad (7) \qquad (1,2,10)$$

Concentration of suspended particles in the lake: S [g$_s$ m^{-3}]; if the partition coefficient refers to POC, S_{oc} can be used instead (see Eq. 16):

$$\frac{\partial S}{\partial t} = \frac{K_z}{A}\frac{\partial}{\partial z}\left(A\frac{\partial S}{\partial z}\right) - v_s\frac{\partial S}{\partial z} + J_s \qquad \text{(M2)}$$

$$\qquad\qquad (4) \qquad\qquad\qquad (6) \qquad (1,2,5)$$

Concentration of compound on particles (or on POC) in mixed layer of sediment column as a function of lake depth: C_s [g g$_s^{-1}$ or g g$_{oc}^{-1}$]

$$\frac{\partial C_s}{\partial t} = \frac{v_s}{M}(1 - f)C_t - \beta\frac{v_s}{M}SC_s + F_s - R_sC_s \qquad \text{(M3)}$$

$$\qquad\qquad (6) \qquad\qquad\qquad (8) \qquad (7) \qquad (11)$$

Processes and Definitions

See Eqs. 15 and 16 for the definition of f and the relation between C_t and the dissolved and particulate fraction.

(1,2,10) J_t: Net input of compound into water column [g m^{-3}d^{-1}] (includes removal by outlet and gas exchange at water surface)

(1,2,5) J_s: Net input of particles into water column [g$_s$ m^{-3}d^{-1}] (includes *in situ* production of particles and removal through outlet)

(4)   Vertical mixing by turbulence

      K_z: Vertical eddy diffusion coefficient [m^{2}d^{-1}]
      A: Lake cross section, depth dependent [m^2]
      z: Vertical coordinate, positive upwards, zero at deepest point of lake [m]

(6)   Settling of particulates by gravitation either into deeper water layers or onto the sediment surface at the side of each water layer

                v_s: settling velocity of particles   [md^{-1}]

Table 1.3. (*continued*)

---

(7)   Boundary flux between sediment surface and overlying water by particle resuspension or diffusive exchange

$$F_s = R_{ex}(K_p f C_t - C_s) \qquad [\text{g g}_s^{-1}\text{d}^{-1}]$$

$$F_l = -MF_s \qquad [\text{g m}^{-2}\text{d}^{-1}]$$

R_{ex}:   resuspension or adsorptive exchange rate $[\text{d}^{-1}]$

K_p: partition coefficient of compound (see Eq. 14) $[\text{m}^3 \text{ g}_s^{-1}]$

M: particle mass in mixed sediment layer $[\text{g}_s \text{ m}^{-2}]$

(8)   Input of particles from mixed sediment into "permanent" layer assuming $M = $ const (see text)

β: preservation factor, that is, fraction of particulate matter reaching the permanent sediment layer. If total solids are used then $\beta \sim 1$; for POC, $\beta < 1$.

(9)   *In situ* reaction of dissolved phase of chemical (photolysis, hydrolysis, etc., see Section 4)

R_l: reaction rate $[\text{d}^{-1}]$

(11)   Reaction of chemical in mixed sediment layer

R_s: reaction rate in sediments $[\text{d}^{-1}]$

Processes and symbols are summarized in the lower part of the table; details have been outlined in the preceding paragraphs.

Equation Ml applies to all compounds, whether they have a particulate phase ($f < 1$) or none ($f \sim 1$). Equation M2 describes the particle (or POC) concentration in the lake. It is used as input into the other equations but is independent of them. Thus, the variable S could as well be introduced into the model as a quantity derived from field data. It is for this reason that we have kept the particle submodel on a relatively primitive level, disregarding processes like coagulation or differential settling as described in Chapter 10 by C. O'Melia.

Treatment of Model Equations.   The equations have to be solved numerically. The simplest approach for a one-dimensional scheme is to replace the continuous profile by a series of horizontal boxes. This leads to the finite difference scheme in which the derivatives are substituted by concentration differencies between adjacent boxes.

4.   APPLICATION OF THE MODEL TO HYDROPHOBIC ORGANIC COMPOUNDS

In this section, the model will be applied to hydrophobic organic compounds to demonstrate its usefulness for assessing the interplay of the various compound- and system-specific processes and their influence on the distribution and residence time of reactive species in a lake. To this end, a set of model compounds and Lake Z will be used.

4.1.   Model Compounds

A major part of the xenobiotic organic chemicals that are continuously introduced into the environment by human activities is hydrophobic. Hydrophobic compounds are defined as being readily soluble in nonpolar organic solvents but only sparingly soluble in water. Important groups of such environmental pollutants include halogenated hydrocarbons, fuel and mineral oil components, polycyclic aromatic hydrocarbons, polychlorinated phenols, plasticizers, and many more.

The transfer and reaction processes to which hydrophobic compounds are subjected in lakes are sorption/desorption, water–air exchange, and chemical and biological transformations. Table 1.4. contains the names, solid (organic carbon)/water partition coefficients, and first-order rate constants for gas exchange, direct photolysis, and hydrolysis of some selected model compounds. These compounds cover a wide range of physicochemical properties and reactivities. The concepts and equations used to specify the various reactions are described in a previous paper (5) and summarized in Table 1.5. For simplicity, it is assumed that gas exchange, hydrolysis, and photolysis act only on dissolved species, and that the kinetics of the sorptive processes are fast compared to all other processes. Furthermore, biologically mediated transformations are not considered, although they could be important for some of the compounds listed in Table 1.4., for example, compounds 5 and 7. However, with the present knowledge of the mechanisms and kinetics of biotransformation of xenobiotic chemicals under natural conditions, it is not possible to reliably predict reaction rates using general concepts in a similar way as for the physical and chemical processes. Thus, in many cases, a comparison between model calculations and actual field data is required to assess the importance of biological transformations.

4.2.   Transfer and Reaction: Rate Constants

The rate at which a given transfer or reaction process occurs in a lake is not only dependent on the compound-specific properties but may also be strongly influenced by a series of environmental parameters such as tem-

Table 1.4. Model Compounds: Transfer and Reaction Parameters (See Also Table 1.5)
[s = Summer (T = 293 K); w = Winter (T = 278 K); n.r. = Not Relevant]

Number	Name	K_{oc}^a [m³ g_oc⁻¹]	v_{tot} (u_{10} = 1 m s⁻¹) [m d⁻¹]	$k_p(0)$ (24-h Average) [d⁻¹]	k_h [d⁻¹]
(1)	Bromomethane	1.9×10^{-5}	s 2.4×10^{-1} w 1.7×10^{-1}	n.r. n.r.	$1.6 \times 10^{-2\ f}$ $1.2 \times 10^{-3\ f}$
(2)	Tetrachloroethylene	3.5×10^{-4}	s 1.8×10^{-1} w 1.2×10^{-1}	n.r. n.r.	n.r. n.r.
(3)	1,4-Dichlorobenzene	8.7×10^{-4}	s 1.3×10^{-1} w 9.1×10^{-2}	n.r. n.r.	n.r. n.r.
(4)	Hexachlorobenzene	7.6×10^{-2}	s 9.8×10^{-2} w 7.2×10^{-2}	$2.4 \times 10^{-2\ b,c}$ $2.4 \times 10^{-3\ b,c}$	n.r. n.r.
(5)	Phenanthrene	1.3×10^{-2}	s 8.4×10^{-2} w 6.2×10^{-2}	$8.6 \times 10^{-1\ c}$ $1.6 \times 10^{-1\ c}$	n.r. n.r.
(6)	2,4-D-2-Butoxyethylester	2.2×10^{-3}	s 4.3×10^{-4} w 4.1×10^{-4}	$2.9 \times 10^{-2\ d}$ $2.9 \times 10^{-3\ e}$	$1.0 \times 10^6 \times 10^{(pH\ -\ 14)\ d}$ $1.3 \times 10^5 \times 10^{(pH\ -\ 14)\ d}$
(7)	n-Heptadecane	4.3	s 7.2×10^{-2} w 4.8×10^{-2}	n.r. n.r.	n.r. n.r.
(8)	Benzo[a]pyrene	5.8×10^{-1}	s 1.2×10^{-3} w 1.1×10^{-3}	12^b 2.9^b	n.r. n.r.

[a] Calculated from K_{ow}.
[b] Estimated value.
[c] Reference 10.
[d] Reference 11.
[e] Estimated from "summer value."
[f] Reference 13.

perature, wind speeds above the surface, particle concentrations, concentrations of dissolved and particulate organic carbon, settling velocities of particles, pH, and many more. Table 1.6 contains calculated overall first-order reaction rate constants for four model compounds in the test Lake Z together with the size of the process-specific rates which contribute to the total reaction. A typical summer and winter situation is given. For these examples, the lake is simply described by two vertically connected boxes during the summer (epilimnion, hypolimnion), and by one (completely mixed) box in the winter. Note that the rate constants refer to the change of the total concentration, C_t. Thus, the reaction rate k_r which only affects the dissolved phase (see Table 1.5) is multiplied by f and the settling rate by 1-f. The reduced rates are denoted as k'.

A comparison between the rate constants given in Table 1.6 shows that usually only one or two reaction processes are important. Compound 2, for example, is representative for a variety of environmental organic pollutants (e.g., chlorinated solvents) which are solely affected by gas exchange—a "surface reaction." Thus, the residence time of this compound is strongly dependent on where its input to the lake occurs (5). Compounds 4 and 8 are eliminated from the water column by both surface reaction and sedimentation. This type of compound will therefore accumulate in the sediments. Finally, compound 6 represents a chemically highly reactive compound which, depending on pH and temperature, may react faster than typical horizontal mixing times. In such cases, the vertical model may have to be replaced by a more sophisticated mixing model.

4.3. Examples

To illustrate the use and limitation of the developed concepts, two examples are discussed. The first represents the class of compounds without an elimination mechanism below the water surface and for which the particulate phase is not relevant. The second demonstrates the long-term memory effect of the lake sediments for a compound with significant elimination via settling particles.

1. *Tetrachloroethylene (PER)*.PER, a compound widely used as a solvent for dry cleaning and metal degreasing, is introduced into lakes primarily by domestic sewage effluents. Schwarzenbach et al. (21) have found significant concentrations in Lake Zurich. The only important elimination mechanisms occur at the surface by gas exchange and through the outlet [0.018 and 0.012 d^{-1}, respectively, if the surface mixed layer is 10 m deep (see Tables 1.2 and 1.6)]. Because of the linearity of the model, the rate constants allow the conclusion that the relative size of the two major elimination processes is 60% to the atmosphere versus 40% into the outlet.

The epilimnion-to-hypolimnion concentration ratio during the summer, as calculated by a two-box model and compared to measured values (5), led to the conclusion that about 25% of PER is entering the lake below the thermocline. The two-box model also showed that for this input distribution, the mean lake concentration (measured value about 55 μg m^{-3}) would be 57% of the mean input concentration; thus C_{in} = 96 μg m^{-3}. This corresponds to a total loading to the lake of 260 kg PER per year. Schwarzenbach et al. (21) have estimated 160 kg yr^{-1} as a lower limit and found a higher value to be quite possible.

The two-box model with its completely mixed hypolimnic box is not able to explain the vertical structure in the concentration profiles as found for many compounds in Lake Zurich. This structure results from incomplete vertical mixing of the water column during the winter period. Also, the two-box model is not sensitive to the exact depth of the hypolimnic input. Therefore, the one-dimensional continuous model described in Section 3 was applied using input, gas exchange, and water exchange rates as explained above. Vertical eddy diffusivity and mixing depth were calculated from water temperature and other standard limnological parameters measured by the Zurich Water Works (unpublished). For instance, in the winter of 1977 to 1978 the level of complete mixing did not reach below 80 m, although at greater depth a significant increase of the vertical diffusivity could be observed.

In Figure 1.5 calculated and measured profiles are compared. The model calculations begin in November 1977 with an idealized profile shaped after the measurements. The best fit was obtained with the following input distribution: 40% into 0–5 m; 30% into 5–10 m; 30% into 10–15 m. Since the average depth of the thermocline is about 10 m, this is in perfect agreement with the two-box results where 25% of the loading was assumed to occur into the hypolimnic box. Also, the absolute concentration level, that is, the

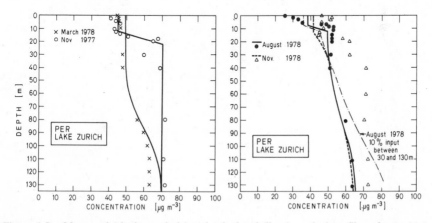

Figure 1.5. Measured (single symbols) and calculated (lines) vertical profiles of tetrachloroethylene (PER) in Lake Zurich. Measurements are from (21) and calculations are from the 1DV model using the characteristic data of Lake Z. Model simulation starts with a schematic profile representing the measurements of November 1977.

Table 1.5. Equations Used to Describe Various Transfer and Reaction Processes to which Hydrophobic Organic Chemicals are Subjected in Lakes

Process	Parameter	Equation(s)	References
Sorption/Desorption	Equilibrium partition coefficient based on organic carbon [$m^3 g_{oc}^{-1}$]	$K_{oc} = b(K_{ow})^a$ $a = 0.72, b = 3.5 \times 10^{-6}$ (chlorinated hydrocarbons)	(7)
		$a = 1.00, b = 4.9 \times 10^{-7}$ (polycyclic aromatic hydrocarbons)	(14)
	(K_{ow} = octanol/water partition coefficient [−])		
Gas exchange[a]	Average first-order rate constant for surface mixed layer [d^{-1}]	$k_g = v_{tot}/h_{mix}$	
	Overall mass transfer coefficient [$m\ d^{-1}$]	$(v_{tot})^{-1} = \dfrac{1}{v_l^0(D_l/D_l^0)} + \dfrac{RT}{H v_g^w(D_g/D_g^w)}$	(15,16)
	Mass transfer coefficient of oxygen [$m\ d^{-1}$]	$v_l^0 = 3.62 \times 10^{-1} \times 1.024^{(T^w - 293\ \text{K})} u_{10}^{1/2}$, for $u_{10} < 5.5\ \text{ms}^{-1}$ $v_l^0 = 2.76 \times 10^{-2} \times 1.024^{(T^w - 293\ \text{K})} u_{10}^2$, for $u_{10} > 5.5\ \text{ms}^{-1}$	(17)
	Mass transfer coefficient of water [$m\ d^{-1}$]	$v_g^w = 253 + 102u_{10} + 11.3(T^w - T^A)$	(18)

(D_l, D_g = molecular diffusivity, in liquid and gaseous phase [$cm^2\ s^{-1}$]; H = Henry coefficient [m^3 atm mole^{-1}]; u_{10} = wind speed, at height 10 m above water surface [$m\ s^{-1}$]; and T = temperature [K].)

Direct photolysis[a]   Near-surface first-order rate constant[b] [d^{-1}] $k_p(o) = 3.9 \times 10^{-16} \phi \sum_{\lambda} \epsilon_{\lambda} W_{\lambda} \Delta\lambda$ (19,20)

Average first-order rate constant for surface mixed layer [d^{-1}] $k_p = \dfrac{1 - e^{-\alpha h_{mix}}}{\alpha h_{mix}} k_p(o)$

Average bulk attenuation coefficient [m^{-1}] $\alpha = \epsilon^{DOC}[DOC] + \epsilon^s S$ (5)
$\epsilon^{DOC} = 0.2{-}0.4 \text{ m}^2\text{g}^{-1}$;
$\epsilon^s = 0.3{-}0.8 \text{ m}^2\text{g}^{-1}$

(W_{λ} = irradiance immediately below water surface [photons cm^{-2}s^{-1}]; ϕ = quantum yield for reaction of compound [$-$]; [DOC] = dissolved-organic-carbon concentration [g m^{-3}]; and S = suspended particle concentration [g m^{-3}].)

Hydrolysis[a]   First-order rate constant [d^{-1}] $k_h = k_A[H_3O^+] + k_B[OH^-] + k_N$

(k_A, k_B, and k_N are the specific rate constants for the acid catalyzed process, the base-promoted process, and the neutral process, respectively.)

[a] Assumed to act only on dissolved species.
[b] An average (wavelength independent) distribution coefficient D of 1.2 is assumed and the integral has been replaced by a sum over finite λ ranges for which mean values of the characteristic parameters are employed.

23

Table 1.6. Transfer and Reaction Processes: First-Order Rate Constants k' of Total Compound Concentrations C_t in Test Lake Z for a Typical Summer and a Typical Winter Situation (se = Summer Epilimnion; sh = Summer Hypolimnion; ww = Winter Whole Lake; and n.r. = Not Relevant, $k' < 10^{-4}$ d^{-1}) (See Also Table 1.5)

Compound			First-Order Rate Constants [d^{-1}]					Fraction in Dissolved Form^f f [-]
Number	Name		Removal on Settling Particles $k_s'^{\,a}$	Gas Exchange $k_g'^{\,b}$	Photolysis $k_p'^{\,c}$	Hydrolysis $k_h'^{\,d}$	Total $k_r'^{\,e}$	
(2)	Tetrachloroethylene	se^g	n.r.	1.8×10^{-2}	n.r.	n.r.	1.8×10^{-2}	$\geqslant 0.99$
		sh^h	n.r.	n.r.	n.r.	n.r.	n.r.	
		ww^i	n.r.	2.4×10^{-3}	n.r.	n.r.	2.4×10^{-3}	
(4)	Hexachlorobenzene	se	1.8×10^{-2}	9.1×10^{-3}	1.1×10^{-3}	n.r.	9.2×10^{-3}	0.93
		sh	4.5×10^{-3}	n.r.	n.r.	n.r.	n.r.	
		ww	3.6×10^{-3}	1.3×10^{-3}	n.r.	n.r.	1.7×10^{-3}	
(6)	2,4-D-2-Butoxyethylester	se	5.0×10^{-4}	n.r.	1.5×10^{-3}	1.0	1.0	>0.99
		sh	1.3×10^{-4}	n.r.	n.r.	1.3×10^{-1}	1.3×10^{-1}	
		ww	1.0×10^{-4}	n.r.	n.r.	1.3×10^{-1}	1.3×10^{-1}	
(8)	Benzo[a]pyrene	se	9.3×10^{-2}	n.r.	3.8×10^{-1}	n.r.	3.8×10^{-1}	0.63
		sh	2.3×10^{-2}	n.r.	n.r.	n.r.	n.r.	
		ww	1.9×10^{-2}	n.r.	1.8×10^{-2}	n.r.	1.8×10^{-2}	

^{a} $k_s' = (1 - f)v_s/h_{mix}$; $v_s = 2.5$ m d^{-1}. ^{b} $u_{10} = 1$ m s^{-1}. ^{c} $\alpha = 2$ m^{-1}. ^{d} pH = 8. ^{e} $k_r' = k_g' + k_p' + k_h'$. ^{f} $S_{oc} = 1$ g$_{oc}$ m^{-3}.
^{g} $h_{mix} = 10$ m; T = 293 K. ^{h} $h_{mix} = 40$ m; T = 278 K. ^{i} $h_{mix} = 50$ m; T = 278 K.

24

total PER input, is confirmed. However, additional information is gained with respect to the depth of the hypolimnic input (just below the epilimnion, i.e., into the thermocline). We have also tried to attribute as little as 10% of the total input into the hypolimnion and were always confronted in the model results with changes in the hypolimnic concentrations not confirmed in the measurements (see example for August 1978 in Fig. 1.5). One exception is the 25% increase of PER in the hypolimnion between August and November 1978 which has been previously explained by input into the hypolimnion (21). However, regarding the low vertical mixing intensity, it would require a very constant input per volume in order to avoid the appearance of local maxima.

The model calculations are not only sensitive to the assumptions for hypolimnic input but also to any subsurface removal mechanism. It has been shown that, for example, in ground water under oxic conditions, 1,4-dichlorobenzene (DCB), a compound also determined in Lake Zurich (21), is quite readily degraded by microbial activity (22). When combining our knowledge on the vertical mixing structure incorporated into the model with the measurements (during the summer, the concentration of DCB in the hypolimnion of Lake Zurich remains constant within 5%), a rate of 3×10^{-4} d^{-1} can be assigned as an upper limit on biodegradation, indicating that this process is not very important at the concentration level at which DCB occurs in Lake Zurich ($\sim$ 10–20 $\mu g\ m^{-3}$).

2. *Hexachlorobenzene (HCB).*   HCB, a pesticide, is presumed to enter lakes mainly at the surface. Concentrations of HCB in rivers are about 0.5 $\mu g\ m^{-3}$ (23); measurements from Swiss lakes and sediments are not available.

The two-box model serves to estimate the relative importance of the different processes affecting HCB in lakes. The four removal pathways (Tables 1.2 and 1.6) are (1) flushing through the outlet, (2) gas exchange, (3) photolysis, and (4) removal on particles. The first three processes occur at the lake surface; thus the corresponding rate constants are inversely related to the depth of the mixed layer, that is, the epilimnion during the summer and the whole lake (or a large portion of it) during the winter. Flushing affects the total HCB concentration, whereas gas exchange and photolysis are assumed to act on the dissolved fraction only.

The removal rate on particles depends strongly on the POC concentration (S_{oc}). To demonstrate the effect from the dynamics of the particles on the total HCB concentration, a POC variation is simulated for Lake Z using Eq. M2 of Table 1.3 as shown in Figure 1.6. Beginning with a low POC concentration (0.2 $g_{oc}\ m^{-3}$) in early spring, as the result of an algal bloom POC reaches peak values close to 3 $g_{oc}\ m^{-3}$ which disappear from the lake by settling within a few weeks.

The rates for the removal mechanisms are summarized in Table 1.7 for the different seasons. During the algal bloom, removal on settling particles

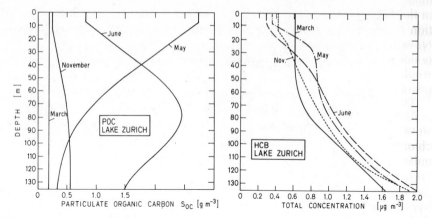

Figure 1.6. Hypothetical POC concentrations in Lake Zurich calculated from Eq. M2 of Table 1.3. Particle production rate at the surface is assumed to be 0.5 $g_{oc}m^{-2}d^{-1}$ from November to February, 1.5 $g_{oc}m^{-2}d^{-1}$ from March to October, interrupted by an algal bloom (8 $g_{oc}m^{-2}d^{-1}$) in May. Settling velocity $v_s = 2.5$ md^{-1}.

Figure 1.7. Hypothetical total concentration of hexachlorobenzene (HCB) in Lake Zurich produced by an input of 0.1 $\mu g\ m^{-2}d^{-1}$ at the lake surface. POC concentrations are shown in Figure 1.6; other parameters are explained in the text. The profiles calculated with the 1DV model are in cyclic steady state, that is, the concentrations assume the same values after one complete annual cycle.

Table 1.7. Removal Rates for HCB in Lake Z for Different POC Concentrations. Approximation for Two-Box Model.[a]

	Winter	Summer during Algal Bloom	
	Whole Lake	Epilimnion	Hypolimnion
h_{mix} (m)	50	10	40
$S_{oc}(g_{oc}\ m^{-3})$	0.2	2.5	2.5
f	0.985	0.840	0.840
Rates (d^{-1})			
Flushing	2.4×10^{-3}	1.2×10^{-2}	—
Gas exchange[b]	1.4×10^{-3}	8.2×10^{-3}	—
Photolysis[c]	9.5×10^{-5}	1.0×10^{-3}	—
Particle settling	7.5×10^{-4}	4×10^{-2}	1.0×10^{-2}

[a] Note that values are slightly different from Table 1.6 due to different POC concentrations.
[b] $u_{10} = 1$ ms^{-1}.
[c] $\alpha = 0.5$ m^{-1} (winter) and 2 m^{-1} (summer).

is the most efficient mechanism followed by flushing from the surface layer. During the winter, the outlet accounts for 53% of HCB removal, gas exchange for 31%, and particle settling for 16%. Photolysis is always a minor mechanism.

Using the lDV model, the temporal and spatial variation of HCB in Lake Z was calculated as it would result from a "standard input" of 0.1 µg m^{-2} d^{-1} at the surface (Fig. 1.7). The lake is in an approximate cyclic steady state, that is, the concentrations are the same at a given time of the year but vary over the course of a year. Note that in spite of the surface input, the concentrations increase with depth during the whole year. This results from the continuous exchange process between the sediments (where C_s is higher due to POC degradation) and the water, a process which affects mainly those layers with the largest sediment surface to volume ratio (see Eq. 13).

Since no adequate measurements are available on the sediment dynamics and the interaction with the overlying water, we have quite arbitrarily chosen the resuspension rate R_{ex} = 0.001 d^{-1}, and the sediment mixed-layer mass M = 600 g$_{oc}$ m^{-2} within which 80% of the POC is degraded (POC preservation factor = 0.2). The particle settling flux is between 0.5 g$_{oc}$ m^{-2} d^{-1} (winter) and 8 g$_{oc}$ m^{-2}d^{-1} (algal bloom); the yearly sum is 600 g$_{oc}$ m^{-2}. Since the preservation factor is 0.2, the mixed layer corresponds to a net POC accumulation of about 5 years.

A substantial fraction of HCB is transported to the lake sediments where, as a result, the HCB concentration reaches values between 0.2 µg g$_{oc}$$^{-1}$ (shallow waters) and 0.6 µg g$_{oc}$$^{-1}$ (deep water). The question arises how the combined lake-sediment system would behave if the external HCB source would be stopped at once. In Figure 1.8 calculations with the lDV model are presented for a period of 2 years after the input of HCB was stopped. Apparently, the long-term memory effect of the lake with respect to HCB is mainly determined by the exchange between sediment and water. It also depends on the transfer rate of sediment material into the "permanent"

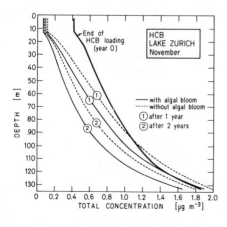

Figure 1.8. Response of total HCB concentration in Lake Z to a total stop of external HCB input, calculated with 1DV model. The profile in November of year 0 (stop of input) is taken from Figure 1.7. The profiles show the situation in November after 1 or 2 years, respectively, with and without an algal bloom in May.

sediment column. In our example the burial rate strongly depends on the particle dynamics during the summer. The suppression of the algal peak in May would approximately double the response time of the system and, at the lake bottom, even temporarily lead to concentrations which are higher than those reached during the HCB loading but with an algal bloom.

The ultimate removal of HCB from the lake is by permanent burial into the sediments and by gas exchange and flushing. Since the latter mechanisms occur at the water surface whereas the largest HCB concentrations are found in the deep part of the lake (in the water and the sediments), the rate of vertical transport within the water column is important for the final removal rate from the lake.

5. CONCLUSIONS

The modeling concepts described in this chapter can be applied to any chemical in order to evelute the relative importance of reaction processes versus transport phenomena in lakes. Hydrophobic organic compounds served as an example to demonstrate the usefulness of a simple one-dimensional vertical mixing model to explain the vertical distribution of such species in lakes. Yet, it should be noted that to date only very scarce field data are available to validate such model calculations. Furthermore, the present knowledge of the kinetics of a number of processes is still very limited. Therefore, it is often difficult to properly describe and quantify compound-specific processes such as sorption/desorption, reactions on solids, and (above all) biologically mediated transformations. On the lake-specific side, open questions refer mainly to the phenomena which deal with the dynamics of particles and the interaction between sediments and water (particle formation, sedimentation, resuspension, etc.) Nevertheless, conceptual models as presented here are indispensable tools to assess the dynamic behavior of chemicals in the aquatic environment. They not only serve to predict concentration levels in lakes resulting from a given loading, but the model calculations—if compared to field data—can also be used to identify and even quantify yet unknown sources or reaction pathways. Keeping this in mind, it is fruitful practice to develop and apply mathematical models as early as possible and to use the results for optimum design of field programs and planning of process-oriented experiments in the laboratory.

REFERENCES

1. W. Stumm, R. P. Schwarzenbach, and L. Sigg, "From Environmental Analytical Chemistry to Ecotoxicology—A Plea for More Concepts and Less Monitoring and Testing," *Angew. Chem.* **95**, 345 (in German); *Int. Ed. Eng.* **22**, 380 (1983).

2. D. M. Imboden and A. Lerman, "Chemical Models of Lakes." In A. Lerman (Ed.), *Lakes: Chemistry, Geology, Physics*, Springer, New York, 1978, p. 363.

3. A. Okubo, "Oceanic Diffusion Diagrams," *Deep Sea Res.* **18**, 789 (1971).

4. H. B. Fischer, E. J. List, R. C. Y. Koh, J. Unberger, and N. H. Brooks, *Mixing in Inland and Coastal Waters*, Academic Press, New York, 1979, p. 483.

5. R. P. Schwarzenbach and D. M. Imboden, "Modelling Concepts for Hydrophobic Organic Pollutants in Lakes." *Ecol. Modelling* **22**, 171 (1984).

6. D. M. Imboden and R. Gächter, "A Dynamic Lake Model for Trophic State Prediction," *Ecol. Modelling* **4**, 77 (1978).

7. R. P. Schwarzenbach and J. Westall, "Transport of Nonpolar Organic Compounds from Surface Water to Groundwater. Laboratory Sorption Studies," *Environ. Sci. Technol.* **15**, 1360 (1981).

8. R. A. Berner, *Early Diagenesis. A Theoretical Approach*, Princeton Univ. Press, Princeton, N.J., 1980.

9. D. M. Imboden, "Interstitial Transport of Solutes in Non-Steady State Accumulating and Compacting Sediments," *Earth Planet. Sci. Lett.* **27**, 221 (1975).

10. W. R. Mabey, J. H. Smith, R. T. Podoll, H. L. Johnson, T. Mill, T.-W. Chou, J. Gates, I. W. Partride, and D. Vandenberg, *Aquatic Fate Process Data for Organic Priority Pollutants*, EPA, Office of Water Regulations and Standards (WH-553), Washington, D.C., 1981.

11. R. G. Zepp, N. L. Wolfe, J. A. Gordon, and G. L. Baugham, "Dynamics of 2,4-D Esters in Surface Waters," *Environ. Sci. Technol.* **9**, 1144 (1975).

12. R. G. Zepp and G. L. Baugham, "Prediction of Photochemical Transformation of Pollutants in the Aquatic Environment." In O. Hutzinger, I. H. Van Lelyveld, and B. C. J. Zoeteman (Eds.), *Aquatic Pollutants: Transformation and Biological Effects*, Pergamon Press, New York, 1978, p. 69.

13. W. R. Mabey and T. Mill, "Critical Review of Hydrolysis of Organic Compounds in Water under Environmental Conditions," *J. Phys. Chem. Ref. Data* **7**, 383 (1978).

14. S. W. Karickhoff, "Semiempirical Estimation of Sorption of Hydrophobic Pollutants on Natural Sediments and Soils," *Chemosphere* **10**, 833 (1981).

15. J. H. Smith, D. C. Bomberger, Jr., and D. L. Haynes, "Prediction of the Volatilization Rates of High-Volatility Chemicals from Natural Water Bodies," *Environ. Sci. Technol.* **14**, 1332 (1980).

16. J. H. Smith, D. C. Bomberger, Jr., and D. L. Haynes, "Volatilization Rates of Intermediate and Low Volatility Chemicals from Water," *Chemosphere* **10**, 281 (1981).

17. R. B. Banks, "Some Features of Wind Action on Shallow Lakes," *Proc. Am. Soc. Civil Engrs.* **101**, 813 (1975).

18. W. Kuhn, "Physikalisch-Meteorologische Ueberlegungen zur Nutzung von Gewässern für Kühlzwecke," *Arch. Meteor. Geophys. Biokl. Ser. A* **21**, 95 (1972).

19. R. G. Zepp and D. M. Cline, "Rates of Direct Photolysis in the Aquatic Environment," *Environ. Sci. Technol.* **11**, 1359 (1977).

20. R. G. Zepp, "Quantum Yields for Reaction of Pollutants in Dilute Aqueous Solution," *Environ. Sci. Technol.* **12**, 327 (1978).

21. R. P. Schwarzenbach, E. Molnar-Kubica, W. Giger, and S. G. Wakeham, "Distribution, Residence Time, and Fluxes of Tetrachloroethylene and 1,4-Dichlorobenzene in Lake Zurich, Switzerland," *Environ. Sci. Technol.* **15,** 1367 (1979).

22. R. P. Schwarzenbach, W. Giger, E. Hoehn, and J. K. Schneider, "Behavior of Organic Compounds during Infiltration of River Water to Groundwater. Field Studies." *Environ. Sci. Technol.* **17,** 472 (1983).

23. M. D. Müller, "Hexachlorbenzol in der Schweiz—Ausmass und Hintergründe der Umweltkontamination," *Chimia* **36,** 437 (1982).

2

CONCEPTUAL MODELS FOR TRANSPORT AT A REDOX BOUNDARY

William Davison

Freshwater Biological Association, The Ferry House, Far Sawrey, Ambleside, Cumbria LA22 OLP, U.K.

Abstract

Fundamental features which are common to the transport of all elements in the vicinity of a redox boundary are discussed. For simplicity, detailed examples are restricted to iron and manganese. Conceptual understanding is based on oxidation or reduction reactions generating a point source, from which newly formed material is transported by random processes. The shapes of concentration–depth profiles of both soluble and particulate components in marine and freshwater sediments and water columns are reviewed. The diverse shapes encountered in nature are due to the particular modifying factors which affect each situation; these factors are discussed in detail.

Generally, transport, either entirely within a water column or within a sediment, may be simply treated because the rate of vertical transport can be regarded as constant. The situation at the sediment–water interface is complicated by the discontinuity in the rate of transport. When the redox boundary coincides with this interface or resides in the sediment, soluble components are transported from the sediment to the water column. However, when the redox boundary occurs in the water column under steady-state conditions, soluble species diffuse from the overlying water to the sediment. The soluble reduced forms of the metals play the major role in transport processes within sediment where particles are relatively immobile. Transport away from particulate peaks within the sediment is therefore usually controlled by chemical processes. In water columns, where advective mixing acts equally on particles and solutions, both components are actively transported and particles are additionally subject to gravitational sinking. The problems associated with using instantaneous measurements of concentration gradients to estimate fluxes are highlighted, and the use of complementary methods is recommended.

1. INTRODUCTION

The transport of elements in the natural environment is governed by the basic thermodynamic principle that substances are transferred from situations of high energy to those of lower energy. Whenever there is a gradient of energy some transport process will operate. The energy may be thermal or kinetic (e.g., derived from sun or wind) with resultant convective movement, or it may relate to the concentration of chemical components. As there can be many chemical components of diverse concentration, it is useful to consider "master variables" (1) such as pH or pe; $pe = - \log \{e\}$, where $\{e\}$ is the electron activity; pe is an assessment of the reducing intensity of an environment. High values of pe correspond to strongly oxidizing situations, usually characterized by the presence of oxygen, whereas low values of pe are diagnostic of reducing conditions which are characterized by the absence of oxygen.

Redox gradients are common in nature: they exist at oxic/anoxic boundaries in marine and freshwater water columns; at the oxyclines (region of large gradient of oxygen versus depth) which are present in sediments, at sediment–water interfaces, or in soils; and in the microenvironments which surround many living organisms. The transport of redox-sensitive elements can be governed by these redox gradients, and this paper considers the processes involved by using two elements, iron and manganese, as illustrative examples.

The transport of iron and manganese in water bodies and sediments has received a great deal of attention because of the central role that these abundant metals can play in the geochemical cycling of other elements. For both elements the higher oxidation state usually occurs as insoluble particles, whereas the lower oxidation state is soluble and relatively free from complexation. Therefore ferrous and manganous ions are the mobile components and the shapes of their concentration profiles both dictate, and are indicative of, the transport of the element.

Several workers have considered the shape of iron and especially manganese profiles (2,3), and mathematical models which simulate the behavior of these elements have been developed (4,5). With some notable exceptions (6,7), such work has concentrated on marine systems. In particular the diversity of profile shapes which are found in the water columns of lakes has received little attention (8). This paper considers the wide variety of differently shaped concentration profiles of reduced and oxidized forms, which occur in the water columns and sediments of marine, estuarine, and fresh-water environments. The shapes are discussed in terms of a single unified model and particular emphasis is placed on understanding the processes which occur in lakes.

2.   INSTANTANEOUS CONCENTRATION PROFILES OF SOLUBLE COMPONENTS

2.1.   Generalized Transport Mechanism

Figure 2.1 illustrates a simple conceptual model for transport of iron and manganese at a redox boundary. The redox gradient is assumed to be dictated by the oxic/anoxic boundary and to be well defined. There is a supply of oxidized metal from the oxic region and at the moment this particulate phase encounters the anoxic region it is reduced to soluble Fe^{2+} or Mn^{2+}. Although other species, for example, $FeOH^+$ and $FeCl^+$, may exist, Fe^{2+} and Mn^{2+} are usually dominant with minimal complexation (9,10). To simplify the model Fe^{2+} and Mn^{2+} will be regarded as the only species. The divalent cation is transported from its site of production by processes, which for the purpose of this discussion, will be designated diffusive. This terminology does not imply molecular diffusion and can embrace enhancement effects (e.g., convection) or retarding effects (e.g., tortuosity). Transport within the oxic and anoxic regions is assumed to be uniform with constant diffusion coefficients of D_O and D_A, respectively.

The model produces divalent metal at a point source and, for a simple one-dimensional example, this soluble component diffuses away from this point in both upward and downward directions. This results in a profile with a sharp peak, on either side of which the concentration declines exponentially. However, in practice, reduction occurs over a finite distance and so the top of the peak is rounded, resulting in the Gaussian shape which is illustrated as the typical profile (Fig. 2.1). The quantitative expression of Figure 2.1 is straightforward as it is simply based on Fick's laws for a sta-

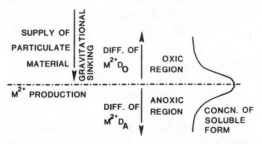

Figure 2.1.   Model for the transport of iron and manganese at a redox boundary, with a schematic representation of the shape of the resultant concentration profile of the dissolved species.

tionary state:

$$D \frac{\partial^2 c}{\partial z^2} + v \frac{\partial c}{\partial z} + J = 0$$

Here D is the effective diffusion coefficient, assumed to be constant with respect to depth z; c is the concentration of the chemical component which is either supplied or removed at a point source with a rate equal to J; and v is a physical transport term which may represent the sedimentation rate in sediments (5,6) or an advective contribution in water columns (3). It is not the purpose of this paper to discuss the detailed application of this equation to particular circumstances as this has already been well documented (2–6). Instead we shall concentrate on conceptual understanding.

2.2. Factors Which Modify Profile Shapes

The model makes only very simple assumptions about the iron and manganese chemistries: the higher oxidation state is insoluble; the lower oxidation state is soluble; and the two oxidation states are readily interconverted at the redox boundary. Six factors which modify profile shapes are listed. The first four factors which modify the profile shape from that shown in Figure 2.1 are related to the physical environment.

1.   The sampling may be inadequate to fully embrace all the environmental conditions implicit in Figure 2.1, and so the resulting data will only represent some portion of the total profile. For example, lake water columns often only show the upward diffusion of cations which are produced in the sediment. Therefore a profile measured in the water column may only represent the top section of the generalized profile shape.

2.   The model assumes that transport is restricted to one dimension. In sediments and deep parts of very large water bodies the horizontal component can be assumed to be the same for all sites, and it is therefore effectively averaged out and need not be considered. Most lake basins are not so large and the horizontal rate of transport is greater than the vertical rate by a factor of 10^2–10^5 (11,12). Therefore interactions between sediment and water can modify or even dominate profile shapes (13,14).

3.   The rate of vertical transport may vary with respect to the position of the concentration profiles. In Figure 2.1 different rates of transport, in oxic and anoxic regions, were permitted by the diffusion coefficients D_O and D_A; but when drawing a profile symmetrically about the redox cline it was assumed that the diffusion coefficients were equal. If D_O and D_A differ greatly, as at a sediment–water interface, the resultant profile shape may be substantially modified.

4.   The redox boundary or redox-cline (region of large gradient of redox intensity versus depth) may not be sharp. The simple model of Figure 2.1 assumed that the redox boundary was distinct and discrete and therefore masked the potential differences in the chemistry of iron and manganese. If the redox environment changes gradually, differences in the shapes of the iron and manganese profiles, due to their different redox potentials and rates of reaction, become apparent.

5.   Steady-state conditions may not apply. This is particularly true for systems controlled by seasonal events, where the position of the redox boundary can move quickly so that the associated chemistry lags behind.

6.   Authigenic (i.e., being produced within the lake) mineral formation depends on the chemistry of the lake. If the solubility product of the divalent cation is exceeded with respect to some anion, such as sulfide or carbonate, the resultant mineral may form with consequent removal of the cation from solution.

To maintain the steady-state profile of Figure 2.1 it is necessary to have some mechanism for removal. Mineral-formation processes often operate in such a capacity below the redox cline. Removal above the redox boundary is usually accomplished by oxidation of the divalent species.

The following sections discuss many of the profile shapes which have been observed for dissolved components in different environments, in relation to the simple model of Figure 2.1, and the modifying factors listed above.

2.3.   Water Columns

Vertical transport in freshwater and marine basins can usually be assumed to take place by eddy diffusion. Figure 2.2 shows some representative examples of profile shapes which have been observed in water columns. These examples have been taken from published works (14–16).

The profiles of dissolved iron and manganese for the Black Sea, a permanently anoxic marine basin, have the basic shape of Figure 2.1 and can be explained by the simple generalized model. Particulate iron and manganese, derived from input to the surface waters and from oxidation of the dissolved fraction at the oxycline, is supplied as a sedimenting component. These particles are reduced when they encounter the anoxic waters and so provide a source of soluble ions which are transported at comparable rates both upwards to be oxidized and downwards towards deeper waters. This two-way transport causes the characteristic maximum, in the concentration–depth profiles, of the soluble species. For the Black Sea, manganese is thought to be still increasing in concentration in the bottom waters and so no removal mechanism is present (16). Dissolved-iron concentrations have

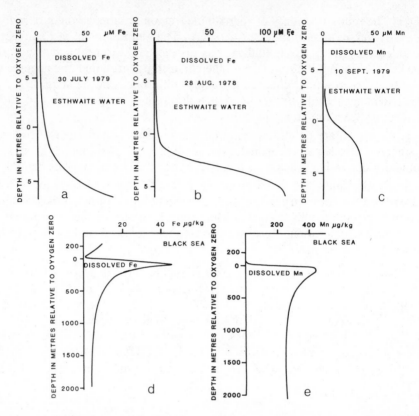

Figure 2.2. Profile shapes of dissolved iron and manganese species in water columns. [Parts (*a*) and (*c*) used with permission of the American Society Limnology and Oceanography, Inc. Adapted from W. Davison et al., *Limnol. Oceanogr.*, **27**, 987–1003 (1982). Part (*b*) used with permission of Birkhäuser Verlag AG. Adapted from W. Davison et al., *Schweiz. Z. Hydrol.*, **42**, 196–224 (1981). Parts (*d*) and (*e*) were adapted from D. W. Spencer and P. G. Brewer, *J. Geophys. Res.*, **76**, 5877–5892 (1971).]

been shown to be influenced by ferrous sulfide precipitation and, since sulfide is in excess, it is this process which controls the profile shape below the iron maximum (17).

Similar peaked manganese profiles were reported for freshwaters in the early part of the century [see Hutchinson (18) for references], and there have also been some recent examples (19,20). Although Einsele (21) thought that such peaks were probably generated by oxidation/reduction processes in the vicinity of the redox cline, Hutchinson (18) commented on the lack of any precise explanation. Indeed quantitative interpretation was pioneered by Spencer and Brewer's (16) study of the Black Sea.

Profiles which increase from the sediment and are either concave or convex upwards [Figs. 2.2(*a*) and 2.2(*b*)] have been observed for both iron and manganese in anoxic freshwater basins, both permanently (22,23) and seasonally (14,15,19,20). They have also been observed in a seasonally anoxic marine basin (24).

These profiles represent segments of the basic shape of Figure 2.1. For the concave profile [Fig. 2(a)] the peak of the soluble component occurs in the interstitial waters of the sediment and so, if the measurements are restricted to the water column, only that region of the profile in which upward transport is taking place is being considered. The redox boundaries between oxic and anoxic water are shown in Figure 2.2. Whereas anoxic water maintains both Fe^{2+} and Mn^{2+} in solution, the active reducing site and the major source of metals is within the sediment. A convex shaped profile [Fig. 2(b)] indicates that some reduction of particulate material takes place in the water column. Although the peak of soluble component straddles the water column and sediment, only that part of the profile which shows upward transport is evident in the water column. When all of the particulate matter is reduced in the water column, the concentration profile in the anoxic water may be vertical [Fig. 2(c)]. The vertical line of invariant concentration actually represents the summit of the peak of dissolved component. In such a small depth of water (a few meters) the peaked shape is not resolved as it is for the Black Sea where the peak extends over many tens of meters.

In a seasonally anoxic basin the entire sequence of profiles shown in Figure 2.2 can be observed as a natural progression in a single season as reducing conditions intensify (19). Because these profiles are still developing, and so do not represent steady-state conditions, there is no need to postulate a removal mechanism for the lower part of the profile. However, in some cases, saturation with respect to ferrous sulfide or manganese carbonate has been observed (17,18). Such authigenic mineral formation will only modify the iron or manganese profile shape if the anion is at a comparable, or excess, concentration to the cation (17).

Because iron and manganese are reduced or oxidized at different rates (1) at any one time the shape of the soluble-iron profile differs from that of manganese. When seasonal anoxia develops, manganese, which is usually reduced before iron, is preferentially released from the sediment (15,20,25,26). Manganese also tends to develop peaked profiles more rapidly. Thus while the concentration of soluble iron may still decrease from the sediment the manganese profile may be vertical or even have a midwater maximum (17,19).

Due to poor definition of the redox boundary there are chemical differences in space as well as time. The iron and manganese chemistries are spatially resolved because sampling intervals are small in comparison to the distances embraced by the relatively rapid rate of vertical transport. Thus dissolved-manganese concentrations are usually maintained up to and through the oxycline, whereas dissolved iron is removed at lower depths where there is no measurable oxygen (15,19,27).

In the stratified waters of most lakes the component of transport in the horizontal plane far exceeds (approximately 100 times) that in the vertical plane (28). It is still possible to interpret vertical profiles; but this horizontal interaction with the sediment must be taken into account, usually by considering the lake as a number of horizontal layers (13,14). The horizontal interaction results in components being supplied to intermediate depths and so the net effect is similar to that of dissolution within the water column: concave-shaped profiles can be transformed into convex ones. In practice the shape of the curve is determined by several factors including: the topography of the lake bottom; the variation with depth in the flux from the sediment; and the depth dependency of the vertical eddy diffusion coefficient. Such complexity makes quantitative modeling very difficult unless some simplifying assumptions can be made. Estuaries are also dominated by the horizontal transport of water because density gradients can be due to the effects of both temperature and salinity. In these dynamic systems steady-state approximations are invalid; but detailed descriptions of the processes which control individual systems are available (29–32).

2.4. Sediments

Berner (33) has discussed the shapes of dissolved-iron and dissolved-manganese profiles in the interstitial waters of sediments. In such environments random transport processes may be approximated by molecular diffusion. Particulate material is supplied by sediment accumulation. This (advective) transport is a pseudotransport originating from the circumstance that the coordinate system is usually fixed at the sediment–water interface. Thus, the coordinates are moving relative to a fixed sediment layer.

In organic-rich coastal marine sediments, it is often impossible to measure reasonable Fe^{2+} and Mn^{2+} profiles because of the immediate formation of minerals which include iron sulfides, such as pyrite, and manganese carbonates, such as rhodochrosite. At these sites the redox gradient is so sharp and so intense that the most reduced species, such as sulfide and methane, are formed simultaneously with ferrous and manganous ions. Transport by diffusion is therefore rendered ineffective by *in situ* mineral-formation processes, and the generalized model of Figure 2.1 (which relies on simple diffusion processes) becomes inoperative. The profile shapes which are obtained are usually best interpreted by considering the sediment–water interface (see next section). Freshwater sediments usually contain much less sulfide and carbonate and so profiles may be obtained even in organic-rich environments. The manganese profile observed by Robbins and Callender (6) in Lake Michigan (Fig. 2.3) is a good example. The profile shape is similar to that shown for the generalized model in Figure 2.1. At this site, particulate manganese is supplied by sediment accumulation, and rhodochrosite formation was shown to account for the manganese consumption with depth (33).

Marine analogs also exist. Murray et al. (34) showed similar profile shapes for iron and manganese in a marine inlet where ferrous sulfide and manganese carbonate precipitation were thought to be controlling at depth. Crerar et al. (35) have reviewed other examples and Emerson (36), observed similar manganese profiles in a lacustrine sediment. However, Emerson (36), Emerson and Widmer (37), and Cook (38) also found iron profiles which were convex upwards and increasing with depth. These shapes do not conform to the simple theory of supply from a point source and are thought to result from the formation of minerals including ferrous sulfides, siderite, and vivianite.

Pelagic deep-sea sediments are very low in organic matter and represent an interesting case. Because of the small supply of organic matter and its relative lack of activity the redox conditions change only gradually with depth in the sediment. Froelich et al. (39) have used the term suboxic diagenesis to describe this process by which the redox boundary is extended. Manganese, being more readily reduced than iron, appears in the interstitial water column at a higher zone in the sediment (Fig. 2.4). Dissolved iron and manganese diffuse towards the sediment surface and are consumed by oxidation. There is a lack of information concerning processes at greater depths in these sediments.

2.5. Sediment–Water Interfaces

Profile shapes appear to be more complicated at sediment–water interfaces. This is because of the marked difference in the rates of transport which operate in the interstitial and overlying waters. Transport within interstitial

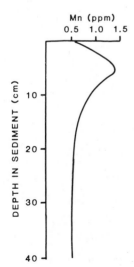

Figure 2.3. Dissolved manganese in the pore waters of medium-reducing Lake Michigan sediments. [Used with permission of the *American Journal of Science*. Adapted from J. A. Robbins and E. Callender, *Am. J. Sci.*, **275**, 512–533 (1975).]

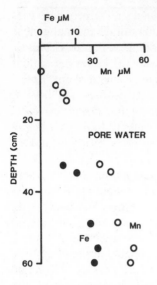

Figure 2.4. Dissolved iron and manganese in core SGCI from the eastern equatorial Atlantic Ocean. [Adapted from P. N. Froelich et al., *Geochim. Cosmochim. Acta*, **43**, 1075–1090 (1979).]

waters can be approximated by molecular diffusion, whereas eddy diffusion, which is usually 1000 times faster, is dominant in water columns. Further complications arise if the redox boundary is also situated at this sediment–water interface.

This latter condition is closely approximated when well-oxygenated water overlies reducing sediments. The profiles of Figure 2.5, which are for manganese at two different sites in Lake Ontario, have shapes which are typical for iron and manganese at many sediment–water interfaces (29,40–42). Particulate material is supplied by sedimentation from the water column and is

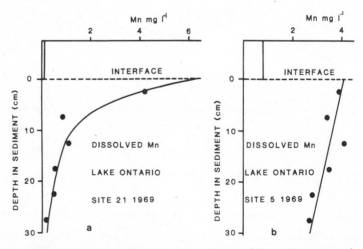

Figure 2.5. Profiles of dissolved manganese in the vicinity of the sediment–water interface. [Plotted from the data of Weiler (40).]

rapidly reduced when it reaches the anoxic sediment. The soluble species which accumulate at the interface are transported rapidly (high D_O) to the overlying waters, but comparatively slowly (low D_A) into the sediment for permanent incorporation. Therefore, when the redox boundary coincides with the sediment–water interface, most of the solubilized metal is returned to the water column. The metal may not show up as a high concentration in the water column because dispersion by the rapid mixing of the water may dominate. The two profile shapes of Figure 2.5 may result. The profile which shows a surface maxima is easily related to that of Figure 2.1 by considering the apparent discontinuity in the vertical scale imposed by a sharp change in the rate of transport. Profiles which do not show a peak can be similarly explained. The surface maximum may be so sharp that the sampling interval may be insufficiently detailed to show it.

Particulate manganese which is supplied to the sediment is usually more readily reduced than iron. Consequently, most of the manganese which is supplied to a highly reducing sediment is solubilized and rereleased to the water column, whereas only a small fraction of the iron is solubilized (15,43). Very little manganese is permanently retained in such a sediment. In contrast most of the iron which reaches the sediment surface becomes incorporated.

The development of a peaked profile in the water column of an anoxic aquatic basin has implications for transport in sediments. To conserve matter between the interstitial waters of the sediment and the overlying waters the concentration of ions in these two distinct environments must be such that fluxes above and below the interface are compatible. Because the concentration in the water column decreases towards the sediment, metal ions must diffuse from the water column into the sediment, and the concentrations in the pore water must therefore also decrease with depth in the sediment so that the downward transport continues. Figure 2.6(a) shows an example of such behavior for an anoxic fjord (44). In this case the sediments clearly act as a sink for iron and manganese which diffuse from the overlying waters.

Diffusion into the sediments has also been observed in lakes. Figure 2.6(b) shows a dissolved-manganese profile for a seasonally anoxic lake. The manganese reaches its maximum concentration in the water which immediately overlies the sediment. In Figure 2.6(b) the scale within the sediment is expanded 1000 times over the scale in the water column. Vertical transport in the water column is governed by eddy diffusion with a diffusion coefficient of approximately 10^{-2} cm^{2}s^{-1} (14); transport in the sediment is controlled by molecular diffusion with a coefficient of approximately 10^{-5} cm^2 s^{-1}. Therefore the scales used in Figure 2.6(b) compensate for a discontinuity in the rate of vertical transport. It is noteworthy that the resultant shape bears a remarkable similarity to that obtained for manganese in the relatively constant transport regime of a permanently anoxic basin [Fig. 2.2(e)]

The proximity of the peak [Fig. 2.6(b)] to the sediment–water interface makes it difficult to decide, solely on the basis of measured concentration profiles, whether manganese is being actively solubilized in the overlying

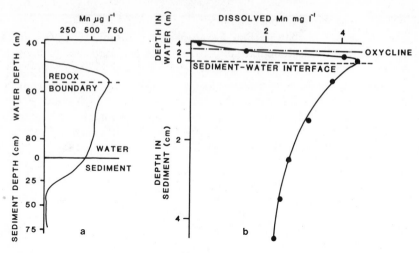

Figure 2.6. Concentration profiles of dissolved manganese extending through the sediment–water interface: (*a*) for a Norwegian fjord, [With permission from J. Hamilton-Taylor and N. B. Price (44). Copyright: Academic Press Inc. (London) Ltd.]; and (*b*) for the seasonally anoxic lake, Esthwaite Water. Water-column data are for October 8, 1971 and pore-water data (kindly supplied by Hamilton-Taylor) are for October 11, 1979. Different depth scales have been used to simulate a constant transport regime.

waters or at the sediment–water interface. Sediment traps (15) used in conjunction with measurements of concentration profiles at 5-cm intervals within the water column (17) were able to show that most of the manganese dissolved in the bottom waters of the lake before it reached the sediment.

3. INSTANTANEOUS CONCENTRATION PROFILES OF PARTICULATE COMPONENTS

3.1. Generalized Transport Mechanism

So far only the soluble species have been considered. However, extension of the simple model of Figure 2.1. permits consideration of concentration profiles of the insoluble form of the respective elements (i.e., iron and manganese) (Fig. 2.7).

We have already seen that soluble metal ions are transported, from their point of production at the redox boundary, in an upward direction by random transport processes. As they progress they encounter oxygen and are consequently oxidized to the insoluble particulate form. At this site of oxidation there is effectively a point source of particulate material. For the simple model we assume that random transport processes operate to distribute the material in both upward and downward directions, and there is an additional gravitational component which only operates in a downward direction. When the particles pass below the redox boundary they are removed by being

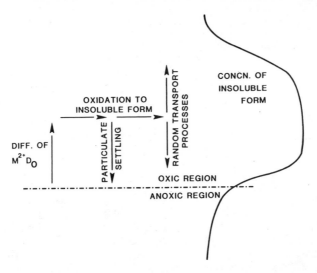

Figure 2.7. Model for the transport of particulate iron and manganese with schematic representation of the resultant concentration profile.

reduced to the soluble phase. The net effect of these processes is to produce a peak in the concentration profile of the insoluble form, analogous to the peak in the concentration of the soluble form (Fig. 2.1). In practice the simplistic ideal of Figure 2.7, in which the particulate peak is immediately above the redox boundary, is rarely encountered. The position of redox boundaries is not usually so well defined either in space or time. When it is, manganese, which is more readily reduced than iron, does tend to develop a particulate peak close to and above the oxycline. The maximum of particulate iron, however, may occur at some considerable distance into the anoxic zone.

The general validity of this model and the relative importance of random transport processes and rates of oxidation will be discussed while considering case examples.

3.2. Water Columns

Peaks of particulate iron or manganese commonly occur in water columns. Figure 2.8(a) shows a particulate-iron profile in a seasonally anoxic lake (14). Similar profiles have been observed for iron and manganese in both marine and freshwater situations (16,27,45–47). The development of such a peak depends upon an appropriate balance between the rate of supply of the oxy/hydroxide and its rate of removal (Fig. 2.7). Whereas pronounced peaks of particulate iron develop systematically during summer stratification in Esthwaite Water (Fig. 2.8(a) peaks of particular manganese are less well defined and of lower concentration (15,48). This is due to the different rates

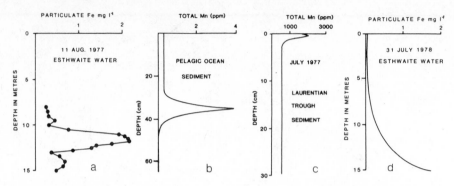

Figure 2.8. Profile shapes of particulate iron and manganese in water columns, (*a*) and (*d*), and sediments, (*b*) and (*c*). [Parts (*a*) and *d*) used with permission of Birkhäuser Verlag AG. Adapted from W. Davison et al., *Schweiz. Z. Hydrol.*, **42**, 196–224 (1981). Part (*b*) used with permission of the *American Journal of Science*. Adapted from D. J. Burdige and J. M. Gieskes, *Am. J. Sci.*, **283**, 29–47 (1983). Part (*c*) was adapted from G. Sundby et al., *Geochim. Cosmochim. Acta.*, **45**, 293–307 (1981).]

of oxidation of the reduced forms. Mn(II) is oxidized more slowly than Fe(II) and so the particulate material is produced at a higher position in the water column. It therefore has further to sink before it can be reduced. Moreover, wind mixing is more prevalent in shallower water and the morphometry is such that the manganese is prone to dilution from the larger volume of water in the upper layers of the lake.

The Black Sea exhibits the reverse phenomena. There is a clear peak of particulate manganese but not of iron (16). Although the particulate iron peak is masked by the high background of iron freshly supplied to the system, it may well be that the peak is too sharply defined at the intense redox gradient to be revealed by the sampling.

Other factors can participate in the formation of particulate peaks in water columns. Removal of particles may be accelerated by coagulation processes (49). In the dynamic regime of small lakes, steady state may not apply. The rate of production of particles may then exceed the rate of removal which will produce a peak as observed by the instantaneous measurement of a concentration profile. Hydrodynamic resuspension of material from sediment at a particular depth in the lake can also provide a point source of particles.

3.3. Sediments and the Sediment-Water Interface

When the redox boundary occurs at an appreciable depth within the sediment a peak of particulate material may be observed in the sediment. In this case, however, the random transport processes which operate on the particulate material are so small that they have an almost negligible effect. The shape of the peak is controlled by burial of fresh material, upward diffusion and subsequently oxidation of the reduced form, and removal of particles by reduction below the redox boundary. Burdige and Gieskes (5) recently provided a mathematical analysis of this model. They were able to simulate the profiles which had been observed by Froelich et al. (39) in pelagic marine sediments [Fig. 2.8(b)]. Note that the shape of this peak is not Gaussian, but sharply pointed with exponential decay on either side of the maximum. The absence of random transport processes allows reduction to occur in such a narrow band that the averaging effect which gives most profiles a bell shape is removed.

In freshwaters the redox boundary more commonly occurs near to or at the sediment–water interface. This can result in large concentrations of particulate iron or manganese accumulating in the surface sediments which may develop a red-brown color (50,51). Many workers have provided data which show elevated concentrations of iron and/or manganese near to the sediment surface [see Callender and Bowser (52) for references]. Demonstration of a peak within the sediment is more difficult, but by using millimeter sampling intervals Sundby et al. (29) have done this for estuarine sediments [Fig. 2.8(c)]. The enriched layer at the sediment surface may consist of a loose association of freshly precipitated material in which case it has an exceptionally high water content. Alternatively, it can form a solid crust. Such solid material, enriched in iron and manganese, has been regarded as a particular type of manganese nodule (52). So the simple model of Figure 2.7 provides one rationale for the formation of manganese nodules.

Peaks of particulate material are not constrained to the sediment. Depending on the proximity of the redox boundary to the physical interfaces, the dissolved species may enter the water column before oxidation takes place. The resultant particulate material is then subject to random transport processes, and within the water column the concentration of particulate material increases towards its source at the sediment–water interface [Fig. 2.8(d)]. The particulate material often sinks through the water to reaccumulate at the sediment–water interface and so the surface sediment can still be enriched. Manganese, which is oxidized more slowly than iron, may be released from the sediment, while iron is retained (53). If the manganese is removed from the system, for example, by being flushed out of the lake basin, the sediment becomes enriched with respect to iron. The concentrations of iron and manganese in surface waters are usually comparable (35) and work on lakes and estuaries has shown that most of the inflowing iron and manganese is transported to the sediment. Therefore without any redox-driven partitioning of the elements the sedimentary ratio would approach unity. In reality the average ratio of iron to manganese in sedimentary rocks is about 50:1, which shows that the preferential release of manganese must be ubiquitous (54).

4. GENERAL DISCUSSION

The foregoing discussion has shown that dissolved and particulate iron and manganese exhibit a wide variety of profile shapes in the interstitial waters and water columns of marine, freshwater, and estuarine environments. Yet in most cases the mechanisms of transport and resulting profile shapes can be explained by the simple composite model of Figure 2.9 which shows that a peak in the particulate component will often occur above a peak in the soluble form. In this schematic diagram concentrations and distances should not be taken literally as they merely serve as a guide to illustrate the processes. Total concentrations will not reflect the simple sum of the components because in sediments especially there can be a large reservoir of nonreactive metal. However, twin peaks have been observed in water columns (16,44), and in sediments (6). One of the most common environmental situations, when oxic water overlies anoxic sediment, should produce a particulate peak in the water column and a soluble peak at the sediment–water interface. The appropriate data have rarely been collected, but unpublished measurements of manganese in Esthwaite Water have shown that the two peaks do exist [Tipping and Woof (53); Hamilton-Taylor and Morris (44)].

Two of the essential ingredients of the model of Figure 2.9 are a well-

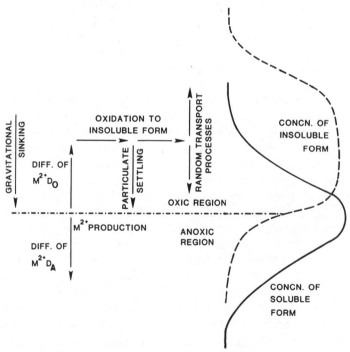

Figure 2.9. Composite model for the redox-driven cycling of iron and manganese showing schematic concentration profiles of the particulate and soluble forms.

defined redox boundary and a regime of random transport controlled by an effective diffusion process. The transport of soluble components entirely within either water columns or sediments usually fulfills these conditions. Although the rates of transport in water are much faster than in sediments, the faster rates are balanced by the greater vertical dimension of the system over which the concentration gradient is developed. Generally, it is useful to remember than when considering fluxes, distance and coefficients of diffusion are related as an inverse function. In sediments distances are considered in millimeters and diffusion coefficients of 10^{-6}–10^{-5} cm^2s^{-1} are typical (33), whereas in water columns distances are considered in meters and diffusion coefficients of 10^{-3}–10^{-2} cm^2s^{-1} are typical (12,55,56). Caution is also necessary to avoid arbitrarily classifying sediments as solids and so distinguishing them from liquid water columns. Recent organic-rich sediments have a high water content (usually $> 90\%$) and could also be regarded as a liquid, containing a high concentration of particles. In some situations it is possible for anoxic basins to serve as surrogate models of highly dispersed sediments, in which the much higher coefficient of vertical transport makes concentration gradients easier to measure. The dispersed nature of the particles also allows them to be more easily differentiated and identified, and so facilitates the study of surface reactions.

The soluble reduced forms of the metals play the major role in transport processes within sediments, where particles are relatively immobile because they are not subject to random transport processes. Transport away from particulate peaks in sediments is therefore more often controlled by chemical rather than physical processes. However, in water columns where advective mixing acts equally on particles and solutions, both components are actively transported by physical movements of water, and particles are additionally subject to transport by gravity.

The transport of both soluble and particulate components at the sediment–water interface is complicated by physical constraints. There are two different rates of transport operating on either side of a physical interface which may or may not coincide with the redox boundary. Generally, however, the same simple reactions occur in the vicinity of all redox boundaries. In seasonally anoxic systems this boundary may move from the sediment to the interface, into the water column, and then back again. As it migrates it takes with it the appropriate redox chemistry which may superficially appear to change owing to the changing time scales dictated by the change in the rates of vertical transport. Because manganese has a slower rate of oxidation than iron, it is more readily transported through oxic environments and as a result it is relatively depleted in sedimentary rocks.

The generalized model of this paper has limitations. Transport cannot always be considered in a single dimension, and when horizontal components are large, as in estuaries, or do not average out, as in the lateral interactions of sediment and water in lake basins, more complicated treatments must be used. Chemical reactions with anions, which result in the formation of minerals, sometimes control the shapes of profiles. In general the profiles of iron are more affected than those of manganese, and more particularly in marine systems which have a greater concentration of salts. As a corollary to the above the profile shapes of other substances may be influenced or controlled by iron or manganese. When dissolved iron is in excess, it can dictate the shape of the sulfide profile (17), and through surface interactions particulate iron can determine the shapes of the vertical profiles of other elements (48,57).

Although the discussion has been restricted to iron and manganese most redox-sensitive elements are transported by a similar mechanism. Nitrogen and iodine species can develop peaked profiles in the vicinity of an oxycline (58). The inherent solubility of the simple chemical components, NO_3^-, NO_2^-, NH_3, IO_3^-, and I^-, does not preclude supply of particulate material from a point source. Decomposition of sedimenting organic matter fulfills this role.

It is appropriate to end this paper, which has focused on the movement of iron and manganese along concentration gradients, on a cautionary note. In practical studies of lake chemistry, concentration gradients are often used to estimate fluxes and so provide quantitative information about the movements of elements in lakes. However, it is unwise to depend solely on such measurement because there are some inherent problems. Concentration profiles reflect only the instantaneous measurement of pseudo-steady-state processes; but lakes are usually dynamic systems and so rapid cycling of elements, or fluxes resulting from transient events, may not be apparent from a concentration profile. In sediments, in particular, there are dangers associated with assuming that a concentration profile is diffusion controlled. Often submillimeter sampling intervals are required to measure fluxes from interstitial waters to overlying water columns.

Sediment traps have been used successfully to measure the flux of particulate material (16). These devices are left in the lake for a period of time and so they give an average flux. This integrated information complements the instantaneous measurement from concentration profiles. Fluxes from sediments can be estimated by measuring concentration profiles in the water which immediately overlies the interface because the higher rate of transport permits the use of reasonable sampling intervals (7). Direct estimation of the fluxes by measuring the time-dependent concentration change in an artificially (43) or naturally (14) enclosed volume of water overlying the sediment is also possible. The best studies of natural systems do not rely on any single technique, but use a variety of complementary methods for estimating fluxes.

5 CONCLUSIONS

One simple conceptual model, incorporating the fundamental features which are common to elemental transport processes at a redox boundary, has been described. It has been used as a unified approach to the understanding of the diverse shapes of concentration–depth profiles which occur in nature. The modifying factors which operate in different environments, such as marine and freshwater water columns, sediments, and the sediment–water interface have been discussed.

Case examples of profiles of iron and manganese have been used to illustrate the universal applicability of this conceptual approach. Transport in water columns is dominated by physical mixing processes. Random convective movements affect soluble and particulate components alike, but the latter are also influenced by gravity. In sediments random transport is brought about by molecular diffusion and so it only applies to the solution phases. Particulate material is effectively transported by sediment accumulation which limits movement to one direction—downwards. The sediment–water interface is characterized by markedly different rates of transport determined by the different physical environments. This interface and the position of the redox boundary controls the net flux of elements. When the redox boundary is in the water column, soluble species have the opportunity of diffusing from the overlying water into the sediment. More commonly, the redox boundary occurs in the sediment and in this case transport is from the sediment to the water column. The closer the redox boundary is to the sediment–water interface, the more rapid is the elemental recycling, as the faster transport regime of the overlying waters becomes dominant.

This work has demonstrated the high level of understanding which can be achieved by a conceptual approach. Quantitative mathematical modeling of a particular case example should only be attempted when the situation is sufficiently well understood so that all the relevant parameters have been measured. This is difficult to achieve in lakes because of two fundamental problems. Changing meteorological conditions usually ensure that coefficients of vertical mixing are markedly dependent upon time (days). It is usually necessary to assume that sediments interact with the overlying water uniformly irrespective of their position and yet their composition will depend on that position. However, mathematical treatments do play a valuable role in formulating general models which, by assessing the relative importance of the various terms, can be used to aid our understanding and complement the approach used here (59).

ACKNOWLEDGMENTS

I wish to thank E. Tipping and D. Imboden for constructively commenting on the manuscript and J. Rhodes for typing the manuscript.

REFERENCES

1. W. Stumm and J. J. Morgan, *Aquatic Chemistry*, 2nd ed., Wiley, New York, 1981.

2. S. E. Calvert and N. B. Price, "Diffusion and Reaction Profiles of Dissolved Manganese in the Pore Waters of Marine Sediments," *Earth Planet Sci. Lett.* **16,** 245 (1972).

3. H. Elderfield, "Manganese Fluxes to the Oceans," *Marine Chem.* **4,** 103 (1976).

4. G. Michard, "Theoretical Model for Manganese Distribution in Calcareous Sediment Cores," *J. Geophys. Res.* **76,** 2179 (1971).

5. D. J. Burdige and J. M. Gieskes, "A Pore Water/Solid Phase Diagenetic Model for Manganese in Marine Sediments," *Am. J. Sci.* **283,** 29 (1983).

6. J. A. Robbins and E. Callender, "Diagenesis of Manganese in Lake Michigan Sediments," *Am. J. Sci.* **275,** 512 (1975).

7. W. Davison, "Supply of Iron and Manganese to an Anoxic Lake Basin." *Nature* **290,** 241 (1981).

8. W. Davison, "Transport of Iron and Manganese in Relation to the Shapes of Their Concentration–Depth Profiles," *Hydrobiologia* **92,** 463 (1982).

9. F. Morel and J. Morgan, "A Numerical Method for Computing Equilibria in Aqueous Chemical Systems," *Environ. Sci. Technol.* **6,** 58 (1972).

10. D. R. Turner, M. Whitfield, and A. G. Dickson, "The Equilibrium Speciation of Dissolved Components in Freshwater and Seawater at 25°C and 1 atm Pressure," *Geochim. Cosmochim. Acta.* **45,** 855 (1981).

11. P. D. Quay, "An Experimental Study of Turbulent Diffusion in Lakes," Ph.D. dissertation, Columbia University, 1977, 194 pp.

12. D. M. Imboden and S. Emerson, "Natural Radon and Phosphorus as Limnologic Tracers: Horizontal and Vertical Eddy Diffusions at Greifensee," *Limnol. Oceanogr.* **23,** 77 (1978).

13. R. H. Hesslein, "Whole Lake Model for the Distribution of Sediment-Derived Chemical Species," *Can. J. Fish. Aquat. Sci.* **37,** 552 (1980).

14. W. Davison, S. I. Heaney, J. F. Talling, and E. Rigg, "Seasonal Transformations and Movements of Iron in a Productive English Lake with Deep-Water Anoxia," *Schweiz. Z. Hydrol.* **42,** 196 (1981).

15. W. Davison, C. Woof, and E. Rigg, "The Dynamics of Iron and Manganese in a Seasonally Anoxic Lake; Direct Measurement of Fluxes Using Sediment Traps," *Limnol. Oceanogr.* **27,** 987 (1982).

16. D. W. Spencer and P. G. Brewer, "Vertical Advective Diffusion and Redox Potentials as Controls in the Distribution of Manganese and Other Trace Metals Dissolved in Waters of the Black Sea," *J. Geophys. Res.* **76,** 5877 (1971).

17. W. Davison, "A Critical Comparison of the Measured Solubilities of Ferrous Sulphide in Natural Waters," *Geochim. Cosmochim. Acta.* **44,** 803 (1980).

18. G. E. Hutchinson, *A Treatise on Limnology*, Vol. I., Wiley, New York, 1957 1015 pp.

19. H. Verdouw and E. M. J. Dekkers, "Iron and Manganese in Lake Vechten: Dynamics and Role in the Cycle of Reducing Power," *Arch. Hydrobiol.* **89,** 509 (1980).

20. J. Liden, "Equilibrium Approaches to Natural Water Systems. A Study of Anoxic- and Ground Waters Based on *In Situ* Data Acquisition," Ph.D. thesis, Amea, Sweden, 1983.

21. E. Einsele, "Versuch Einer Theorie der Dynamik der Manganund und Eisenschichtung in Eutrophen See," *Naturwissenschaften* **17,** 257 (1940).

22. P. Campbell and T. Jorgensen, "Maintenance of Iron Meromixis by Iron Re-deposition in a Rapidly Flushed Moni-Molimnion," *Can. J. Fish. Aquat. Sci.* **37,** 1303 (1980).

23. J. Kjensmo, Jr., "The Development and Some Main Features of 'iron-mer-omictic' Soft Water Lakes," *Arch. Hydrobiol. Suppl.* **32,** 137 (1967).

24. W. Davison and S. I. Heaney, "Determination of the Solubility of Ferrous Sulphide in a Seasonally Anoxic Marine Basin," *Limnol. Oceanogr.* **25,** 153 (1980).

25. C. H. Mortimer, "The Exchange of Dissolved Substances between Mud and Water in Lakes; III and IV," *J. Ecol.* **30,** 147 (1942).

26. H. H. Howard and S. W. Chisholm, Seasonal Variation of Manganese in a Eutrophic Lake," *Am. Midl. Nat.* **93,** 188 (1975).

27. P. Baccini and T. Joller, "Transport Processes of Copper and Zinc in a Highly Eutrophic and Meromictic Lake," *Schweiz. Z. Hydrol.* **43,** 176 (1981).

28. A. Jassby and T. Powell, "Vertical Patterns of Eddy Diffusion During Strat-ification in Castle Lake, California," *Limnol. Oceanogr.* **20,** 530 (1975).

29. G. Sundby, N. Silverberg, and R. Chesselet, "Pathways of Manganese in an Open Estuarine System," *Geochim. Cosmochim. Acta.* **45,** 293 (1981).

30. J. W. Murray and G. Gill, "The Geochemistry of Iron in Puget Sound," *Geo-chim. Cosmochim. Acta.* **42,** 9 (1978).

31. A. W. Morris, A. J. Bale, and R. J. M. Howland, "The Dynamics of Estuarine Manganese Cycling," *Est. Coast. Shelf. Sci.* **14,** 175 (1982).

32. J. C. Duinker, R. Wollast, and G. Billen, "Behavior of Manganese in the Rhine and Scheldt Estuaries. II. Geochemical Cycling," *Est. Coast. Mar. Sci.* **9,** 727 (1979).

33. R. A. Berner, "Early Diagenesis," Princeton Univ. Press, Princeton, N.J., 1980, 241 pp.

34. J. W. Murray, V. Grundmanis, and W. M. Smethie, "Interstitial Water Chem-istry in the Sediments of Saanich Inlet," *Geochim. Cosmochim. Acta.* **42,** 1011 (1978).

35. D. A. Crerar, R. K. Cormick, and H. L. Barnes, "Geochemistry of Manganese: An Overview." In I. M. Varentsov and G. Graselly (Eds.), *Geology and Geo-chemistry of Manganese. Vol. I. General Problems*, pp. 293–334, Academiai Kiado, Budapest, 1980, 463 pp.

36. S. Emerson, "Early Diagenesis in Anaerobic Lake Sediments: Chemical Equi-libria in Interstitial Waters," *Geochim. Cosmochim. Acta.* **40,** 925 (1976).

37. S. Emerson and G. Widmer, "Early Diagenesis in Anaerobic Lake Sediments. II. Thermodynamic and Kinetic Factors Controlling the Formation of Iron Phosphate," *Geochim. Cosmochim. Acta.* **42,** 1307 (1978).

38. R. B. Cook, "The Biogeochemistry of Sulfur in Two Small Lakes," Ph.D. thesis, Columbia University, 1981, 234 pp.

39. P. N. Froelich, G. P. Klinkhammer, M. L. Bender, N. A. Leudkte, G. R. Heath, D. Cullen, P. Dauphin, D. Hammond, B. Hartman, and V. Maynard, "Early Oxidation of Organic Matter in Pelagic Sediments of the Eastern Equatorial Atlantic: Suboxic Diagenesis," *Geochim. Cosmochim. Acta.,* **43,** 1075 (1979).

40. R. R. Weiler, "The Interstitial Water Composition in the Sediments of the Great Lakes. I. Western Lake Ontario," *Limnol. Oceanogr.* **18,** 918 (1973).

41. H. Elderfield, R. J. McCaffrey, N. Leudtke, M. Bender, and V. W. Truesdale, "Chemical Diagenesis in Narragansett Bay Sediments," *Am. J. Sci.* **281,** 1021 (1981).

42. G. P. Nembrini, F. Rapin, J. I. Garcia, and U. Förstner, "Speciation of Fe and Mn in a Sediment Core of the Baie de Villefrance (Mediterranean Sea, France), *Environ. Sci. Technol. Lett.* **3,** 545 (1982).

43. R. C. Aller, "Diagenetic Processes near the Sediment–Water Interface of Long Island Sound. II. Fe and Mn." *Adv. Geophys.* **22,** 351 (1980).

44. J. Hamilton-Taylor and N. B. Price, "The Geochemistry of Iron and Manganese in the Waters and Sediments of Bolstadfjord, S. W. Norway," *Est. Coast. Shelf. Sci.* **17,** 1 (1983).

45. M. Tanaka, "Manganese Dioxide Particulates in Lake Waters." In Y. Miyake and T. Koyama (Eds.), *Recent Researches in the Fields of Hydrosphere, Atmosphere and Nuclear Geochemistry*, Kenkyusha, Tokyo, 1964, p. 285.

46. R. A. Horne and C. H. Woernle, "Iron and Manganese in a Coastal Pond with an Anoxic Zone," *Chem. Geol.* **9,** 299 (1972).

47. A. Hoshika, O. Takimura, and T. Shiazawa, "Vertical Distribution of Particulate Manganese and Iron in Beppu Bay," *J. Oceanogr. Soc. Japan* **34,** 261 (1978).

48. E. R. Sholkovitz and D. Copland, "The Chemistry of Suspended Matter in Esthwaite Water—A Biologically Productive Lake with a Seasonally Anoxic Hypolimnion," *Geochim. Cosmochim. Acta* **46,** 393 (1982).

49. C. R. O'Melia, H. Weisner, U. Weilenmann, and W. Ali, "The Influence of Coagulation and Sedimentation on the Fate of Particles, Associated Pollutants, and Nutrients in Lakes," Chapter 10 in this book.

50. C. H. Mortimer, "Chemical Exchanges between Sediments and Water in the Great Lakes—Speculations on Probable Regulatory Mechanisms," *Limnol. Oceanogr.* **16,** 387 (1971).

51. E. Gorham and D. J. Swaine, "The Influence of Oxidizing and Reducing Conditions upon the Distribution of Some Elements in Lake Sediments," *Limnol. Oceanogr.* **10,** 268 (1965).

52. E. Callender and C. J. Bowser, "Freshwater Ferromanganese Deposits." In K. H. Wolf (Ed.), *Handbook of Strata-Bound and Stratiform Ore Deposits. II. Regional Studies and Specific Deposits*, Elsevier, Amsterdam/New York, 1976, p. 341.

53. W. Davison and C. Woof, "A Study of the Cycling of Manganese and Other Elements in a Seasonally Anoxic Lake, Rostherne Mere, U.K.," *Water Res.* **18,** 727 (1984).

54. K. H. Wedepohl, "Geochemical Behaviour of Manganese." In I. M. Varencov and G. Grasselly (Eds.), *Geology and Geochemistry of Manganese. Vol. I. General Problems*, pp. 335–351, Akademiai Kiado, Budapest, 1980.

55. P. D. Quay, W. S. Broecker, R. H. Hesslein, and D. W. Schindler, "Vertical Diffusion Rates Determined by Tritium Tracer Experiments in the Thermocline and Hypolimnion of Two Lakes," *Limnol. Oceanogr.* **25,** 201 (1980).

56. K. Wyrtki, "The Oxygen Minima in Relation to Ocean Circulation," *Deep Sea Res.* **9,** 11 (1952).

57. E. Tipping and C. Woof, "Seasonal Variations in the Concentrations of Humic Substances in a Soft-Water Lake," *Limnol. Oceanogr.* **28,** 168 (1983).

58. E. Emerson, R. E. Cranston, and P. S. Liss, "Redox Species in a Reducing Fjord: Equilibrium and Kinetic Considerations," *Deep Sea Res.* **6A,** 859 (1979).

59. D. M. Imboden and R. P. Schwarzenbach, "Spatial and Temporal Distributions of Chemical Substances in Lakes: Modeling Concepts," Chapter 1 in this book.

3

AQUEOUS SURFACE CHEMISTRY: ASSESSMENT OF ADSORPTION CHARACTERISTICS OF ORGANIC SOLUTES BY ELECTROCHEMICAL METHODS

Božena Ćosović

Center for Marine Research, Zagreb, "Rudjer Bošković" Institute, Zagreb, Yugoslavia

Abstract

Electrochemical methods, based on the measurement of adsorption phenomena at the mercury electrode, are direct and nondestructive and therefore convenient for the research and control of surface-active material in natural waters. The two different techniques, measurement of polarographic maxima and capacity-current measurement, were developed and applied to the analysis of surface-active substances in freshwater and marine samples. The approximate characterization of predominant groups of surface-active compounds is made through a comparison of the shape and the intensity of the electrochemical response obtained in natural samples with those of different model substances.

Humic substances were found to be predominant surface-active material in river-water and ground-water samples. Field observations of surface-active substances in the Adriatic Sea by electrochemical methods and experiments with phytoplankton culture media demonstrated that the content and the composition of surface-active material in the sea are closely related to the biological activity of the sea. The enrichment factors in the sea surface film were found to be higher for hydrophobic lipid material than for more soluble wet surfactants. Electrochemical studies have strengthened the role of lipid material in adsorption processes in the sea.

Preliminary investigations of the interaction between surfactants and cadmium at the electrode/water interface showed that: (1) most naturally occurring surfactants have little effect on the mass and charge transfer process at the electrode surface; (2) synthetic compounds, like commercial detergents, slow down the kinetics of the processes at interfaces; and (3) some organic coatings at the surface, as, for example, unsaturated fatty acids, may interact with metal ions resulting in an enrichment of metal ions in the organic layer at the surface.

1. INTRODUCTION

The importance of dissolved and colloidally dispersed organic matter in natural waters is recognized in many marine and freshwater sciences and this topic acts as a strong link between subjects which otherwise have very little in common. It is a central problem in microbiology and an important consideration in the study of phytoplankton and zooplankton. It is now also well known that dissolved organic compounds may influence the physicochemical state and processes of inorganic substances in aquatic environments and thus their acceptability for aquatic life, adsorption at interfaces, and biogeochemical cycles.

Dissolved organic compounds (DOC) in the sea originate from several internal and external sources, including excretion by plants and animals, bacterial decomposition, autolysis of dead organisms, inputs by rivers and effluents, and from the atmosphere. The concentrations in the "open sea" vary between 0.3 to 3 mg C dm^{-3} with values higher in surface layers than in deep waters. In coastal waters the concentrations may be significantly elevated because of increased primary production and because of pollution (1).

Rivers, lakes, and estuarine waters contain higher concentrations of organic matter (approximately 1–10 mg dm^{-3}) with a considerable contribution of humic substances derived from weathering of soils (2).

Most biogeochemical processes in natural waters take place at phase discontinuities, namely, at the atmosphere–hydrosphere and at the lithosphere–hydrosphere interface (3,4). Organic compounds with surface-active properties are concentrated by adsorption processes at the phase boundaries of water with the atmosphere, solid particles, sediment, and biota.

Surface-active substances, both natural and pollutants, modify the structure of the interboundary layers and affect the processes of mass and energy transfer through them.

Adsorption processes have important effects on sedimentation and mineralogy, bubble flotation of particles, and enrichment of organic and inorganic material in the sea-surface microlayer. The distribution of chemical elements in natural waters is controlled to a great extent by scavenging or adsorption onto solid surfaces. The role of organic coatings of solid particles in the scavenging processes is still the subject of controversial interpretations because of the scarcity of data on the adsorption effects of different complex mixtures of organic compounds, such as the composition of dissolved and colloidal organic matter in natural waters (5–7).

It was found that significant fractions of copper (6–30%), lead (3–12%), and mercury (4–50%) were associated with surface-active dissolved and colloidal organic matter isolated from the water column of a controlled experimental ecosystem (8). Qualitative and quantitative information on the nature of these associations is still lacking.

Recently, there has been an ever-increasing interest in trying to learn more about the content and composition of organic substances in natural waters and their interactions with other macro- and microconstituents, both in bulk water and at interfaces. The method of investigation should be sufficiently sensitive for direct determination (without pretreatment procedures) in order to avoid changes in the composition of organic substances which are initially present in the sample. The same is valid for the interaction of organic matter with other constituents, especially metal–organic interactions.

In view of these problems one should take into account the fact that the chemistry of natural waters invariably involves reactions at extreme dilution,

Table 3.1. Application of Electroanalytical Techniques in Natural Aquatic Systems

Conductometry	Salinity
Potentiometry	Macroconstituents
Polarographic and voltametric methods:	
1. Oxidation–reduction processes at electrodes	Trace elements (10^{-8}–10^{-11} mole dm^{-3})
	Speciation studies
	Metal–organic interactions (complexing capacity)
2. Adsorption processes at electrodes	Determination of surface-active substances and approximate characterization of predominant classes of compounds
3. Oxidation–reduction processes at the electrode surface covered by the adsorbed layer of organic substances	Permeability of adsorption layer for mass and charge transfer processes; interactions in the adsorbed layer

with possible competition of different components in the complex mixture (9). For seawater, these reactions take place in a strong salty solution. Electroanalytical methods have found many uses in the study of natural aquatic systems because of their simplicity and the possibility of determining trace or even ultratrace amounts of different constituents without laborious analytical procedures (10–12). The main applications of electroanalytical techniques are schematically presented in Table 3.1.

The electroanalytical approach of the investigation of the adsorption process of organic substances at the mercury electrode/water interface and the implications to the adsorption phenomena in natural waters will be presented in more detail.

2.   DETERMINATION OF SURFACE-ACTIVE SUBSTANCES BY ELECTROCHEMICAL METHODS

Electrochemical determination of surface-active substances in natural waters is based on the measurement of adsorption phenomena at a mercury electrode. These include decrease of interfacial surface tension, changes in the capacity of the electrode double layer, and the suppression of polarographic maxima.

The main advantage of the mercury electrode is its uniform, reproducibly renewable surface, smooth (without roughness) and energetically controlled. The potential and the charge of the electrode surface are determined by the excitation signal and the composition of the supporting electrolyte. This electrolyte is either the salt content of the natural seawater sample or sodium chloride added to freshwater samples prior to the measurement. The surface concentration of the adsorbed surface-active substance, expressed either as Γ in mole cm^{-2} or as surface coverage $\theta = \Gamma/\Gamma_{max}$, depends both on the thermodynamic and kinetic parameters of the adsorption process. The first parameters are determined by the affinity of the particular organic compound for the surface in question and correspond to the parameters of the equilibrium adsorption isotherm. Mass transfer of surface-active molecules towards the electrode surface by diffusion and/or convection processes is also a very important factor in determination of surface-active substances by the measurement of adsorption effects at electrodes. In very diluted solutions of strongly adsorbable species, the equilibrium adsorption can be reached only after long accumulation periods. For a shorter time period the apparent adsorption isotherm is obtained, which reveals the equilibration of the electrode surface with the diffusion layer, where the concentration of the surface-active substances is lower than in the bulk solution because of the depletion effect.

In the adsorption processes at the electrode surface, water dipoles are replaced by organic molecules which have different dielectric properties from water (13). The dielectric constant of water is $D = 80$, while for most organic compounds values range between 2 and 30. The capacity of the electrical double layer and thus also of the charging current depend on the character of the ions or other substances present in the solution. The formula for the capacity of a condenser, $C = \epsilon/4\pi d$, where ϵ is the dielectric constant and d the distance between the layers, shows that the dependence on the dielectric constant is such that by changing the dielectric constant in the interface, due to adsorption of surface-active substances, the capacity of the electrical double layer, and hence the charging current, is changed, that is, decreased.

Since 1973 we have developed and used several different electrochemical techniques for direct determination of naturally occurring surface-active substances in various aquatic systems and for monitoring pollution effects at pollution sources. These are schematically presented in Figure 3.1 and can be summarized as follows:

Method A.   Measurement of suppression of polarographic maxima of mercuric ions and dissolved oxygen by surfactants. The large increase of the current above the diffusion limit value (polarographic maximum) is caused by an increased transport of reducible substances (mercuric ions and molecular oxygen) through convective streaming in the mercury drop and adhering solution layers. Adsorption processes of organic substances slow

ELECTROCHEMICAL DETERMINATION OF
SURFACE ACTIVE SUBSTANCES

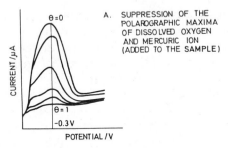

A.   SUPPRESSION OF THE POLAROGRAPHIC MAXIMA OF DISSOLVED OXYGEN AND MERCURIC ION (ADDED TO THE SAMPLE)

B. CAPACITY CURRENT MEASUREMENTS

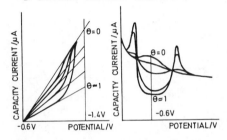

KALOUSEK COMMUTATOR A.C. POLAROGRAPHY
TECHNIQUE

Figure 3.1.   Typical curves which are obtained in electrochemical measurements of adsorbable organic substances by different techniques and increasing values of surface coverage, θ.

down the convective movements resulting in suppression of the polarographic maxima. This widely known phenomena of streaming maxima (14)
were successfully applied in the analysis of surface-active constituents of
natural waters (15–17).

Method B.   Measurement of the capacity current at the hanging mercury
drop electrode either by the alternating current (ac) polarography (18) or by
the Kalousek commutator technique (16,19) with the accumulation of surface-active substances at the potential of maximum adsorption, -0.6 V vs
the Ag/AgCl electrode, the approximate potential of the electrocapillary
maximum. In ac polarography the excitation signal consists of the sinusoidal
potential with the small and constant amplitude which is superimposed on
the linearly increasing potential scan (20,21), while in the Kalousek commutator technique (19,22) the excitation potential consists of a series of
pulses with linearly increasing amplitude from 0 to about 1.5 V. If the frequency of the excitation signal is not too low (we use $f = 64$ Hz for the
Kalousek commutator technique and $f = 230$ Hz for ac polarography), the
capacity-current measurement can be done in the presence of oxygen, that
is, without prior deaeration of the sample.

Typical charging-current curves obtained by ac polarography and by the
Kalousek commutator technique for increasing values of the surface coverage of adsorbed surface-active substances are schematically presented in
Figure 3.1. At potentials near to the electrocapillary maximum of the mercury electrode, the charging current is decreased owing to adsorption of
organic substances, while at more positive or more negative potentials, the
coulombic attractions between the charged-electrode surface and the op-
positely charged ions of the supporting electrolyte cause desorption of organic molecules from the electrode surface. In ac polarography desorption
processes produce more or less pronounced peaks, while in the measurement
by the Kalousek commutator technique the charging current continuously
increases with increasing desorption of organic molecules until the current
finally reaches the same value that is obtained for the supporting electrolyte
without surfactant. The high sensitivity of the capacity-current measurement, needed for the determination of natural surface-active substances in
bulk seawater and freshwater samples with very low organic content, is
reached by enhancing the accumulation of surfactants at the electrode surface by stirring the solution.

The two methods (A and B) differ in the type of electrical excitation,
measured response, time scale, and the potential region. These facts result
in different specific sensitivities to various classes of surface-active molecules. Calibration curves for various types of surface-active compounds are
presented in Figure 3.2. It is evident that the capacity-current measurement
at an uncharged and negatively charged electrode are more sensitive to hydrophobic and heavily soluble lipid material and to synthetic detergents than
determinations made by the polarographic maximum method. On the other

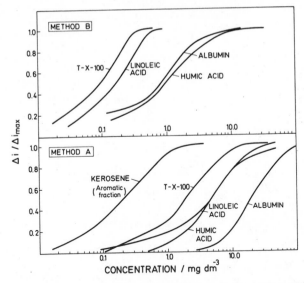

Figure 3.2. Calibration curves of selected model substances for two different electroanalytical techniques.

hand, the polarographic maximum method is more sensitive to aromatic hydrocarbons (23) and to some polysaccharides, such as, for example, alginate (16).

The adsorption phenomena at the electrodes represent the gross effect of a complex mixture of surfactants present in the natural sample. For a quantitative determination of the surfactant activity of natural samples, the calibration curve prepared with an arbitrary standard surfactant, Triton-X-100, is used. The approximate characterization of the predominant group of surface-active substances is possible through a comparison of the shape and the intensity of the electrochemical response obtained in the natural sample with those of different model substances and their mixtures.

3. SURFACE-ACTIVE SUBSTANCES IN FRESHWATERS

Surface-active substances have been determined by electrochemical methods in samples of river waters, groundwaters, and tap water. River-water samples were collected from three Yugoslavian rivers in different seasons and at different flow rates. The results will be presented in more detail elsewhere (24). The average concentrations of Ca^{2+} and Mg^{2+} salts and the range of the DOC values of the investigated freshwaters are given in Table 3.2. As shown in curve 1 Figure 3.3, all data of surfactant-activity values of river-water samples, obtained by the two electrochemical methods, fit fairly well the positive linear correlation both in the logarithmic as well as in the linear scale ($r = 0.44$; 76 pairs of data, linear scale). The same is valid for

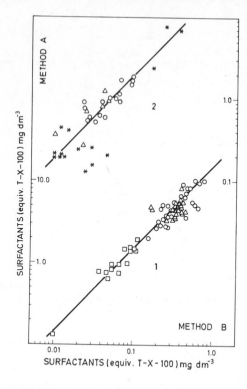

Figure 3.3. Composite plot comparing surfactant contents (equivalent Triton-X-100) obtained by two electrochemical methods (A and B) for: curve 1, river-water samples from the Sava (O), Drava (△), and Kupa (□); and curve 2, ground-water samples from the Sava (O) and Drava (△) aquifers, and tap waters (*).

ground-water and tap-water samples (Figure 3.3, curve 2; $r = 0.95$; 36 pairs of data, linear scale). However, the content of organic surface-active substances in ground water is generally lower than in surface waters. The infiltration of organic substances into the subsurface decreases on increasing distance of the observation well from the river bank.

Although we have not measured systematically the DOC content of all

Table 3.2.   Average Concentrations of Ca^{2+} and Mg^{2+} and the Approximate Range of the DOC Values in the Investigated Freshwaters

	Ca^{2+} (mmole dm^{-3})	Mg^{2+} (mmole dm^{-3})	DOC (mg dm^{-3})
The Sava			
River water	1.48	0.56	7–30
Ground water	1.79	0.60	
The Drava			
River water	0.85	0.28	10–15
The Kupa			
River water	1.58	0.45	<5

investigated freshwater samples, the approximate ranges of DOC values that were obtained from several DOC analyses show the same trend as the surfactant activity values for the three investigated rivers. Hunter and Liss (17) arrived at the same conclusion on the basis of a large number of analyses of estuarine samples on the DOC content and this paralleled the surfactant content measured by the polarographic maximum method. These authors proposed that the surfactant-activity measurements be used to obtain a simple measure of the DOC concentration. Where geochemical interest centers on the interfacial effects of organic matter, the determination of the surface-active component of dissolved organic matter is more appropriate than the determination of DOC.

The characterization of predominant groups of surface-active substances in freshwaters is possible on the basis of the comparison of the shape of corresponding curves and the intensity of adsorption effects of natural samples and different model substances. Since it is well known that soil humic acids are surface active (25) and that terrestrial humics represent predominant organic substances in fresh waters (2), it is not surprising that the electrochemical determinations reveal the presence of humic material. Figure 3.4(a) presents capacity-current curves obtained from ac polarographic measurements of different concentrations of humic acid which were isolated from Rupia detritus. Figure 3.4(b) presents typical capacity-current curves

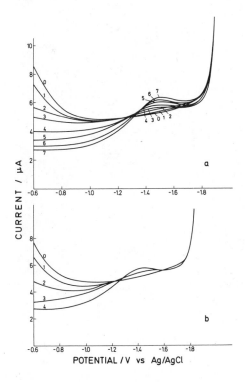

Figure 3.4. Capacity-current curves obtained by ac polarography ($E = 10$ mV, $f = 230$ Hz, accumulation period 60 s with stirring). (a) Humic acid in 0.55 mole dm^{-3} NaCl, 0.03 mole dm^{-3} NaHCO$_3$: (0)0, (1)0.2, (2)0.8, (3)1.6, (4)3, (5)10, (6)30, and (7)50 mg dm^{-3}. (b) Surface-active substances present in different freshwater samples: (0) electrolyte without surfactants, (1, 3) tap water, (2) ground water, and (4) river water.

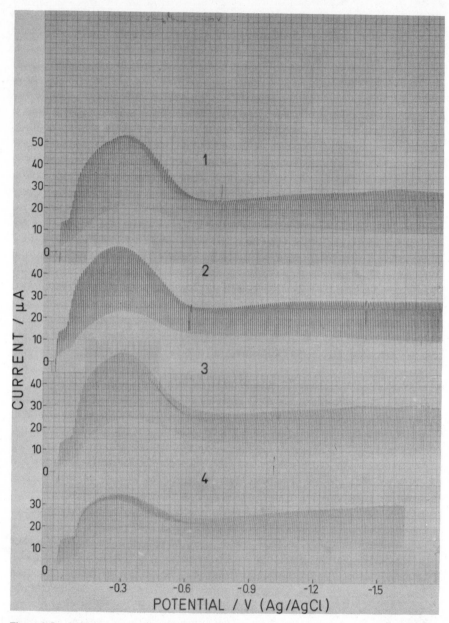

Figure 3.5. Polarograms (actual recordings) of (1) 5 mg dm^{-3} humic acid, (2) 5 mg dm^{-3} fulvic acid, and (3) 5 mg dm^{-3} and (4) 10 mg dm^{-3} sodium lignosulphonate, in 0.55 mole dm^{-3} NaCl and 5 × 10^{-3} mole dm^{-3} NaHCO$_3$.

of different samples of freshwaters. The same results were obtained also by the polarographic maximum method, as illustrated in Figures 3.5 and 3.6.

We should point out that in the electrochemical measurements the investigated humic and fulvic acids (predominant molecular weights between 20,000 and 50,000) and sodium lignosulphonate give very similar curves with

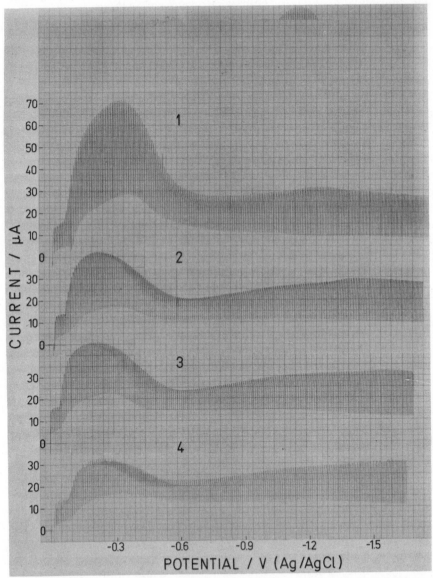

Figure 3.6.   Polarograms (actual recordings) of different freshwater samples: (1) and (2) ground water, and (3) and (4) the Sava river. Concentration of added sodium chloride 0.55 mole dm^{-3} and 5×10^{-3} mole dm^{-3} NaHCO$_3$.

regard to the shape and the intensity of the response, and thus all data fit very well the same calibration curves for humic material. These curves, as obtained by methods A and B, are presented in Figure 3.7 together with the calibration curves of Triton-X-100, which are given for comparison. The surfactant-activity values of different freshwater samples, determined by using the calibration curves of humic material, range between 0.5 and 15 mg dm^{-3} of humic material for both electrochemical methods.

The simple relationship found by Hunter and Liss (17):

surfactant activity (mg dm^{-3}) of Triton-X-100 = 1.3 DOC (mg dm^{-3}) is valid for natural samples where humic surface-active substances form a relatively constant and predominant fraction of dissolved organic matter. We confirmed that by comparing the calibration curves of humic material and Triton-X-100 for the measurement by the polarographic maximum method, and by taking into account that the investigated humic have approximately 50% of the carbon content.

Because of the fact that the capacity-current measurement (method B) is highly sensitive, particularly to Triton-X-100, the surfactant-activity values of most natural samples are about one order of magnitude lower for this method than for the polarographic maximum method if they are given in equivalent effect of Triton-X-100. The capacity-current measurement is also more sensitive for detection of pollution effects by detergents in natural waters than the polarographic maximum method. This is well illustrated in

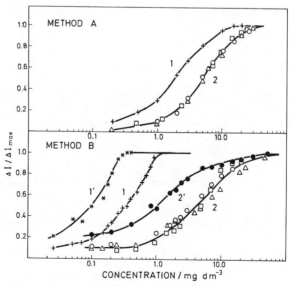

Figure 3.7. Calibration curves (adsorption isotherms) of selected model substances. Method A: (1) Triton-X-100; (2) humic acid (○), fulvic acid (△), and sodium lignosulphonate (□). Method B: (1) and (1′) Triton-X-100; (2) humic acid (○), fulvic acid (△), and sodium lignosulphonate (□); and (2′) humic acid. Accumulation period: (1) and (2) 60 s and (1′) and (2′) 15 s with stirring.

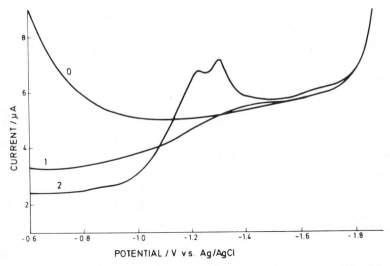

Figure 3.8.   Capacity-current curves (ac polarography) of two freshwater samples of the river Drava with different content of anionic detergents: (1) 10 and (2) 260 μg dm^{-3}. Curve (0) is supporting electrolyte without surfactants.

Figures 3.8 and 3.9. The two different river-water samples, both high in organic content, but only one containing a high content of anionic detergents (260 μg dm^{-3}), gave practically the same suppressions of the polarographic maxima, while the form of the capacity-current curve and the increased adsorption effect fully reveal the presence of detergents in one of the samples.

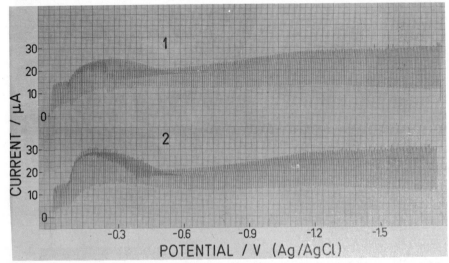

Figure 3.9.   Polarograms (actual recordings) of the same freshwater samples as presented in Figure 3.8. Concentration of anionic detergents: (1) 10 and (2) 260 μg dm^{-3}.

4. SURFACE-ACTIVE SUBSTANCES IN THE SEA

Concentrations of surface-active material in the sea are generally lower than in freshwaters. The concentration ranges of surfactant-activity values of different natural waters, which were obtained on the basis of electrochemical measurements of a large number of samples, are given in Table 3.3. Only bulk seawater and sea-surface microlayer samples, collected by the screen sampler, from locations remote from direct pollution sources, have been considered. However, in coastal waters, at places which are under the influence of sewage and industrial effluents, the surfactant-activity values can be highly elevated. Some typical surfactant-activity values of seawater samples taken at pollution sources are presented in Table 3.4.

The main source of natural surface-active substances in the sea is the excretion of phytoplankton species (9,26,27). The surfactant-activity values of bulk seawater samples show seasonal variations that have the same trend as the variations of biological activity of the sea. Higher surfactant-activity values, usually from spring to autumn, correspond to seasons of higher biological activity, while the lowest values are obtained in the winter season and coincide with the period of the lowest biological activity. The seasonal variations of surfactant-activity values in the North Adriatic Sea, determined by the polarographic maximum method (26), are presented in Figure 3.10. The most remarkable peak of the surfactant-activity values observed in 1977 for the whole North Adriatic, coincided with an intensive phytoplankton bloom in the upper (5 m) layer of the stratified water column. However, the most pronounced peak of the surfactant-activity values, determined by ac polarography, in the North Adriatic Sea, was observed in 1978 (18), as is shown in Figure 3.11. At the same time, the surfactant-activity values obtained by the polarographic maximum method showed only a moderate seasonal increase.

We have obtained no correlation between the surfactant-activity values, determined by the two electrochemical methods, for a larger number of seawater samples. The most pronounced differences were observed in measurements of different phytoplankton culture media (26). The typical results

Table 3.3.   Surfactant-Activity Values (Equivalent Triton-X-100 in mg dm^{-3}) of Different Natural Waters as Obtained by Two Electrochemical Methods

Samples	Method A	Method B
River water	1–10	0.05–1
Ground water	0.2–2	0.01–0.2
Seawater	0.2–2	0.01–0.2
Sea surface Microlayer	1–3	0.1–2

Table 3.4. Surfactant Activity and Content of Anionic Detergents in Seawater Samples Collected at Pollution Sources Along the Adriatic Coast

| Station | Date | Surfactant Activity | | Anionic Detergents (μg dm^{-3}) |
| | | Method A | Method B | |
		($mg\ dm^{-3}$)		
Rovinj Harbour	3/9/76	6	2.15	620
	7/8/76	1.1	0.29	8
	1/11/77	2.5	1.7	83
Rovinj	3/9/76	1.55	0.28	8
Fish Cannery	7/8/76	0.84	0.07	9
	1/11/77	1.9	0.13	6
Pula Harbour	6/18/77[a]	10	0.27	20
Rijeka Harbour	12/17/76	1.35	0.15	25
	3/4/77	4.3	0.42	102
Oil Refinery of	12/17/76	3.1	0.29	27
Rijeka	3/3/77	4.8	0.38	6
Split Plastics	12/28/78	8.7	57	200
Industry	2/1/79	9.5	41	140

[a] The phytoplankton bloom was intensified in the harbor by local pollution.

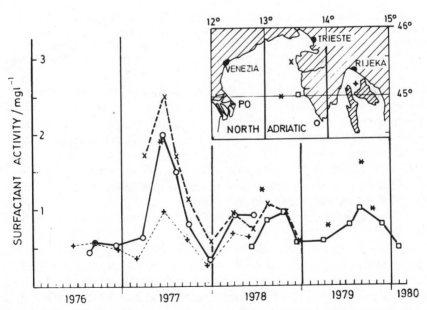

Figure 3.10. Surfactant activity of seawater samples (0.5 m) measured by method A at selected stations in the Northern Adriatic (26).

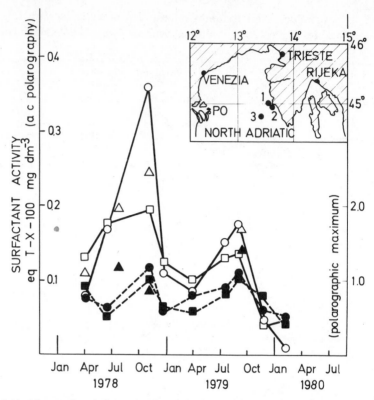

Figure 3.11. Seasonal variations of surfactant-activity values by ac polarography (open symbols) and by polarographic maximum method (closed symbols) in seawater samples taken at offshore stations in the Northern Adriatic at 0.5-m depth: station 1 (O●); station 2 (□■); and station 3 (△▲). [From B. Ćosović and V. Vojvodić, *Limnol. Oceanogr.* **27**, 361–369 (1982).]

for a diatom Skeletonema and a dinoflagellate Cryptomonas are presented in Figure 3.12. It is known that healthy cells excrete surface-active material during growths. The two experimental techniques measure increased content for cultures of higher density. For the polarographic maximum method the surfactant-activity values (expressed in equivalents of the model substance, Triton-X-100) of both culture media fell in the same concentration range; the differences between cultures were mainly qualitative with respect to the shape of the obtained polarographic curves.

Using capacity-current measurement in the same series of experiments, the observed surfactant-activity values of Cryptomonas culture were more than two orders of magnitude higher than for the Skeletonema culture of approximately the same cell densities. However, the effect of surfactants in the Cryptomonas culture was beyond the saturation of the method for all samples investigated and therefore the determination of the surfactant activity was performed after dilution with the sodium chloride solution to reach a measurable concentration range of surfactants. The loss of surface-active

material during the filtration procedure (1.2 μm millipore filter) was so tremendous that the surfactant activity of filtered samples decreased for one
or two orders of magnitude in comparison with unfiltered samples and became comparable with the surfactant-activity values of the Skeletonema
culture.

Qualitative and quantitative adsorption characteristics of the dinoflagellates culture media can be summarized as follows: (1) irregular oscillations
in the signal of the polarographic maximum; (2) extremely high specific sensitivity of the capacity-current measurements to the present surfactants; and
(3) significant loss of surfactants in the filtration procedure. A very similar
behavior was observed also with artificial dispersions of different unsaturated lipid material, which, in our investigations, was represented by oleic
and linoleic acid.

Although the unsaturated lipid substances are produced in significant
amounts in the sea, they are to a great extent eliminated from seawater by
adsorption processes or changed into more soluble associations. Recently,
it was reported that unsaturated aliphatic compounds, most probably trig-
licerides, have a major role in the formation of marine humic and fulvic acids
(28). There are still numerous controversial discussions about the concentration and surface chemistry of fatty acids in the sea. As Garrett very early
proposed (29), the higher polar and water-insoluble lipid material might be
the most important and durable family of compounds at the air/sea interface.
The lack of direct analytical techniques of sufficient sensitivity and the
changes the fatty acids undergo during the pretreatment, including both a
loss of material and eventual leaching from microscopic organisms, are the
main reasons why this subject was not treated appropriately.

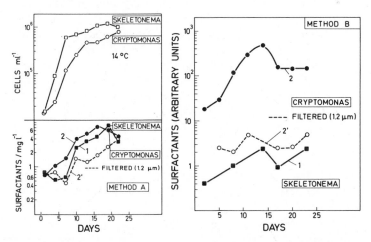

Figure 3.12. Surfactant-activity values of the culture medium vs culture age for (1) Skeletonema costatum, and (2) Cryptomonas sp. Filtered sample represented by (2′). Comparison of
results obtained from parallel determinations with the two electrochemical methods (26).

Later experiments by Baier et al. (30), using germanium plate for collecting the sea-surface microlayer, presented evidence that "wet" surfactants, such as glycoproteins and proteoglycans, may be common on the surface of the ocean. These compounds are essentially hydrophylic, but attached to the surface by the occasional hydrophobic groups. In the same way as is known that a mercury electrode surface is not sensitive to simple hydrocarbons, amino acids, and sugars, (although they tend to get enriched at the air/water interface), it might be also questionable whether, in the case of Baier's sampling, the germanium plate collects and retains a complete surface film, or only specifically adsorbed molecules.

Electrochemical investigations of surfactants in the sea-surface-microlayer samples imply the presence of both types of surface-active substances—hydrophobic and hydrophilic—at the sea surface (16,18). With respect to adjacent subsurface waters, the enrichment factors of surfactants in the sea-surface microlayer, determined by the polarographic maximum method, ranged between 2 and 8, the same as many authors had found for DOC values. On the contrary, the values obtained by the capacity-current measurement, which is more sensitive to hydrophobic surfactants in particular, ranged between 2 and 100. Surface-microlayer samples from polluted areas, with petroleum hydrocarbons as prevalent organic substances were the only one exception where the enrichment factors for the polarographic maximum method were higher than for the capacity-current measurement (31). This is, however, quite understandable, because the polarographic maximum method is specifically more sensitive to petroleum aromatic hydrocarbons (23).

The concentration and composition of surface-active material in the sea and the fate of different types of organic substances depend on their physicochemical properties, biodegradability, and input and output mechanisms. A great part, approximately 70%, of dissolved organic matter in the sea is more or less adsorbable substances. The electrochemical experiments demonstrated that significant adsorption effects may be observed at low concentrations of strongly adsorbable substances and even in the presence of higher concentrations of weakly adsorbable substances. The competition and interaction of different organic substances in the adsorption process from a complex mixture, such as the composition of organic matter in natural waters, remains an open problem that requires further investigation.

5. INTERACTION OF SURFACE-ACTIVE SUBSTANCES WITH METALS STUDIED AT THE ELECTRODE/WATER INTERFACE

The study of adsorption phenomena and mass and charge transfer processes at charged electrodes covered, partly or completely, with the adsorbed layer of different surface-active substances, which are representative for composition of natural and polluted waters, opens up new possibilities for simulation of interfacial phenomena and processes at natural phase boundaries.

The adsorbed layer of organic molecules generally presents a barrier to the transport of ions and electrons at the interface. In the electrochemical measurement, adsorbed surface-active substances influence oxidation-reduction processes of ions and molecules present in solution, mainly by reducing the rate of the electrode reaction (32). The effects have been studied so far with relatively high bulk concentrations of surfactants. To study interfacial phenomena at low surfactant concentrations, similar to those of the constituents of natural and polluted waters, it is necessary to use a suitable and sensitive technique. This technique should enable registration of the smallest changes in the kinetics of the electrode reaction at the stationary electrode after prior formation of the adsorbed layer of surface-active substances at a conveniently selected potential. The use of differential pulse polarography is advantageous for three principal reasons (33):

1. The high sensitivity of this technique enables measurements at relatively low concentrations of electroactive species.
2. A special technique of performing the measurement in pulse polarography provides a high level of separation between the faradaic (oxidation reduction) and nonfaradaic (capacity) current.
3. A relatively wide range of electrode kinetic parameters can be deduced from experiments.

In differential pulse polarography the decrease of the electrode reaction rate is usually followed by a lowering and broadening of the polarographic wave. We have used differential pulse polarography to study oxidation reduction processes of cadmium(II) in the presence of the absorbed layer of surface-active substances at the hanging mercury drop electrode. Cadmium was chosen because of its very convenient polarographic characteristics, which result from its well-defined reversible and two-electron polarographic wave. This wave appears at potentials near to the electrocapillary maximum of the mercury electrode, that is, at potentials of maximum adsorption. Besides, cadmium is a very toxic metal and one of the potential pollutants in the aquatic environment. Moveover, this method can be applied for investigation of other metals. The kinetic parameters, the rate constant, and the transfer coefficient α of the electrode reaction, are estimated from the shape and the height of the corresponding differential pulse polarograms using theoretical curves obtained by digital simulation of the electrode process (33). The investigations with model substances (34,35) showed that the synthetic compounds, such as Triton-X-100, which is presented in Figure 3.13, inhibit the electrode reduction of cadmium at the mercury electrode at very low concentrations of surfactant. Model substances representative for biopolymers, such as albumin in Figure 3.13 (the same is valid for geopolymers

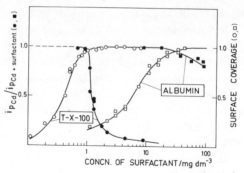

Figure 3.13.   Adsorption isotherms capacity-current measurement of Triton-X-100 (○) and albumin (□) in seawater, and the dependence of the peak height of the differential pulse polarogram of cadmium (10^{-4} mole dm^{-3}) on the concentration of Triton-X-100 (●) and albumin (■). Accumulation period 5 min by the diffusion process.

humic and fulvic acids) create adsorbed films which are very porous for cadmium ions.

One should generally expect an interaction, most likely complexation or coordination, of trace metals with the adsorbed layer of organic molecules at various interfaces. Our experiments with different surface-active compounds and with cadmium demonstrated the unique behavior of unsaturated fatty acids (oleic and linoleic) in this respect (36). The increase of the peak height of cadmium, up to 10-fold of the value obtained in the electrolyte without fatty acid, is caused by the adsorption of metal ions at the so-modified electrode surface. The accumulation of the metal increases with an increase in the adsorption time (at potentials more positive than the reduction potential of cadmium), as shown in Figure 3.14, and with an increase in salinity. The same was also obtained for lead and zinc.

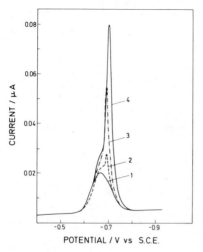

Figure 3.14.   Differential pulse polarograms of 10^{-5} mole dm^{-3} cadmium(II) in the presence of 36.3 mg dm^{-3} linoleic acid in 0.55 mole dm^{-3} NaCl and 3×10^{-3} mole dm^{-3} NaHCO$_3$. Adsorption time: (1) 0, (2) 2, (3) 4, and (4) 8 min. Adsorption potential -0.4 V vs S.C.E., pH = 8.7 (36).

We have not observed the accumulation of cadmium in the adsorbed layer of humic and fulvic acids. However, one should not neglect the probability of the interaction between the adsorbed layer of humic and fulvic substances with copper ions which are known to form stronger complexes with humic material than cadmium does (37). The river-water samples with a high content of surface-active material showed no influence on the kinetics of the electrode reduction of cadmium (Table 3.5), the same as was obtained with the adsorbed layer of humic and fulvic acids.

The inhibition effects of surface-active substances, present in seawater samples taken from polluted sites and in effluents, increased with increasing concentrations of synthetic compounds in the total content of surface-active material (Table 3.6).

Electrochemical investigations with model substances and with polluted waters and effluents, imply that the adsorbed layer of synthetic surfactants represents a barrier for the interaction between the surface and the metal ions. The mechanism of the inhibitory effect of the electrode process is based on the concept of the competitive adsorption of electroactive species (metal ions), solvent molecules, and inhibitor molecules (38). This results in a strong

Table 3.5.   The Influence of Surface-Active Substances Present in River-Water Samples on the Rate Constant of the Electrochemical Reduction of 10^{-5} mole dm^{-3} Cadmium (II)

Sample	Surfactant Activity[a] (Equivalent T-X-100) (mg dm^{-3})	$k_{\theta=0}/k_{\theta}$ [b]
The river Sava (May 20, 1983)		
1	0.51	1
2	0.53	1
3	0.45	1
4	0.52	$\geqslant 1.5$
5	0.44	$\geqslant 1.6$
6	0.50	$\geqslant 1.3$
The river Drava (June 21, 1983)		
1	0.39	1
2	0.43	1
3	0.43	$\geqslant 1.3$
4	0.47	1
5	0.41	1
6	1.5	$\geqslant 1.2$

[a] Method B.

[b] $k_{\theta=0} \geqslant 0.112$ cm s^{-1} in 0.55 mole dm^{-3} NaCl.

Table 3.6.   The Influence of Surface-Active Substances, Present in Seawater Samples Taken at Pollution Sources and in Effluents, on the Rate Constant of the Electrochemical Reduction of 10^{-4} mole dm^{-3} Cadmium (II)

Seawater Sample	Surfactant Activity[a] (Equivalent T-X-100) (mg dm^{-3})	$k_{\theta=0}/k_\theta$ [b]
Rovinj Harbour	0.84	$\geqslant 1.44$
Rovinj Fish Cannery	2	$\geqslant 2.49$
Split Plastics Factory	0.6	$\geqslant 4.48$

Effluents	Surfactant Activity[a] (Equivalent T-X-100) (mg dm^{-3})	$k_{\theta=0}/k_\theta$ [b]
Emulsion polymerization	31	$\geqslant 56$
Phtalate esters	1000	$\geqslant 22$

[a] Method B.
[b] $k_{\theta=0} \geqslant 0.112$ cm s^{-1} in 0.55 mole dm^{-3} NaCl.

dependence of the rate constant of the electrode reaction on the inhibitor concentration.

If the adsorbed layer of organic molecules at the surface of the mercury electrode represents an adequate model for organic coatings of the particles in natural waters, these preliminary electrochemical studies on cadmium indicate that: (1) most of the naturally occurring substances (biopolymers and geopolymers) may have little effect in the interaction between metal ions and the surface; (2) synthetic compounds, like commercial detergents, may slow down to a great extent the kinetics of the processes that occur at interfaces; and (3) some organic coatings may specifically interact with metal ions which results in the enrichment of the metal in the organic layer at the surface.

6.   CONCLUSIONS

1.   Electrochemical methods, based on measurements of the adsorption phenomena at the mercury electrode, are simple, rapid, direct and nondestructive and therefore convenient for the research and control of surface-active material in natural waters.

2.   The adsorption phenomena at the electrodes represent the gross effect

of a complex mixture of surfactants present in natural samples. For a quantitative determination of the surfactant activity of natural samples, the calibration curve of an arbitrary standard surfactant, Triton-X-100 is used.

3.   The approximate characterization of the predominant groups of surface-active substances is possible through a comparison of the shape and the intensity of the electrochemical response obtained in natural samples with those of different model substances.

4.   Humic substances are predominantly surface-active material in freshwaters. For natural samples, where humic substances form a relatively constant and predominant fraction of dissolved organic matter, the surfactant-activity values are a simple measure of the DOC concentration.

5.   Field observations of surface-active substances in the Adriatic Sea by electrochemical methods and experiments with phytoplankton culture media, demonstrated that the content and the composition of surface-active material are closely related to the biological activity of the sea. Electrochemical studies have strengthened the importance and contribution of lipid material to the surfactant activity of the sea.

6.   The investigations of electrochemical processes of cadmium(II) at the charged mercury electrode covered with the adsorbed layer of surface-active substances showed that: (1) most naturally occurring substances have little effect on the interaction between metal ions and the surface; (2) synthetic compounds, like commercial detergents, slow down the kinetics of the processes at interfaces; and (3) some organic coatings, like unsaturated fatty acids, interact with metal ions resulting in an enrichment of metal ions in the organic layer at the surface.

ACKNOWLEDGMENTS

The author wishes to express her thanks to Marko Branica, who inspiringly introduced her to the field of electroanalytical environmental analysis; to Vera Žutić and Zlatica Kozarac for a longstanding collaboration on this work; and to Vjeročka Vojvodić and Tinkica Pleše for their valuable participation in the analysis of samples.

The generous gift of the humic material from Gustave Cauwet is also gratefully acknowledged.

This work has been supported by the Self-Managed Authority for Scientific Research, SR Croatia.

REFERENCES

1. D. W. Menzel, "Primary Productivity, Dissolved and Particulate Organic Matter, and the Sites of Oxidation of Organic Matter," In E. D. Goldberg, (Ed.), *The Sea*, Vol. 5, Wiley, New York, 1974, pp. 659–679.

2. W. Stumm and J. J. Morgan, *Aquatic Chemistry* (2nd ed.), Wiley–Interscience, New York, 1981.

3. G. A. Parks, "Adsorption in the Marine Environment." In J. P. Riley and G. Skirrow (Eds.), *Chemical Oceanography*, Vol. 2, Academic Press, New York, 1975, pp. 241–301.

4. K. A. Hunter and P. S. Liss, "Organic Sea Surface Films." In E. K. Duursma and R. Dawson, *Marine Organic Chemistry*, Elsevier Oceanographic Series, Elsevier, New York, 1981, pp. 259–295.

5. I. G. Loeb and R. A. Niehof, "Marine Conditioning Films," *Adv. Chem. Ser.* **145,** 319–335, (1975).

6. L. Balistrieri, P. G. Brewer, and J. W. Murray, "Scavenging Residence Times of Trace Metals and Surface Chemistry of Sinking Particles in the Deep Ocean," *Deep Sea Res.* **28A,** 101–121, (1981).

7. Yuan-Hui Li, "Ultimate Removal Mechanisms of Elements from the Ocean," *Geochim. Cosmochim. Acta* **45,** 1659–1664, (1981).

8. Gordon T. Wallace, Jr., "The Association of Copper, Mercury and Lead with Surface-Active Organic Matter in Coastal Seawater, *Marine Chem.* **11,** 379–394, (1982).

9. P. J. Wangersky and R. G. Zika, *The Analysis of Organic Compounds in Sea Water*, NRCC No. 16566, 1978, MACSP, Atlantic Regional Laboratory, NRC of Canada, Halifax, N.S.

10. H. W. Nürnberg and P. Valenta, "Polarography and Voltametry." In E. D. Goldberg (Ed.), *The Nature of Seawater*, Dahlem Konf., Berlin, 1975, pp. 87–136.

11. W. Davison and M. Whitfield, "Modulated Polarographic and Voltametric Techniques in the Study of Natural Water Chemistry, *J. Electroanal. Chem.* **75,** 763–789, (1977).

12. M. Whitfield and D. Jagner, *Marine Electrochemistry*, Wiley-Interscience, New York, 1981.

13. B. B. Damaskin and O. A. Petrii, *Adsorption of Organic Compounds on Electrodes*, Plenum Press, New York, 1971.

14. H. H. Bauer, "Streaming Maxima in Polarography," *Electroanal. Chem.* **8,** 169–279, (1975).

15. T. Zvonarić, V. Žutić, and M. Branica, "Determination of Surfactant Activity of Seawater Samples by Polarography, *Thalassia Jugoslav.* **9,** 65–73, (1973).

16. B. Ćosović, V. Žutić, and Z. Kozarac, "Surface Active Substances in the Sea Surface Microlayer by Electrochemical Methods," *Croat. Chem. Acta* **50,** 229–241, (1977).

17. K. A. Hunter and P. S. Liss, "Polarographic Measurement of Surface Active Material in Natural Waters," *Water Res.* **15,** 203–215, (1981).

18. B. Ćosović and V. Vojvodić, "The Application of a.c. Polarography to the Determination of Surface Active Substances in Seawater," *Limnol. Oceanogr.* **27,** 361–369, (1982).

19. B. Ćosović and M. Branica, "Study of the Adsorption of Organic Substances at a Mercury Electrode by the Kalousek Commutator Technique," *J. Electroanal. Chem.* **46,** 63–69, (1973).

20. D. E. Smith, "A.C. Polarography and Related Techniques: Theory and Practice," *Electroanal. Chem.* **1,** 1–148, (1966).
21. H. Jehring, *Elektrosorptionanalyse mit der Wechselstrom-Polarographie,* Akademie Verlag, 1976.
22. J. Radej, I. Ružić, D. Konrad, and M. Branica, "Instrument for Characterization of Electrochemical Processes," *J. Electroanal. Chem.* **46,** 261–280, (1973).
23. V. Žutić, B. Ćosović, and Z. Kozarac, "Electrochemical Determination of Surface Active Substances in Natural Waters. On the Adsorption of Petroleum Fractions at Mercury Electrode/Seawater Interface," *J. Electroanal. Chem.* **78,** 113–121, (1977).
24. B. Ćosović, V. Vojvodić, and T. Pleše, Electrochemical Determination and Characterization of Surface Active Substances in Freshwaters, *Water Res.* (1984) (in press).
25. T. Tschapek and C. Wasowski, "The Surface Activity of Humic Acid," *Geochim. Cosmochim. Acta* **40,** 1343–1345, (1976).
26. V. Žutić, B. Ćosović, E. Marčenko, N. Bihari, and F. Kršinić, "Surfactant Production by Marine Phytoplankton." *Marine Chem.* **10,** 505–520, (1981).
27. P. J. Williams, "Biological Aspects of Dissolved Organic Material in Seawater" In J. P. Riley and G. Skirrow (Eds.), *Chemical Oceanography*, Vol. 2, Academic Press, New York, 1975, pp. 301–363.
28. G. R. Harvey, D. A. Boran, L. A. Chesal, and J. M. Tokar, "The Structure of Marine Fulvic and Humic Acids," *Marine Chem.* **12,** 119–132, (1983).
29. W. D. Garrett, "The Organic Chemical Composition of the Sea Surface," *Deep Sea Res.* **14,** 221–227, (1967).
30. R. E. Baier, D. W. Goupil, S. Perlmutter, and R. King, "Dominant Chemical Composition of Sea Surface Films, Natural Slicks and Foams," *J. Res. Atm.* **8,** 571–600, (1974).
31. B. Ćosović and V. Žutić, "Surface Active Substances in the Rijeka Bay," *Thalassia Jugoslav.* **17,** 197–209, (1981).
32. J. Lipkowski and Z. Galus, "On the Present Understanding of the Nature of Inhibition of Electrode Reactions by Adsorbed Neutral Organic Molecules," *J. Electroanal. Chem.* **61,** 11–32, (1975).
33. Z. Kozarac, S. Nikolić, I. Ružić, and B. Ćosović, "Inhibition of the Electrode Reaction in the Presence of Surfactants Studied by Differential Pulse Polarography. Cadmium(II) in Seawater in the Presence of Triton-X-100," *J. Electroanal. Chem.* **137,** 279–292, (1982).
34. Zlata Kozarac, "The Influence of Surface Active Substances on Electrochemical Processes of Cadmium(II) at the Mercury Electrode," Ph.D. thesis, University of Zagreb, 1980.
35. Z. Kozarac and B. Ćosović, "On Interaction of Cadmium(II) with Surfactants. Model Studies at Electrode/Seawater Interface," *Rapp. Comm. Int. Mer. Medit.* **28,** 155–158, (1983).
36. D. Krznarić, B. Ćosović, and Z. Kozarac, "The Adsorption and Interaction of Long-Chain Fatty Acids and Heavy Metals at the Mercury Electrode/Sodium Chloride Solution Interface," *Marine Chem.* **14,** 17–29 (1983).

37. R. A. Saar and J. H. Weber, "Fulvic Acid: Modifier of Metal-Ion Chemistry," *Environ. Sci. Technol.* **16**, 510A–517A (1982).

38. G. Pyzik and J. Lipkowski, "On the Mechanism of Ion Transfer Across the Monolayers of Organic Surfactants. The Mechanisms of Cu(II) Deposition at Hg Electrodes Covered by Monolayers of Aliphatic Alcohols, *J. Electroanal. Chem.* **128**, 351–364, (1981).

4

STRATEGIES FOR MICRO-BIAL RESISTANCE TO HEAVY METALS

John M. Wood and Hong Kang Wang

Gray Freshwater Biological Institute, University of Minnesota, Navarre, MN 55392, USA

Abstract

We have witnessed huge changes in the distribution of elements at the surface of the earth. Microorganisms are adapting to these changes by evolving strategies to maintain low intracellular concentrations of toxic pollutants. First, this adaptation to resist toxic substances may have been inherited from organisms that lived in extreme environmental conditions. Second, some bacteria have acquired resistance relatively easily through the acquisition of extrachromosomal DNA molecules (i.e., plasmids).

The biochemical basis for resistance to metal-ion toxicity is emerging and it is complicated by the different resistance mechanisms. Several strategies for resistance to metal-ion toxicity have been identified:

1. The development of energy-driven efflux pumps which keep toxic-element levels low in the interior of the cell. Such mechanisms have been described for Cd(II) and As(V).

2. Oxidation (e.g., AsO_3^{2-} to AsO_4^{3-}) or reduction (e.g., Hg^{2+} to Hg^0) can enzymatically and ultracellularly convert a more toxic form of an element to a less toxic form.

3. The biosynthesis of intracellular polymers which serve as traps for the removal of metal ions from solution, such as the traps which have been described for cadmium, calcium, nickel, and copper.

4. The binding of metal ions to cell surfaces.

5. The precipitation of insoluble metal complexes (e.g., metal sulfides and metal oxides) at cell surfaces.

    6. Biomethylation and transport through cell membranes by diffusion-controlled processes.

    Each of the mechanisms for resistance to toxicity described above requires inputs of cellular energy, and as such represent a nonequilibrium component for the distribution of elements at the earth's surfaces.

1. INTRODUCTION

Microorganisms have evolved a number of strategies to maintain low intracellular concentrations of heavy metals. These strategies are influenced greatly by whether the metal being considered is essential or nonessential for cell growth and cell division. Essential metal ions are defined as those elements which have an important biological function, and for these elements elaborate transport systems have evolved which are very specific and therefore carefully regulated by the cell. Many essential metals are toxic at high concentrations and therefore homeostatic mechanisms operate in cells to maintain a proper balance for biological function. These mechanisms can vary from species to species with important differences between prokaryotes and eukaryotes. These differences are reflected by the evolution history of these two cell types. Before examining a number of examples of the strategies employed by microorganisms in resisting metal-ion toxicity, let us consider the following basic questions:

1. Which elements are essential for the growth and cell division of microorganisms?
2. Why were these elements selected in the evolution of microorganisms?
3. What is the role of the geosphere in the uptake of essential elements?
4. What is the role of the biosphere in the selection of these elements?
5. What is known about the mechanisms for transport of essential metal ions into microbial cells?

    Obviously, the uptake of essential elements depends on their chemical and physical properties. Of the 92 elements in the Periodic Table, 30 have been found to be essential to the sum total of microbial life. In addition to the bulk elements carbon, nitrogen, hydrogen, and oxygen, 26 other elements are required in intermediary to trace amounts. The reason for the selection of these elements in the evolution of microorganisms appears to have been determined by abundance in the earth's crust and solubility in water under anaerobic conditions. Table 4.1 presents a list of elements in order of their crustal abundances. Twenty-two of the 26 trace elements listed are found to be essential for microbial growth. Nonessential elements are generally of low abundance and availability in the earth's crust and therefore should not be effective in competing with essential elements in cells through their spe-

Table 4.1. Crustal Abundance of Elements

O, Si, Al[a], Fe, Na, Ca, Mg, K, H, Mn, P, S, C, V, Cl, Cr, Zn, Ni, Cu, Co, N, Pb[a], Sn[a], Br, Be[a], As, F, Mo, W, Tl[a], I, Sb[a], Cd[a], Se . . . all the remaining elements are less than 0.1 μg/g.

[a] These elements have no known biological function.

cific transport systems. However, this situation can change markedly as a result of anthropogenic inputs of nonessential elements which can elevate concentrations by many orders of magnitude in local situations so that nonessential toxic elements compete for transport systems and toxic effects become evident. Williams (1) has listed a number of chemical parameters which must be considered when competition between nonessential and essential metal ions occurs in an environment rich in heavy metals. These parameters are: (1) charge; (2) ionic radius; (3) preference for the coordination of metals to certain organic ligands; (4) coordination geometry; (5) spin pairing between metal ions to provide more stability; (6) available concentrations of metal ions in solution; (7) kinetic controls which are pertinent to metal-ion transport; and (8) the chemical reactivity of the chemical species of the metal in solution (i.e., whether the reactive species functions as an acid or a base).

Some of the principles can now be formulated for the uptake of essential metal ions by microorganisms. The availability of metal ions for transport into cells is restricted by natural abundance and by solubility. The solubility is profoundly influenced by pH and by the standard reduction potential (E^0) of the metal ion under consideration. Both the pH and E^0 can vary widely from outside the cell to inside the cell. For example, many essential transition metals, such as iron, copper, cobalt, chromium, and nickel, occur in high oxidation states outside the cell and low oxidation states inside the cell. Such changes in oxidation state and pH affect the aqueous-ion chemistry as well as steric factors which are important to the mode of binding of essential metals to intracellular active sites. Williams (1) has recognized those factors which can cause the failure of a cell to transport sufficient essential elements. These factors are:

1. Low availability.
2. Excessive competition from other elements (e.g., in a polluted environment).
3. Inadequate synthesis of carrier molecules by the cell.
4. Excessive excretion of the elements by the cell.
5. Failure of the energy-driven uptake system.

Excessive element uptake can occur through the reverse of the above five factors.

With the exception of extensive studies on the transport of iron into microbial cells (2,3) and also the transport of the alkali and earth metals, such as Na^+, K^+, and Ca^+ (1), there is very little basic understanding of the processes involved for the cellular uptake of essential metal ions, such as copper, nickel, and chromium. This area has been neglected in biochemistry, and therefore, until recently, we have not understood some of the mechanisms which are operative in keeping intracellular concentrations of such transition metals below toxic levels. Nonetheless, from our extensive knowledge of the iron and calcium transport systems, it is possible to construct a model which can be tested for a variety of metal ions. At least this model can be used to demonstrate why the cellular uptake of metal ions is not at equilibrium with the external environment. Since most attempts to model the fate of heavy metals assumes that the polluted environment is at equilibrium, we should not be shocked by our lack of understanding of the uptake of metals by biota, which is a very important nonequilibrium component of any pollution situation.

2. A GENERAL MODEL FOR METAL TRANSPORT BY MICROORGANISMS

Let us examine here, in a general way, those parameters which are likely to control metal transport in both prokaryotes and eukaryotes. In what follows we examine this basic problem in terms of both thermodynamics and kinetics.

Figure 4.1 illustrates those parameters which must be taken into consideration in both the transport and the control of intracellular concentrations of essential trace elements. Let us consider a general trace metal M and ligands L_1 through L_4 which can coordinate to M.

The free external concentration of M is determined by environmental factors and so are the stability constants for ML_1 (where L_1 is a variety of ligands, e.g., humic acids, etc.). The internal concentrations are:

1. Supplied by selective carrier ligands L_2.
2. Supplied by ligands which can remove M from solution at the cell surface L_3.
3. Supplied by concentration gradients that are established between the exterior and the interior of the cell through binding to L_2, or by energy-coupled channels and pumps.
4. Controlled by removal of M from solution by special biomolecules, that is, L_4 (e.g., the removal of cadmium by metallothionen).

It is apparent that uptake is controlled by metabolic activity, and of crucial importance is the specific design of L_2 for transport of each specific metal ion (e.g., the transport of iron into bacteria and fungi by siderophores (2,3). L_2 must be able to compete effectively with the external ligand L_1; therefore

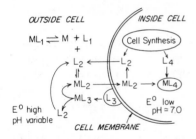

Figure 4.1. Ligands L_2, L_3, and L_4 are synthesized by the cell. Ligand L_1 represents extracellular complexes.

stability constants are of critical importance to element transport. Also ML_2 must have an affinity for cell membranes by reacting at a specific membrane receptor site. This is well established for iron transport in prokaryotes (3). Therefore, there will be a distribution coefficient K_D, for each element in this membrane interaction. Competition between similar elements (e.g., Co^{II} and Ni^{II}) must be taken into account, and the kinetics for ML_2 interactions outside the cell, at the membrane, and inside the cell, is very important in determining intracellular concentrations of M.

It is clear from the above concepts that the accumulation of a trace metal by a cell is not at equilibrium, because metabolic activity is responsible for the synthesis of L_2, L_3, and L_4, as well an energy-coupled channels, pH gradients, and redox-potential differences. Clearly, uptake requires the investment of cellular energy. Therefore the thermodynamics of trace-metal-ion transport depends on the following parameters:

$\log K_{aq}$ = stability constants for ligand/metal-ion interactions in aqueous solution.

ΔpH = internal versus external pH.

ΔE° = internal versus external redox potentials.

Free (M) = outside cell.

Free (L_1) through (L_4) = inside, outside, and in-cell membranes.

The problem becomes even more complicated if one considers these principles for uptake in terms of kinetics. (For example, the rates of input from the environment, the rates for transport through cell membranes, the rates for cellular exclusion or inclusion on cellular traps, and the rates of exit from the cell.) Outside the cell, the availability of a trace metal ion is determined by its abundance and solubility in aqueous solution. Inside the cell, redox potentials and pH gradients are determined by different environmental situations. For example, high pH and E° favor higher oxidation states for trace metals and it follows that low solubility and availability result. Transfer through membranes requires a combination with a carrier or the presence of special channels in membranes (e.g., Ca^+ channels, etc.). Carrier molecules can be specific small molecules or proteins. It will become apparent from the content of this review that there is a great lack of understanding of the selectivity principles involved for trace-metal-ion transport and of those mechanisms which energize inward transport.

3. MICROBIAL LIFE IN METAL-RICH ENVIRONMENTS

Microorganisms which live in extreme environmental conditions are quite common (4). Therefore it is not unusual to find naturally occurring populations of bacteria and algae which tolerate high concentrations of toxic metal ions.

The microbiological leaching of ore is rapidly gaining preference over other technologies for the removal of uranium, strategic metals, and precious metals by the mining industries (5–8). Microorganisms which live in hot springs tolerate not only high temperatures, but low pH and high metal-ion concentrations (4,9). Recent work on microbial life in the deep-sea hydrothermal vents indicates that a number of primitive bacteria are using iron and manganese oxidation as a major energy source for growth and cell division (10). Analysis of vent water shows that it contains very high concentrations of heavy metals in solution, including such trace metals as beryllium (11). It would be expected that the primordial shallow seas, which covered the earth four to five billion years ago, would have a low pH and would be rich in heavy metals. Therefore, the most successful primitive organisms developed strategies to prevent intracellular accumulation of heavy metals to toxic levels.

4. BIOMETHYLATION

Recent analyses on oil shale shows that primitive organisms were actively synthesizing organometals and organometalloids (12). Therefore, biomethylation must have given certain microorganisms selective advantages for the elimination of heavy metals, such as mercury and tin, and for metalloids, such as arsenic and selenium. The synthesis of less-polar organometallic compounds from polar inorganic ions has certain advantages for cellular elimination by diffusion-controlled processes (13,14). The microbial synthesis of organometallic compounds from inorganic precursors is well understood in both the terrestrial environment and in the sea. Mechanisms for B_{12}-dependent synthesis of metal-alkyls have been discovered for the metals Hg, Pb, Tl, Pd, Pt, Au, Sn, Cr, and for the metalloids As and Se (14–19). Also, pathways for the synthesis of organoarsenic compounds have been shown to occur by a mechanism involving S-adenosylmethionine as the methylating coenzyme (20). To date two different mechanisms have been determined for methyl transfer from methyl B_{12} to heavy metals. These mechanisms are: (1) electrophilic attack by the attacking metals on the Co–C bond of methyl B_{12} (Fig. 4.2); and (2) methyl-radical transfer to an ion pair between the attacking metal ion and the corrin-macrocycle (Fig. 4.3). Metal ions which displace the methyl group by electrophilic attack are Hg(II), Pb(IV), Tl(III), and Pd(II). Examples of free-radical transfer are Pt(II)/Pt(IV), Sn(II), Cr(II), and Au(III). The latter mechanism (Fig. 4.3)

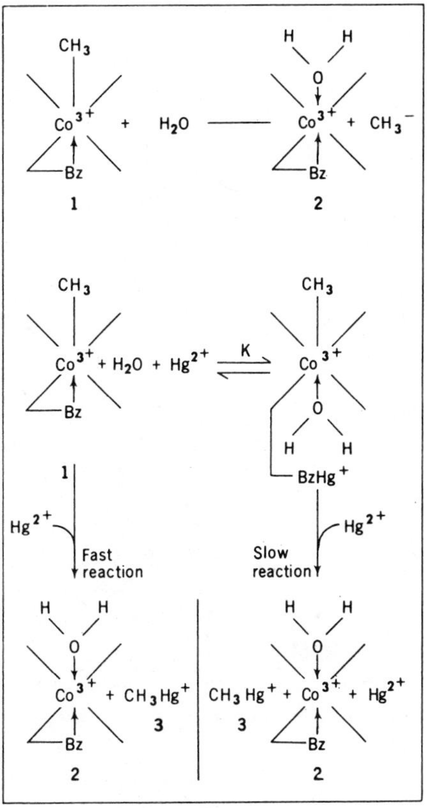

Figure 4.2. Electrophilic attack on the Co–C bond by Hg(II) on both sides of the "base-on" to "base-off" equilibrium (Bz = the benzimidazole axial base).

provides experimental support for Kochi's ideas on charge transfer complex formation and electron transfer (21).

The ecological significance of B_{12}-dependent biomethylation is best illustrated by B_{12}-dependent and B_{12}-independent strains of *Clostridium cochlearium*. The B_{12}-dependent strain is capable of methylating Hg(II) salts to $CH_3Hg(II)$, whereas the B_{12}-independent strain is incapable of catalyzing this reaction. Both strains transport Hg(II) into cells at the same rate, but the B_{12}-independent strain is inhibited by at least a 40-fold lower concentration of Hg(II) than the B_{12}-dependent strain. This result clearly demonstrates that *Clostridium cochlearium* uses biomethylation as a mechanism

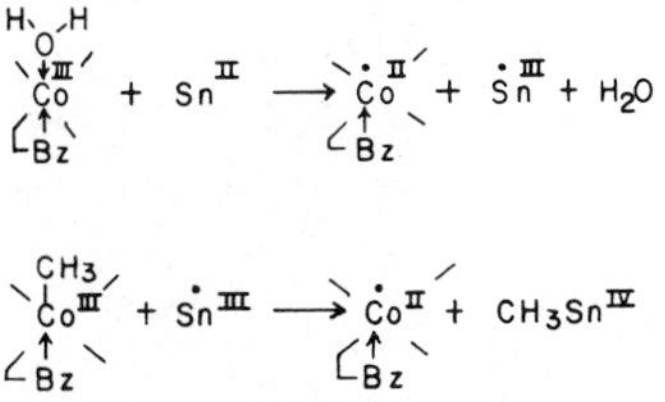

Figure 4.3. Free-radical attack by Sn(II) on the Co–C bond where the resulting methyl-tin radical product is oxidized to give stable CH_3Sn^{IV} in solution. This oxidation step will proceed with single-electron oxidants under strictly anaerobic conditions, or aerobically in the presence of molecular oxygen.

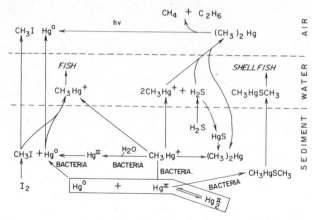

Figure 4.4. The mercury cycle showing those reactions catalyzed by bacteria, the chemical disproportionation by H_2S, and the photochemistry of organomercury compounds.

for detoxification giving the organism a clear advantage in mercury-contaminated systems. This biomethylation capability was shown to be plasmid mediated (22).

Once methyl-mercury is released from the microbial system, it enters food chains as a consequence of its rapid diffusion rate. In the estuarine environment, the reduction of sulfate by *Desulfovibrio* species to produce hydrogen sulfide is quite important in reducing $CH_3Hg(II)$ concentrations by S^{2-}-catalyzed disproportionation to volatile $(CH_3)_2Hg$ and insoluble HgS. It should be pointed out here that there is overwhelming evidence to support the notion that membrane transport of methyl mercury is diffusion controlled. Fluorescence techniques and high-resolution NMR show that diffusion is the key to $CH_3Hg(II)$ uptake (14). Also, a field study of the uptake of $CH_3Hg(II)$ by tuna fish in the Mediterranean fits perfectly the diffusion model for biota in tuna fish food chains (23). An update view of the mercury cycle is presented in (Fig. 4.4).

5. INTRACELLULAR TRAPS

The biosynthesis of intracellular traps for the removal of metal ions from solution represents a temporary measure adopted by cells to prevent metals from reaching toxic levels. This temporary measure precedes mechanisms for the release of these metals from vacuoles by exocytosis. However, such temporary traps can be very effective, for example, the biosynthesis of metallothionin and the removal of cadmium (1), or copper (24) by this sulfhydryl-containing protein. The strategy adopted for the biosynthesis of intracellular traps fits quite closely the predicted partitioning for elements in organic or inorganic matrices. For example, Na, K, Mg, Cu, Al, P, Si, and

B prefer to react with an oxygen-donor matrix; but Cu, Zn, Fe, Ni, Co, Mo, Cd, and Hg prefer a nitrogen- and sulfur-donor matrix. A recent example which we discovered in our laboratory came about through the selection of nickel-tolerant mutants of the cyanobacterium *Synechococcus*. Mutants which would tolerate up to $20 \times 10^{-6} M$ NiSO$_4$ were selected (25) and were found to synthesize large quantities of an intracellular granule (26). Nickel analysis and electron microscopy of sections of these nickel mutants showed that this polymer effectively removes nickel from solution providing an intracellular mechanism to prevent nickel toxicity. Figure 4.5 shows an electron micrograph of a nickel mutant of *Synechococcus* showing the presence of large granules. We failed to isolate similar mutants of *Synechococcus* showing tolerance to $10 \times 10^{-6} M$ CuSO$_4$. However, we found that our

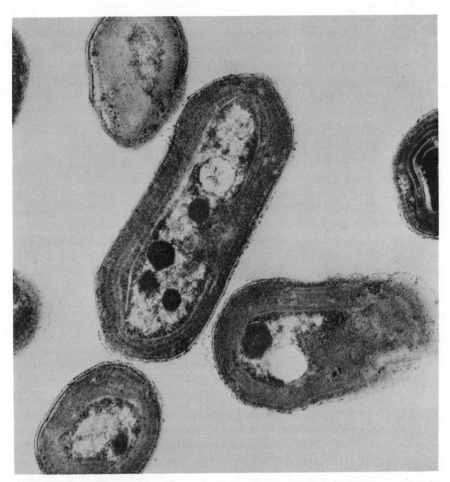

Figure 4.5. Nickel-tolerant *Synechococcus* with electron-dense cyanophycin granules. The organism was grown in $20 \times 10^{-6} M$ NiSO$_4$ (magnification 62,200).

nickel-tolerant mutants were also resistant to copper. This was presumably due to the stronger coordination of copper. Mutants with intracellular trapping mechanisms tend to bioconcentrate the toxic metal intracellularly approximately 200 times over the external concentration, and while this strategy works quite well for some organisms, it does not compare favorably with those organisms which bind, or precipitate, metals extracellularly.

Intracellular concentrations of metal ions can be controlled by deposition on a solid surface as in the crystallization of calcium salts in blood platelets or the removal of copper and nickel by intracellular granules. The concentration of free metal ions can also be controlled by the biosynthesis of ligands in the form of small molecules with high-stability constants. For example, the removal of iron by siderophores fits this category. Energy may be expended by the cell to pump the metal ion out of the cell (e.g., the sodium/ potassium ATPase pump). The cell may synthesize ligands which bind metals strongly at the cell surface or use the activities of surface-bound enzymes to precipitate metals extracellularly.

6. THE BINDING OF METAL IONS TO CELL SURFACES AND TO EXTRACELLULAR LIGANDS

Microorganisms, including the algae, synthesize extracellular ligands which complex metals and prevent their cellular uptake. The research groups of Francois Morel and Pamela Stokes, at MIT and the University of Toronto, respectively, have carried out extensive work on the complexation of copper with a variety of extracellular substances (27–33). Also a considerable literature exists on the toxicity of nickel and copper to algae (34–39). Brown algae, cyanobacteria, and green algae all bioconcentrate nickel (39–44). Recently, we used the isotope nickel-63 in a study of nickel binding by seven different strains of nickel-tolerant algae (45). The cyanobacteria were found to be more sensitive to nickel toxicity than the green algae, which points out differences in transport mechanisms for prokaryotes versus eukaryotes. Both cyanobacteria and green algae could concentrate nickel primarily at the cell surface 3000-fold over the concentration in the culture medium. Both nickel binding and nickel toxicity was shown to be very pH dependent. The optimum pH for binding was between 8 and 8.5. Binding was shown to be rather specific with charge, ionic radius, and coordination geometry being the predominant factors. The only significant competing cation for nickel Ni(II) binding was Co(II). The orientation of ligands at the cell surface must be important, because only surface-active substances, such as humic acids, could effectively compete for nickel binding (45). Table 4.2 shows the ability of these seven strains of nickel-tolerant algae to bioconcentrate nickel simply through proton exchange with surface ligands. Figure 4.6 shows the binding isotherm for one of these strains, *Scenedesmus* ATCC 11460; these data

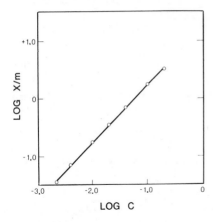

Figure 4.6. The adsorption isotherm for ^{63}Ni(II) by *Scenedesmus* ATCC 11460.

indicate that this organism functions rather similarly to synthetic ion exchange media. Research of this kind on the ability of resistant strains and metal-tolerant mutants to bioconcentrate metals with a high degree of specificity may prove to be of appreciable practical importance. The chemical-processing industries are faced with the problem of removing toxic metals, such as Cd, Cr, Ni, Cu, Hg, and so on, which are present in low concentrations in industrial effluents. This problem can now be addressed by using biotechnology to select algae which will specifically remove metal ions even in the ppm range. Algae can also be used to monitor the presence of toxic metals in effluents due to their ability to rapidly bioconcentrate specific metals and therefore algae could be used as biological indicators.

Table 4.2. Concentration Factor (CF)^a for ^{63}Ni(II)^b by Different Strains of Algae at Different pH Conditions

	pH					
Algae	4	5	6	7	8	9
Scenedesmus ATCC 11460	NS^c	NS	7.9×10^2	1.8×10^3	2.2×10^3	2.0×10^3
Scenedesmus B-4	NS	9.0×10	2.8×10^2	6.6×10^2	1.0×10^3	4.0×10^2
Synechococcus ATCC 17146	NS	NS	4.2×10^2	3.0×10^3	3.3×10^3	3.1×10^3
Synechococcus Nic^7	NS	NS	2.2×10^2	4.4×10^2	5.5×10^2	NS
Oscillatoria UTEX 1270	NS	2.9×10	9.5×10^2	1.1×10^3	1.1×10	NS
Chlamydomonas UTEX 89	2.7×10	3.8×10	5.4×10^2	NS	NS	NS
Euglena UTEX 753	NS	NS	NS	1.7×10	6.9×10^2	NS

^{a} CF = μg of ^{63}Ni(II) removed per gram of alga per μg of ^{63}Ni(II) in the culture medium.

^{b} [^{63}Ni(II)] = 0.02 μg ml^{-1} (0.34 μM), incubated for 6 h at 20°C in the light.

^{c} NS = no significant uptake at the 95% confidence level.

7. THE PRECIPITATION OF METALS BY THE ACTIVITIES OF ENZYMES AT THE CELL SURFACE

The precipitation of insoluble metal complexes occurs through the activities of membrane-associated sulfate reductases (6), or through the biosynthesis of oxidizing agents, such as oxygen or hydrogen peroxide (6). The reduction of sulfate to sulfide, and the diffusion of O_2 and H_2O_2 through the cell membrane, provides a highly reactive region where metals can be complexed and precipitated. This process depends on the metabolic activity of the cell and is closely tied to heavy-metal resistance. Several bacteria have been found which precipitate silver as Ag_2S at the cell surface. Also, certain fungi are very efficient at recovering uranium (5). However, the most interesting organisms are certain strains of green algae which grow in acidic conditions and at high temperatures (4). *Cyanidium caldarium* is such an organism which has the remarkable ability to grow in media containing either 1 N H_2SO_4 or 1 N HCl. A strain of *Cyanidium caldarium* has been adapted to grow at 45° on acid mine water from a copper–nickel mine. This thermophilic alga grows extremely well in sulfuric acid at pH 2.0, and removes toxic metal ions from solution by their precipitation at the cell surface as metal sulfides. Batch cultures of *Cyanidium caldarium* have been grown which remove very high concentrations of metals from solution, for example, 68% of the iron, 50% of the copper, 41% of the nickel, 53% of the aluminum, and 76% of the chromium.

Since *Cyanidium caldarium* is both thermophilic and acidophilic, it is easily grown free of other contaminating autotrophs without having to use sterilization procedures for the acid mine-water medium. This alga has great promise as a biological agent for the recovery of metals and for cleaning metal-polluted waste waters.

In the early 1950s, Allen (8) was the first to report the isolation of *Cyanidium caldarium* from acidic hot springs in California. Since that time substantial research has been done on the microbiology, life cycle, structure, and ecology of this unusual alga (4). This organism has an optimum growth temperature of 45°C and grows very well in the temperature range 33–55°C. Cells grow equally well in the pH range 1.0–4.0. *Cyanidium* adapts to fit a whole range of temperatures and pH without any apparent selection of specialized strains for specific ecosystems. Cultures grow faster when provided with 5% CO_2, and excellent cell yields are obtained under strictly autotrophic conditions (4). However, more than double the cell yield is obtained if cultures are provided with 1% of a soluble carbon source such as glucose. Therefore, *Cyanidium caldarium* grows very well heterotrophically in the dark on a variety of organic substrates (e.g., monosaccharides, disaccharide, mannitol, glycerol, ethanol, succinate, glutamate, lactate, and acetate.) The organism is easily maintained on acid medium with viabilities of stock cultures lasting longer than 1 year (4).

A culture of *Cyanidium caldarium* isolated from the Waimangu Caldron

Outlet, North Island, New Zealand, was slowly adapted to growth in acid mine water by adding 10%-increment increases of mine water to a culture grown under the conditions described by Allen (8). Using this procedure it is possible to select for strains of *Cyanidium caldarium* which grow very well on a culture medium consisting of unfiltered acid mine water provided with 5% CO_2. Table 4.3 presents data on the elemental composition of acid mine water at pH 2.1, and shows the removal efficiencies for stationary-phase cultures of *Cyanidium caldarium* in the presence and absence of glucose plus ammonium sulfate. Figure 4.7 shows a section of *Cyanidium caldarium* examined under the electron microscope; microcrystals of metal sulfides are found to adhere to the external cell membrane. Cells contain approximately 20% metal as the basis of dry weight. Clearly, toxic metals, such as copper, nickel, and chromium, are prevented from entering the cell through an extracellular precipitation mechanism. This result suggessted to us that *Cyanidium caldarium* possess a membrane-associated sulfate reductase system. Heterotrophic cultures which are allowed to attain anaerobic conditions in the dark, do produce hydrogen sulfide gas quite efficiently. Therefore, sulfide precipitation of metals can be regarded as a cellular detoxification mechanism.

Of special interest is the removal efficiencies of chromium and nickel which are present in very low concentrations in the acid mine water. These metals are known to have potential as carcinogens (46) and their selective

Table 4.3.   Composition of Acid Mine Water and Removal Efficiencies for Stationary-phase Cultures of *Cyanidium Caldarium* in the Presence and Absence of Glucose Plus Ammonium Sulfate.

Element[a]	Acid-Mine Water (pH 2.1) (mg liter^{-1})	Culture Supernatant + 1% Glucose + 1% $(NH_4)_2 SO_4$ + 5% CO_2 (mg liter^{-1})	Percent Removed	Culture Supernatant + 5% CO_2 (mg liter^{-1})	Percent Removed
Ca	342	219	36%	277	19%
Mg	456	228	50%	342	25%
Fe	632	205	68%	386	39%
Cu	119	60.6	50%	95	20%
Al	329	155	53%	245	25%
Cr	1.31	0.31	76%	0.38	71%
Na	28.4	13.2	54%	20.9	26%
Ni	4.32	2.55	41%	2.69	28%
P	27.0	13.5	50%	19.1	29%

[a] Analyzed by plasma emission spectroscopy.

Cells grown for 1 week from 1-liter inoculum into 8 liters of acid mine water.

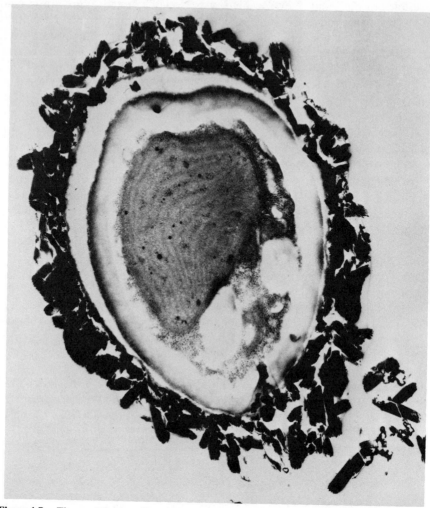

Figure 4.7.   The precipitation of inorganic complexes at the cell surface of *Cyanidium caldarium* grown on acid mine water. The organism was grown on 5% CO_2 in acid mine water at pH 2.1 and 45°C.

removal from waste waters is highly desirable. *Cyanidium caldarium* has several advantages in the development of biotechnology for the recovery of metals from waste waters. The organism can be grown under controlled conditions since it is both acidophilic and thermophilic. Growth on mine water simply requires a source of light and 5% carbon dioxide. Cultures of the organism are very stable, with innocula having a shelf life in excess of 12 months. Sterile microbiological techniques need not be used in growing the organism in continuous culture. Organisms such as *Cyanidium caldarium* may well be effective in treating polluted waters so that effluents can meet federal standards for toxic elements.

8.   CONCLUDING REMARKS

It is clear that microorganisms have evolved a number of mechanisms for regulating the uptake of metal ions. Even today we find microorganisms which are extremely tolerant to heavy metals in extreme environments, such as hot springs, volcanic lakes, deep-sea vents, and metal-contaminated soils. In this paper four strategies for resistance to metal-ion toxicity have been identified. These include (1) biomethylation and transport through cell membranes of the resulting metal alkyl by diffusion-controlled processes; (2) the biosynthesis of intracellular polymers which serve as traps for the removal of metal ions from solution; (3) the binding of metal ions to cell surfaces; and (4) the precipitation of insoluble metal complexes (e.g., metal sulfides and metal oxides) at cell surfaces.

An appreciation of the biochemical processes which are involved in metal-ion transport through membranes helps to explain why it is difficult to model the movement of metals in complex environmental situations. Several mechanisms for transport are available to microbial populations, and some organisms are more resistant to toxic metal ions than others. Physical parameters such as pH, temperature, redox conditions, and the presence of competing cations and anions have a profound effect on toxicity and on the selection principles involved. Chemical speciation is very important also because of toxicity differences between individual complexes. Where we have living organisms there is a nonequilibrium situation at work, and therefore a kinetic approach to metal-ion biomagnification is often more meaningful than a thermodynamic one. After all, the rates at which toxic substances penetrate cells are much more important than the actual concentrations of the toxic substance in solution. This point is often missed by those analysts who put great stock in monitoring toxic substances without any appreciation for the subtleties of the various biological uptake systems. With this in mind it is clearly advantageous to use living organisms as indicators of pollution. Such systems work extremely well when applied to the dynamics of heavy-metal uptake. Coal miners in Wales recognized how useful canaries could be as indicators for natural-gas leaks. When the birds stopped singing and fell off their perches, it was time to get out of the mine. This was one of the first kinetic models for environmental safety, and we need many more such indicator systems for this expanding industrial world.

ACKNOWLEDGMENTS

We wish to acknowledge the NIH Grant AM 18101-90 and Atlantic Richfield Company in support of this research.

REFERENCES

1. R. J. P. Williams, "Physicochemical Aspects of Membrane Transport through Membranes," *Philos. Trans. Roy. Soc. London* **57** (1981) and Dahlem Konferenzen (May 1983), Berlin.

2. J. B. Neilands, "Biomedical and Environmental Significance of Siderophores." In N. Kharasch (Ed)., *Trace Metals in Health and Disease*, Raven Press, New York 1979, pp. 27–43.

3. J. B. Neilands, "Continuous Synthesis of Iron-Regulated Membrane Proteins," *Chemica Scripta* (Proceedings of a Nobel Symposium on Inorganic Biochemistry) **21**, 123–129 (1983).

4. T. D. Brock, *Thermophilic Microorganisms and Life at High Temperatures*, Springer-Verlag, 1978.

5. B. Z. Seigal, "The Bioaccumulation of Uranium by Molds," *Science* **219**, 285–286 (1983).

6. J. M. Wood, F. Engle, and R. Nice (unpublished results) (1983).

7. J. M. Wood, "Selected Biochemical Reactions of Environmental Significances," *Chemica Scripta* **21**, 157–162 (1983).

8. M. B. Allen, "Studies with *Cyanidium caldarium*, a Pigmented Chlorophyte," *Arch. Mikrobiol.* **32**, 270–277 (1959).

9. J. Seckbach, F. A. Baker, and P. M. Shugarman, "Algae Thrive under Pure Carbon Dioxide," *Nature* **227**, 744–745 (1970).

10. J. M. Edmond and K. VanDamm, "Hot Springs on the Ocean Floor," *Sci. Am.* **248**, 4–10 (1983).

11. K. Van Damm, speech at a Gordon Research Conference on Chemical Oceanography, Ventura, California, February 1983.

12. Dr. R. Fish, Lawrence Radiation Laboratory, Berkeley, California has discovered organoarsenic compounds in oil shale (personal communication).

13. J. M. Wood, "Biological Cycles for Elements in the Environment," *Naturwissenschaften* **62**, 357–364 (1975).

14. J. M. Wood, A. Cheh, L. J. Dizikes, W. P. Ridley, S. Rackow, and J. E. Lakowicz, "Biochemical Pathways for Toxic Elements," *Fed. Proc.* **37**, 16–21 (1978).

15. W. P. Ridley, L. J. Dizikes, and J. M. Wood, "Biomethylation of Toxic Elements in the Environment," *Science* **197**, 329–332 (1977).

16. W. P. Ridley, L. J. Dizikes, A. Cheh, and J. M. Wood, "Recent Studies on Biomethylation and Demethylation of Toxic Elements," *Environ. Health Pers.* **19**, 43–46 (1977).

17. J. M. Wood, "The Biochemistry of Toxic Elements." In A. Ehrenberg, R. D. Keynes, and G. Felsenfield, (Eds.), *Quarterly Review of Biophysics*, Vol. 9, No. 2, pp. 467–479 (1978).

18. J. M. Wood, Y.-T. Fanchiang, and W. P. Ridley, "Biochemical Pathways for Toxic Elements." In F. E. Brinckman and J. M. Bellama (Eds.), *ACS Monograph*, No. 82, 56–64 (1978).

19. P. J. Craig and J. M. Wood, "The Biological Methylation of Lead." In D. R.

Lynman, L. E. Piantanida, and J. F. Cole (Eds.), *Environmental Lead*, Academic Press, New York, 1981, pp. 333–357.

20. B. C. McBride, H. Merilees, W. R. Cullen, and W. Pickett, "Anaerobic and Aerobic Methylation of Arsenic." In F. Brinckman and J. M. Bellama (Eds.), *ACS Monograph*, No. 82, 94–116 (1978).

21. J. K. Kochi, "Mechanisms for Alkyl-Transfer in Organometals." In, F. Brinckman and J. M. Bellama (Eds.), *ACS Monograph*, No. 82, 205–235 (1978).

22. H. S. Pan-Hou and N. Imura, "Involvement of Mercury Methylation in Microbial Detoxification," *Arch. Microbiol.* **131**, 176–177 (1982).

23. R. Buffoni, "A Model for the Accumulation of Mercury in Tuna Fish in the Mediterranean." In Proceedings of the Seventh International Conference on the Chemistry of the Mediterranean, Primosten, Yugoslavia, May 6th–12th; *Thalassia Jugoslav.* (1982).

24. K. Lerch, "Metallothionen, Some Aspects of Its Structure and Function in Binding Copper," *Chemica Scripta* **21** (1983) (in press).

25. R. Smith and F. K. Gleason, "Copper and Nickel Mutants of *Synechococcus* sp.," *Archiv. Microbiol.* (1983) (in press).

26. R. D. Simon and P. Weathers, "Determination of the Structure of a Novel Polypeptide Containing Aspartic Acid and Arginine Which is Found in Cyanobacteria," *Biochim. Biophys. Acta* **420**, 165–173 (1976).

27. D. M. McKnight and F. M. M. Morel, "Release of Weak and Strong Copper-Complexing Agents by Algae," *Limnol. Oceanogr.* **24**(5), 823–832 (1979).

28. K. C. Swallow, J. C. Westall, D. M. McKnight, and F. M. M. Morel, "Potentiometric Determination of Copper Complexation by Phytoplankton Exudates," *Limnol. Oceanogr.* **23**(3), 538–542 (1978).

29. N. M. L. Morel, J. G. Rueter, and F. M. M. Morel, "Copper Toxicity for *Skeletonema costatum*," *J. Phycol.* **14**(1), 43–48 (1978).

30. P. M. Stokes and S. I. Dreier, "Copper Requirement of a Copper-Tolerant Isolate of *Scenedesmus* and the Effect of Copper Depletion on Tolerance," *Can. J. Bot.* **59**(10), 1817–1823 (1981).

31. P. M. Stokes "Copper and Nickel-Tolerant Algae," 13th International Botanical Congress, *Proc. Int. Bot. Congr.* **13**, 84 (1981).

32. G. Mierle and P. M. Stokes, "Heavy Metal Tolerance and Metal Accumulation by Planktonic Algae," Proceedings of University Missouri Annual Conference on Trace Substances and Environmental Health, Vol. 10, pp. 113–122 (1976).

33. D. F. Spencer, In J. O. Nriaga (Ed.), *Nickel in the Environment*, Wiley–Interscience, New York, 1980, pp. 339–349.

34. J. M. Hassett, J. H. Jennett, and J. E. Smith, "Microplate Technique for Determining Accumulation of Metals by Algae," *Appl. Environ. Microbiol.* 1097 (1981).

35. J. S. Fezy, "Effect of Nickel on the Growth of the Freshwater Diatom *Navicula pelliculosa*," *Environ. Pollut.* **20**(2) 131 (1979).

36. G. W. Stratton and C. T. Corke, "Effect of Mercuric, Cadmium and Nickel Ion Combinations on a Blue Green Alga," *Chemosphere* **8**(10), 731 (1979).

37. D. F. Spencer and R. W. Green, "Effects of Nickel on Seven Species of Freshwater Algae," *Environ. Pollut. SERA* **25**(4), 241 (1981).

38. P. T. S. Wong et al., "Toxicity of a Mixture of Metals on Freshwater Algae," *Can. Fishers Res. Board J.* **35**(4), 479 (1981).

39. P. Foster, "Concentrations and Concentration Factors of Heavy Metals in Brown Algae," *Environ. Pollut.* **10**(1), 45 (1976).

40. R. Fuge, "Trace Metal Concentation in Brown Seaweeds from Cardigan Bay, Wales," *Marine Chem.* **1**(4) 281 (1973).

41. P. M. Sivalingam and I. Rodziah. "*Cladophora fascicularis* as a Prominent Global Algal Monitor for Trace Metal Pollutants." *JPNJ Phycol.* **29**(3), 171 (1981).

42. Akira, Kurata, Yoshida Yoichi, and Taguchi, "Accumulation of Metals by Marine Algae," *MER (Tokyo)* **18**(2), 1 (1980).

43. Antonio Ballester and Josefina Castelvi, "Bioaccumulation of Vanadium and Nickel by Marine Organisms and Sediments," *INVEST PESQ.* **44**(1), 1 (1980).

44. H. Hischberg, H. Skane, and E. Thorsby, "Nickel Accumulation in Chlorophytes," *Plant Cell Physiol.* **16**, 1167 (1977).

45. Hong-Kang Wang and J. M. Wood, "The Bioaccumulation of Nickel by Algae," *Environ. Sci. Technol.* (in press).

46. *Trace Metals in Health and Disease*, N. Kharasch (Ed.) Raven Press, New York, 1978.

5

CARBON ISOTOPES AND PRODUCTIVITY IN THE LACUSTRINE AND MARINE ENVIRONMENT

Judith A. McKenzie

Geology Institute, Swiss Federal Institute of Technology (ETH), Zurich, Switzerland

Abstract

The carbon-isotope composition of the dissolved inorganic carbon in lacustrine and marine waters is primarily controlled by the photosynthesis–respiration cycle. Carbon-12 is essentially transferred from the surface waters to the deeper waters by a photosynthesis–respiration pathway; photosynthetically produced organic matter is enriched in carbon-12, and as it sinks, it is oxidized, releasing carbon-12 enriched CO_2 to underlying waters. This isotope fractionation produces a similar $\delta^{13}C$ profile in both lakes and oceans; the $\delta^{13}C$ values of the surface waters are more positive, while the values of intermediate to deep waters are relatively more negative. As carbonate precipitates are basically in isotopic equilibrium with the dissolved inorganic carbon, calcium carbonate (which is precipitated in situ within the water column either as a biogenically induced precipitate or the test* of a microorganism) contains a carbon-isotope composition characteristic of the water depth. Significant changes in the rate of surface-water productivity are reflected by fluctuations in the $\delta^{13}C$ values incorporated into the surface-water carbonates. For example, an increased nutrient supply to an aquatic basin promotes increased productivity and more and more carbon-12 is removed from the surface waters by the sinking organic matter,

* Test = hard part.

which may be subsequently buried as the oxygen of the bottom waters becomes depleted. Simultaneously, the $\delta^{13}C$ value of the surface-water carbonates trends towards more and more positive values. Carbon-isotope stratigraphy in lacustrine and marine sediments is therefore a measure of the paleoproductivity of the basin.

1. INTRODUCTION

The carbon cycle in both the lacustrine and marine environments is intrinsically related to surface-water productivity. During photosynthesis, the phytoplankton incorporate the dissolved inorganic carbon into organic matter. This process leads to carbon-isotope fractionation as the lighter carbon isotope (^{12}C) is preferentially assimilated into the organic matter resulting in an enrichment of the heavier isotope (^{13}C) in the remaining dissolved inorganic carbon. In the case of lacustrine environments, photosynthesis has a second effect on the carbon balance. The removal of dissolved CO_2 produces a disequilibrium in the bicarbonate–carbonate system, the pH rises, and calcium carbonate precipitates. Since the precipitate is in apparent isotopic equilibrium with the dissolved carbonate ions, it records the carbon-isotope composition of the surface waters. Therefore, variations in the carbon-isotope ratio of the surface waters due to changes in the rate of productivity should be recognizable in the isotopic composition of the authigenic carbonate component in the underlying sediments. A study of the carbon-isotope stratigraphy in cored sediments provides a history of the productivity within a lake basin.

In the marine environment, calcium carbonate sediments are oozes comprising biogenically secreted tests of nannoplankton, foraminifera, and, sometimes, pteropods. Isotopic analysis of the fossil remains of these phyto- and zooplankton from the pelagic realm reveals that they contain a record of carbon-isotope variations related to changes in the carbon cycle throughout geologic time. These natural variations are frequently related to changes in the rate of surface-water productivity. An evaluation of the processes causing carbon-isotope fluctuations in the marine record becomes extremely interesting in light of modern man's perturbation of the carbon cycle by increasing atmospheric CO_2 through the combustion of fossil fuels. An understanding of the causes of past geologic events may aid in predicting the consequences of modern perturbations of the oceanic atmosphere system. For this reason, the study of productivity changes in lakes using the carbon-isotope signal in the sediments has immediate relevancy as biogeochemical processes in the lacustrine and marine environment are often analogous.

Lakes are natural "beakers" in which geochemical processes can be effectively studied. In general, they are much quicker to respond to environmental pressures than ocean basins, and, because of the smaller size of the reservoirs, the geochemical signals of such perturbations are amplified.

Therefore, lakes serve as an ideal medium in which to evaluate the effects of productivity changes on the carbon-isotope balance in a dynamic system. Lake models can be developed and subsequently used to interpret isotope data obtained from the marine environment.

This chapter proposes to introduce the reader to the process of carbon-isotope fractionation in the lacustrine environment and to demonstrate its relationship to productivity. This will be accomplished by using actual examples from studied lakes. A simple model for productivity-induced, carbon-isotope fractionation evolves from this interpretation, which can be likewise applied to carbon-isotope data from the marine record. Further, the recognition of comparable biogeochemical controls on carbon-isotope fractionation in both lacustrine and marine environments adds new insight to our understanding of how these processes operate; in this case, the overall effect of organic productivity on the carbon cycle is discernible.

2. CARBON ISOTOPES

Carbon is one of the most abundant elements in nature and is undoubtedly the most important element in the biosphere. Carbon has two stable isotopes with the following abundances: $^{12}C = 98.89\%$ and $^{13}C = 1.11\%$. Although carbon isotopes have chemically identical behavior, their mass difference results in different vibrational frequencies or reaction rates for the isotopic molecules. This mass difference leads to the natural fractionation of carbon isotopes. In the aquatic environment, two fractionation processes are most important: (1) photosynthesis which leads to the enrichment of ^{12}C in organic matter and (2) isotope exchange reactions between CO_2 gas and the aqueous carbonate species which leads to the enrichment of ^{13}C in the carbonate phase. In this discussion of carbon isotopes and productivity, photosynthesis is the dominant fractionation process and, along with the subsequent respiration of the organic matter, produces a distinctive carbon-isotope signal, which is recorded in the carbon-isotope content of the dissolved inorganic carbon and calcium carbonate precipitates. [See Faure (1) or Hoefs (2) for a more comprehensive discussion of carbon isotopes and their fractionation in nature, and Pearson and Coplen (3) for a discussion of stable isotopes and lakes.]

Using a mass spectrometer, the stable-isotope ratios of dissolved carbon dioxide or carbonate precipitate are measured on CO_2 gas previously released from the analyzed samples. The methodology for the analysis of the isotopic composition of lacustrine carbonates is discussed by the author (4). The isotope ratios are relative measurements and are reported as such in the "delta" notation relative to the international standard, Peedee Belemnite (PDB):

$$\delta^{13}C_{PDB} \ (\permil) = \left(\frac{(^{13}C/^{12}C)_{sample}}{(^{13}C/^{12}C)_{standard}} - 1 \right) \times 10^3.$$

When the oxygen-isotope values are reported, the ratio is then a measurement of ^{18}O to ^{16}O. Oxygen-isotope equilibrium between carbonate precipitates and water is temperature dependent. The $\delta^{18}O$ value of the carbonate is determined by the temperature and the $\delta^{18}O$ value of the water from which it was precipitated.

3.   DISSOLVED INORGANIC CARBON

3.1.   Lacustrine Environment

The carbon-isotope budget in a lake is controlled by three processes: (1) the isotopic composition of the inflow; (2) CO_2 exchange between the atmosphere and water; and (3) photosynthesis and respiration. The dominance of the latter process can be dramatically demonstrated by monthly, carbon-isotope profiles of the dissolved inorganic carbon in lake waters. A good example of such profiles collected during a single year from Lake Greifen, a small freshwater lake in northeastern Switzerland, illustrates this phenomenon (4). During the winter and early spring when photosynthetic activity is very low, the $\delta^{13}C$ value of dissolved inorganic carbon remains relatively constant throughout the water column, with an average value of about -11.5 ‰ (Fig. 5.1). With the beginning of mass blooms in late spring–early summer,

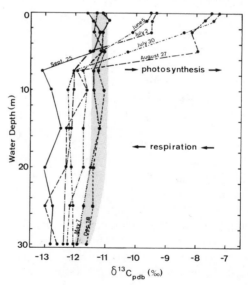

Figure 5.1. Monthly $\delta^{13}C$ profiles of the dissolved inorganic carbon in Lake Greifen. The shaded area represents the range of $\delta^{13}C$ values found between December and May. The carbon-13 increase in surface waters due to photosynthesis and the carbon-13 depletion in deep waters resulting from the respiration of the sinking organic matter are depicted for the summer months from May to September. [Modified from McKenzie (4).]

a significant increase in the $\delta^{13}C$ value of dissolved inorganic carbon in the surface waters is particularly noticeable and continues to increase throughout the summer months reaching a maximum of about -7.5 ‰. Simultaneously, the $\delta^{13}C$ value of water below 5 m becomes increasingly more negative approaching -13.0 ‰.

The partition of the carbon isotopes between the surface and bottom waters is readily explained by carbon-isotope fractionation during photosynthesis and subsequent respiration of the sinking organic matter (5). The onset of the mass blooms occurs after the early spring overturn which brings a fresh supply of nutrients to the surface waters. The phytoplankton, diatoms, and other planktonic algae preferentially incorporate carbon-12 into organic matter, and this process depletes the dissolved inorganic carbon of carbon-12 atoms. On an average, lacustrine plankton have $\delta^{13}C$ values of about -30 ‰ (5). As the decaying plant material sinks below the photic zone, carbon-12 enriched CO_2 is released to the underlying waters. This regeneration of CO_2 can be traced by the utilization of dissolved oxygen (Fig. 5.2). During the summer months, the dissolved oxygen is essentially exhausted in waters below 5 m, due to the oxidation of the large amounts of organic matter in Lake Greifen. Photosynthesis and respiration can also be traced by changes in concentration of dissolved inorganic carbon in near-surface and bottom waters of Lake Greifen throughout the year (6). As shown in Figure 5.2, the concentration of H_2CO_3 at 1 m decreases 50-fold due to

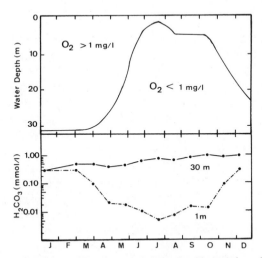

Figure 5.2. Monthly dissolved oxygen (above) and dissolved inorganic carbon (below) profiles from Lake Greifen. The oxygen profile (unpublished data from EAWAG, Switzerland) illustrates the depletion of the oxygen content of waters below 2.5 m due to respiration during the summer and early fall months. Not illustrated by this graph is the oxygen supersaturation of the surface waters during peak periods of photosynthesis. The dissolved inorganic carbon profiles, calculated as H_2CO_3 by Weber (6), depict the utilization of CO_2 in the surface waters (1 m) during photosynthesis and the regeneration of CO_2 to the bottom waters (30) during respiration.

the consumption of carbon dioxide during the summer months, while a less drastic but significant increase in concentration occurs at 30 m due to the oxidation of organic matter.

The Lake Greifen example can serve as a model for assessing the relationship between productivity and carbon-isotope fractionation. It shows clearly that significant fractionation can occur between organic matter and dissolved inorganic carbon. Further, it can be postulated that changes in the magnitude of surface-water productivity will be reflected by increasing or decreasing $\delta^{13}C$ values for the dissolved inorganic carbon. An important corollary to this proposed relationship between productivity and carbon isotopes is the prerequisite of a sufficient nutrient supply to promote the production of organic matter. Increasing productivity requires an increase in the flow of nutrients to the photic zone. In the case of Lake Greifen, modern pollution has supplied sufficient nutrients, such as phosphate and nitrate ions, to result in eutrophication.

3.2. Marine Environment

Analogous to the lacustrine environment, variations in carbon-isotope content of marine waters are primarily controlled by the photosynthesis–respiration cycle (7). Accordingly, during marine photosynthesis, organic matter enriched in carbon-12 is produced and subsequently descends into deeper water where it is oxidized. Carbon-12 is essentially transferred from the surface into the deeper waters. Although temperature-dependent exchange of CO_2 between the atmosphere and sea can also regulate the carbon-isotope composition of the dissolved inorganic carbon, the biological mechanism apparently dominates.

As in lakes, the effects of photosynthesis and respiration on the carbon-isotope balance in the sea are readily illustrated by profiles of geochemical measurements versus water depth (8). Figure 5.3 is a composite of $\delta^{13}C$, dissolved oxygen, and dissolved inorganic carbon profiles from a typical midlatitude Pacific Ocean site. The surface $\delta^{13}C$ value is +2.1 ‰, but becomes increasingly more negative with depth, approaching a minimum value of about +0.2 ‰ at 2.5 km. Over this same depth interval, there is a corresponding decrease in dissolved oxygen plus an increase in dissolved inorganic carbon. Apparently, these parameters indicate the control of photosynthesis and respiration on the distribution of carbon isotopes in marine waters. Extensive geochemical data indicate that there is an inverse relationship between $\delta^{13}C$ and phosphate concentration of marine waters, Δ $\delta^{13}C/\Delta$ PO_4 = 0.93 ‰ μm^{-1} kg^{-1} (7). In other words, the utilization of phosphate during photosynthesis corresponds to increasing $\delta^{13}C$ values for the dissolved inorganic carbon, while respiration of organic matter releases both phosphate ions and carbon-12 enriched CO_2 to seawater. Numerous carbon-isotope profiles from the Atlantic, Pacific, and Antarctic Oceans

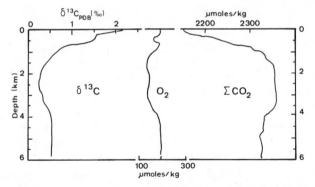

Figure 5.3. Profiles of carbon-13 content of dissolved inorganic carbon ($\delta^{13}C$), dissolved oxygen (O_2), and total dissolved inorganic carbon (ΣCO_2) from a typical midlatitude Pacific Ocean station (17°S, 172°W). This graph illustrates the control of photosynthesis and respiration on these three parameters in the marine environment; that is, carbon-12 and CO_2 are removed from the surface waters during photosynthesis and transferred to the intermediate waters by the oxidation of the sinking organic matter. [Redrawn from Kroopnick et al. (8).]

show a distinct pattern, whereby the $\delta^{13}C$ values of warm surface waters are 1 to 2 ‰ more positive than the underlying deep waters (7). In areas of high productivity (e.g., upwelling zones), the $\delta^{13}C$ values of the intermediate waters, as well as other geochemical parameters, reflect an increased input of CO_2 derived from organic matter. For example, this trend is seen in N–S horizontal sections of $\delta^{13}C$ in the southeastern Atlantic Ocean in the region of the highly productive upwelling waters near the west coast of Africa, where shallow dissolved O_2 minima and nutrient maxima mark the zone of respiration (9).

4.   CARBONATE PRECIPITATES

4.1.   Lacustrine Environment

During the seasonal periods of mass blooms, the CO_2 content of the surface water can be depleted to the point of supersaturation with respect to calcium carbonate (10). In freshwater lakes, the biologically induced precipitate is calcite, which apparently precipitates in isotopic equilibrium with the dissolved inorganic carbon. The $\delta^{13}C$ values of calcite collected over two-day intervals in sediment traps set in Lake Greifen indicate the attainment of carbon-isotope equilibrium with the surface-water bicarbonate (McKenzie, unpublished data), as predicted by experimental studies of carbon-isotope fractionation during the precipitation of calcium carbonate. The calcite is more enriched in carbon-13 than the dissolved bicarbonate, but the carbon-isotope signal sealed in the calcite at the time of precipitation corresponds to the $\delta^{13}C$ value of the dissolved inorganic carbon. For example, at 20°C the $\delta^{13}C$ value of calcite is 1.85 ‰ more positive than the corresponding bicarbonate (11). Any change in the value of the dissolved inorganic carbon will be reflected by a parallel change in the precipitate. Therefore, if the carbon-13 content of the dissolved inorganic carbon increases due to increased productivity, as in Figure 5.1, the $\delta^{13}C$ value of the calcite increases. The carbon-isotope ratio of the precipitate records the degree of productivity in the surface waters.

As mentioned previously, Lake Greifen is eutrophic, a modern example of the effects of human pollution. Its lacustrine sediments are composed predominantly of calcite which was precipitated inorganically in the surface waters during periods of peak biological activity. Since about 1940, the bottom waters remain anoxic throughout most of the year, which prevents bioturbation and produces varved sediments. These annual layers of calcite precipitate can be individually counted, separated, and analyzed for their stable-isotope composition (4). Figure 5.4 presents the data for varves formed between 1955 and 1978. From about 1952 to 1965, the phosphate content of the lake waters increased at least fivefold (12). This corresponds to an increase of $\delta^{13}C$ values from about -7 to -5 ‰. As the average preeutrophic $\delta^{13}C$ value of the sediments is -7.1‰, the trend towards more positive values is interpreted as the direct consequence of increased productivity related to the increased nutrient supply (4). Yearly fluctuations in the carbon-isotope content of the varves could be related to variable production rates dependent on other factors, such as weather. It should be noted

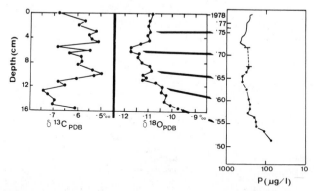

Figure 5.4. Stable-isotope data (McKenzie, unpublished) from the calcite laminae of varved sediments deposited between 1955 and 1978 in Lake Greifen compared with the phosphate content of the waters (12) during the same period. The coresponding variation of these three annual parameters can be interpreted as follows: the increased influx of nutrients (PO_4^{2-}) between 1955 and 1972 promoted increased productivity as reflected by the increased carbon-13 content of the calcite precipitated from the surface waters. Because phosphate ions can inhibit calcite precipitation (13), the decrease in $\delta^{18}O$ between 1955 and 1972 is apparently related to the temperature of the surface waters, or the time of summer in which the average calcite precipitated—that is, the more phosphate, the greater its inhibiting effect and the later in summer precipitation occurred. After 1972, the phosphate flux decreased resulting in decreased productivity and more negative $\delta^{13}C$ values and a reversion in the $\delta^{18}O$ curve.

that attempts to reduce pollution in the lake have resulted in a decreasing phosphate content since 1973, which corresponds to an isotopic trend towards more negative $\delta^{13}C$ values.

Although oxygen isotopes are not the theme of this paper, the data for the varved sediments are included in Figure 5.4 because of a possible relationship between the oxygen signal and productivity. From 1955 to 1970 the $\delta^{18}O$ value becomes increasingly negative by about 2 ‰. It is reasonable to assume that the $\delta^{18}O$ value of the lake water did not change during this period, so the isotope change must represent a temperature increase of 10°C. Since the air temperature in Switzerland did not increase by 10°C over this period, the temperature effect must lie in the lake water itself. Surface-water temperatures in Lake Greifen increase steadily from 4°C in March to 26°C in July. The oxygen-isotope content of the calcite will then depend upon the time of year it precipitates. Also, note in Figure 5.4 that the $\delta^{13}C$ and $\delta^{18}O$ values tend to diverge with increasing phosphate concentrations; with greater amounts of productivity, the calcite has a tendency to precipitate from warmer waters later in the summer. This delayed precipitation may be related to phosphate inhibition of calcite precipitation (13) as the record shows a turn around in the $\delta^{18}O$ ratio after the antipollution measures began to reduce the phosphate concentration.

If the $\delta^{13}C$ values of the surface-water precipitate is a true reflection of the value of the dissolved inorganic carbon and shows changes in the rate of photosynthesis, calcite precipitated in the bottom waters should reflect carbon-isotope changes related to respiration. An elegant example of this phenomenon is an isotopic study of bivalves from Lake Zürich sediments (14). The isotopic composition of a single species of bivalves was measured for samples taken across the last glacial–interglacial boundary. As seen in Figure 5.5, the glacial bivalves have a $\delta^{13}C$ value of about -5 ‰, which steadily decreases after the boundary to -9 ‰. This decrease corresponds to a gradual sixfold increase in the percent organic carbon in the interglacial sediments (15). Although the question of isotope equilibrium has not been established for these bivalves, the significant decrease in their $\delta^{13}C$ value of up to 4 ‰ surely represents an increased influx of carbon-12 enriched organic matter, which was subsequently oxidized and built into their shells. The trend towards more negative $\delta^{13}C$ values (more respiration) indicates a postglacial increase in surface-water productivity, as also suggested by the organic content of the sediments.

4.2. Marine Environment

The calcium carbonate sediments in the deep-sea environment are ooozes composed of biogenically secreted tests of plants and animals living in the pelagic realm or at the sediment–water interface. There are two major types of carbonate oozes: nannofossil and foraminiferal. Nannofossil oozes are a

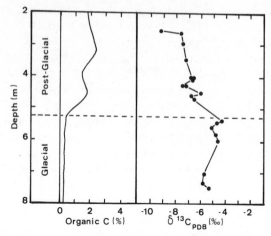

Figure 5.5. Carbon-isotope content of *in situ* bivalves, *Pisidium conventus,* from deep-water sediments of Lake Zürich (Lister, unpublished data) and the corresponding organic carbon content (15). The increase in the carbon-12 content of the postglacial bivalves indirectly reflects increased surface-water productivity with the climatic amelioration. As shown by the increased organic carbon content of the sediments, more organic matter entered the bottom waters causing increased respiration and the release of carbon-12 enriched CO_2. This CO_2 was subsequently assimilated by the bivalves, progressively increasing the carbon-12 content of their shells as the surface-water productivity increased.

fine-grained sediment (<30 μm) dominated by the calcareous tests of nannoplankton of the recent algae family coccolithophoridae and the extinct group called discoasters. These phytoplankton live within the photic zone where there is sufficient light to promote photosynthesis. Foraminiferal oozes are dominated by the calcareous tests of foraminifera, single-celled animals largely of sand size (>61 μm). Planktic foraminifera live in the surface waters and also at intermediate depths, while the benthic foraminfera dwell on or in the bottom sediments. In general, a foraminiferal ooze contains abundant nannofossils.

With both nannoplankton and foraminifera, the carbon-isotope content of the dissolved inorganic carbon is incorporated into the secreted calcareous material. It is debatable whether carbon-isotope equilibrium is obtained between the deposited carbonate and the dissolved inorganic carbon. Unlike inorganic lacustrine precipitates, a biological or vital effect influences the isotope fractionation during the precipitation of the carbonate. In spite of this effect, the $\delta^{13}C$ value of the fossils does have a relationship to the $\delta^{13}C$ value of the dissolved inorganic carbon varying as the $\delta^{13}C$ of the water changes. As seen in Figure 5.3, the photosynthesis–respiration cycle results in a wide range of $\delta^{13}C$ values within the water column. Nannoplankton and foraminifera living in the surface waters, the zone of productivity, exhibit $\delta^{13}C$ values more positive than those for foraminifera living in intermediate waters where respiration occurs. Benthics have even more negative $\delta^{13}C$

values than the planktics as more organic-derived CO_2 exists within the
bottom waters of the oceans.

The partitioning of the carbon isotopes in the water column by photosynthesis–respiration processes can be illustrated by a $\delta^{18}O$ vs. $\delta^{13}C$ plot.
Sea water becomes progressively colder with depth, having in the open ocean
a possible temperature range from about 27 to 2°C. Since oxygen-isotope
fractionation has a significant temperature dependency, the $\delta^{18}O$ value of
the fossil is related to the water depth in which it lived; that is, the heavier
the $\delta^{18}O$ value, the deeper (colder) the water. A plot of $\delta^{18}O$ vs. $\delta^{13}C$ shows
the variation of the $\delta^{13}C$ value of the dissolved inorganic carbon with depth.

In a recent work, Biolzi (16) illustrates this phenomenon using the isotopic
composition of fossils from Oligocene–Miocene (~30 million years ago) sediments from equatorial Atlantic Ocean, DSDP Site 354. As seen in Figure
5.6, the nannofossils and planktic foraminifera (*Globigerinoides spp.* and
Globorotalia mayeri) living in the surface waters have the most positive $\delta^{13}C$
values (+ 1.0 and 2.5 ‰), an indication of the effect of photosynthesis on
the carbon-isotope composition of the dissolved inorganic carbon. Planktic
foraminifera (*Globigerina venezuelana*) living in the intermediate waters
have a more negative $\delta^{13}C$ value (+ 0.5 to 1.5 ‰), an indication of the effect
of respiration on the carbon-isotope composition of the dissolved inorganic
carbon. The very deeply dwelling planktic foraminifera (*Catapsydrax dis-
similis*) and the benthic foraminifera (*Cibicidoides pseudooungerianus*) have
even more negative $\delta^{13}C$ values (−0.5 to 1.0 ‰), a consequence of the

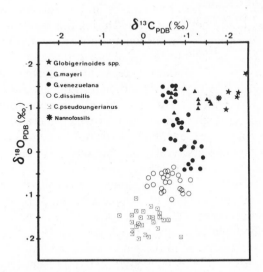

Figure 5.6. A plot of $\delta^{18}O$ vs $\delta^{13}C$ for representative planktic and benthic microorganisms
from middle Oligocene–early Miocene equatorial Atlantic sediments, DSDP Site 354. The graph
illustrates the variation of the carbon-13 content of the microfossils as a function of their depth
habitat and indirectly the control of photosynthesis and respiration on the carbon-13 content
of the dissolved inorganic carbon in the water column. [Modified from Biolzi (16).]

availability of abundant amounts of CO_2 derived from organic matter in the bottom waters and the sediments.

As in the lacustrine environment, the effect of the photosynthesis–respiration cycle on the marine carbon-isotope budget is recorded in the calcareous sediment, as illustrated in Figure 5.6. Likewise, it is proposed any changes in the rate of productivity should also be recorded in the isotopic signal of the surface-water dwellers; that is, increases in the carbon-13 content correspond to increased productivity and vice versa.

5. CARBON ISOTOPES AND PALEOPRODUCTIVITY

5.1. Lacustrine Environment

The carbon-isotope stratigraphy of lacustrine sediments can be a useful tool for evaluating paleoproductivity in a lake. The Lake Greifen example illustrated the relationship between the carbon-isotope content of the lacustrine calcite and productivity changes resulting from changes in the nutrient supply (Fig. 5.4). Looking at a Lake Greifen short core with sediments from the past 190 years (Fig. 5.7), the evolution from a mesotrophic to an eutrophic lake can be traced visually and chemically (4). Between 1788 and 1878, the sediments were homogeneous, bioturbated marl with an average carbonate content of 65% and a $\delta^{13}C$ value of -7.1 ‰. Between 1878 and 1928, the lake became gradually more productive as indicated by the presence of an increasing amount of black sulfide streaks in the sediment, a sign of partial oxygen depletion in the bottom waters. During this transitional phase from mesotrophic to euthrophic, the carbonate content increased to 85%, while the $\delta^{13}C$ value became slightly more negative (-7.5 ‰). These changes represent the onset of the modern increase in the nutrient flux to the lake promoting increased photosynthesis and, hence, the quantity of biogenically induced calcite precipitation. The shift towards more negative $\delta^{13}C$ values can be interpreted as increased respiration which increases the carbon-12 content of the deeper waters eventually to be refluxed into the surface water. After 1928, varved couplets, yearly alternations between white calcareous layers and black gelatinous layers rich in organic matter and diatoms, developed. After 1955, the $\delta^{13}C$ value became more positive approaching -5.0 ‰, while the carbonate content of the calcareous layers approached 90%. The preservation of varves occurs when the oxygen content of the bottom waters is severely depleted during most of the year (Fig. 5.2). These anoxic conditions are not conducive to the bioturbating activities of bottom dwellers. The more positive $\delta^{13}C$ values after 1955 represent increased productivity in the surface waters plus the removal by sedimentation of organic matter from the lake's carbon cycle. The organic carbon content of the varved section ranges from 3 to 8% (17). The removal and storage of organic carbon in sediments is undoubtedly one of the principal factors required to

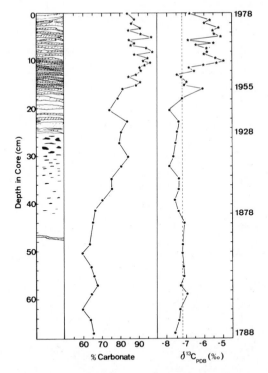

Figure 5.7. The percentage of carbonate and carbon-isotope stratigraphy of the Lake Greifen short core. The progressive eutrophication of the lake, beginning about 1878, is depicted by the change from homogenous marls to varved sediments, white chalks alternating with black organic and diatom-rich layers. With the increasing bottom-water anoxia, the sediments are pigmented by black, unstable iron sulfides. Data for samples taken over 0.5-cm intervals are represented by dots, while those from single chalk (calcite) laminae within the varves are represented by stars. The dashed line represents the average δ ^{13}C value (-7.1‰) for preeutrophic sediments. Between 1878 and 1955, increased productivity resulted in the recycling of more carbon-12 enriched CO_2 within the lake as reflected by the trend toward more negative δ ^{13}C values for the sediments. After 1955, continued increasing productivity combined with organic carbon preservation in the sediments removed carbon-12 enriched CO_2 from the system as reflected by the trend towards more positive δ ^{13}C values for the laminae. [Modified from McKenzie (4).]

drastically increase the carbon-13 content of the dissolved inorganic carbon in any aqueous reservoir, lacustrine or marine.

Other methods besides water pollution, such as increased soil erosion or evaporative concentration, could increase the available nutrient content of a lake and be recorded in the carbon-isotope content of the authigenic carbonate sediments. For example, a piston core from Lake Greifen shows an increase in carbon-13 content of the lake carbonate which curiously corresponds to a decrease in the percent carbonate between 2.0 and 1.0 m (Fig. 5.8). The core can be dated as 2000 B.P. at 1.4 m with the first appearance of characteristic pollen of agricultural plants introduced to the region by

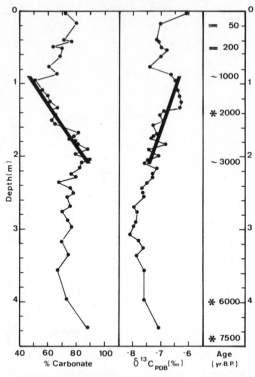

Figure 5.8. The percentage carbonate and carbon-isotope stratigraphy of the Lake Greifen piston core 76-16 (McKenzie, unpublished data) correlated with a time scale determined from a combination of varve, ^{210}Pb, and pollen ages by Weber (6). The pollen date (∗) 2000 B.P. represents the first sedimentary appearance of pollen from plants introduced to the area by the Romans. Between 3000 and 1000 B.P., the δ ^{13}C values, which have been corrected for the contribution of detrital carbonate (4), progressively increased by more than 1‰, which corresponded to a decrease of 40% in the percentage carbonate. Deforestation and agricultural development of the area surrounding Lake Greifen could account for an increased influx of detrital material, which would dilute the authigenic carbonate content of the sediments. Greater erosion of the watershed would also increase the nutrient input allowing for increased rates of productivity, which, then, could explain the trend towards more positive δ ^{13}C values. The reverse trend in both parameters after 1000 B.P. could be associated with forest restoration during the Middle Ages, while the changes of the last 200 years are associated with the modern eutrophication of the lake (Fig. 5.7).

Roman conquerors. The decrease in carbonate content could be interpreted as an increased detrital influx to the lake with deforestation of the watershed as more land was brought into agricultural use. This erosion could have been accompanied by an increased nutrient supply which facilitated higher rates of photosynthesis. The increasing δ^{13}C value makes this scenario quite plausible.

The Great Salt Lake, Utah, a hypersaline lake in a closed basin, provides an example of nutrient concentration with evaporation. Water flowing into the Great Salt Lake can only leave through evaporation. The lake level rises and falls as the ratio of freshwater input versus evaporation varies. The biogenically induced precipitate in the Great Salt Lake is aragonite—the stable phase of calcium carbonate in hypersaline waters. An extensive stable-isotope study of the Great Salt Lake sediments (18) demonstrates that lake-level changes are reflected by variations in the oxygen-isotope composition of the sediments. The lake level rises when there is an increase in the fresh-water input, which correlates to a more negative $\delta^{18}O$ value in the sediments. As water evaporates, its oxygen-18 content increases. Thus, during periods of lowered lake levels, the $\delta^{18}O$ value of the sediments is more positive. Variations in the aragonite composition and the $\delta^{13}C$ value parallel the ox-ygen-isotope stratigraphy (Fig. 5.9). This synchronization is interpreted as follows: During periods of decreasing inflow or conversely more evapora-tion, the lake waters become more saline resulting in more positive $\delta^{18}O$ values and a greater concentration of nutrients which would support a large standing stock of phytoplankton biomass. The increased productivity facil-

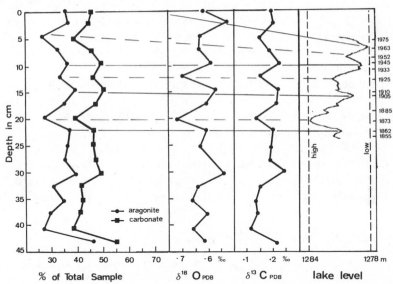

Figure 5.9. The percentage aragonite and carbonate and stable-isotope stratigraphy of the Great Salt Lake, Utah, gravity core 79-05G2 (McKenzie and Eberli, unpublished data) correlated with the historic lake-level curve from 1855 to 1975 (19). The exceedingly close correspondence between the increasing percentage aragonite and $\delta^{18}O$ and $\delta^{13}C$ values during periods of lowered lake levels is probably related to evaporative concentration. More evaporation produces an increased oxygen-18 and nutrient content of the waters, which results in increased productivity and more biologically induced precipitation of aragonite with more positive $\delta^{13}C$ values. During periods of higher lake levels, the reverse situation occurs. For the Lake Lisan–Dead Sea system in Israel, where biological activity is normally nonexistent, there is no correspondence between the oxygen- and carbon-isotope contents of the authigenic aragonite (20). This fact further emphasizes the importance of the effect of productivity on the carbon-isotope cycle in the very biologically active Great Salt Lake.

itates the precipitation of greater quantities of aragonite imprinted with a more positive $\delta^{13}C$ signal.

Although the examples presented in this paper are isotope stratigraphy of Holocene lacustrine sediments, the methods and interpretations for these recent lakes can be applied to the study of ancient lakes. A careful evaluation of the carbon-isotope content of fossil carbonate sediments, many of which are valuable source rocks for petroleum, will undoubtedly provide information about paleoproductivity and the types of lacustrine environment promoting the preservation of organic matter.

5.2. Marine Environment

The carbon-isotope record for marine sediments reveals the presence of major increases in the carbon-13 content of widely distributed pelagic carbonates, which are associated with periods of increased preservation of organic matter in black shales (21). One such event is illustrated by a large $\delta^{13}C$ perturbation of about 4‰ during the early Cretaceous (135–100 million years ago) (Fig. 5.10). This same time period is associated with some of the world's major petroleum reservoirs, which were deposited globally in anoxic basins. It is interesting to speculate on the geologic conditions surrounding the deposition of these economic reserves.

As has been frequently stated in this paper, the changes in the $\delta^{13}C$ content of carbonate sediments are related to variations in the carbon-isotope content of the dissolved inorganic carbon, which is primarily controlled by the photosynthesis–respiration cycle. Increasing productivity can increase the $\delta^{13}C$ content of the authigenic carbonates; but the removal of the carbon-12 enriched organic matter from the carbon cycle by burial is undoubtedly required to maintain very positive $\delta^{13}C$ values for several million years as seen in the isotope record for the early Cretaceous. The factors controlling the

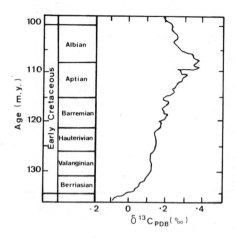

Figure 5.10.   The carbon-isotope stratigraphy of the early Cretaceous, pelagic marine limestones of the Peregrina Canyon, Mexico, section showing a progressive increase in the carbon-13 content between 135 and 110 million years. The positive carbon-13 signal at the Aptian–Albian boundary coincides with a period in which extensive organic-rich shales were deposited worldwide. The carbon-isotope trend of this curve could be explained by along-term, gradual increase in productivity beginning about 135 million years and culminating with extensive burial of organic matter around 110 million years. [Redrawn from Scholle and Arthur (21).]

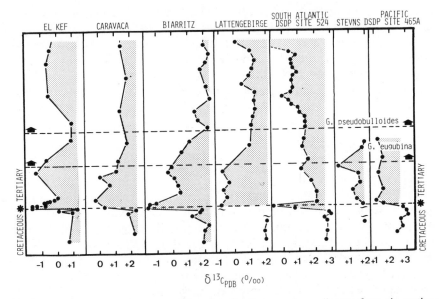

Figure 5.11. The carbon-isotope stratigraphy of pelagic marine sediments from six sections across the Cretaceous/Tertiary boundary. The rapid decrease in the δ ^{13}C value at the C/T boundary is interpreted as a sharp decrease in plankton productivity associated with a catastrophic mass mortality. The negative trend quickly reversed itself after the boundary, as best documented by the El Kef section. A second decrease in δ ^{13}C values occurred tens of thousands of years after the boundary with the first appearance of the foraminifera *Globigerina eugubina* and is associated with the mass extinction (decreased productivity) of the Cretaceous planktic microorganisms, which had never fully recovered from the catastrophic C/T boundary event. [Modified from Perch-Nielsen et al. (23).]

preservation of sedimented organic matter remain controversial; but high primary production and bulk sedimentation rates appear to be most important (22).

Productivity in the marine environment can also decrease. This apparently occurred at the Cretaceous/Tertiary boundary (~63 million years ago) when there was a mass mortality followed by a mass extinction of the marine planktic biomass. If surface-water productivity decreased, an increase in the carbon-12 content of the dissolved inorganic carbon should follow. An examination of the carbon-isotope record from many sedimentary sections across the Cretaceous/Tertiary boundary shows that, indeed, there was a very sharp decrease in the δ^{13}C value of about 3 ‰ exactly at the boundary (Fig. 5.11). The δ^{13}C value returned to a nearly normal value soon afterwards as photosynthetic activity recovered, only to be followed by a more gradual decrease over a period of the next 15,000–50,000 years. The first, rapid perturbation was interpreted as a consequence of a catastrophic mass mortality at the boundary (23,24). This "negative" example of decreased productivity, the so-called "Strangelove Effect" (7,25) emphasizes further the role played by the marine biosphere in controlling the carbon-isotope distribution by photosynthetic activity. The large $\delta^{13}C$ decrease could be a manifestation of the elimination of the surface-to-bottom, carbon-isotope gradient in ocean waters (Fig. 5.3) at a time when carbon-isotope fractionation by a photosynthesis-respiration mechanism became ineffective. This single parameter, the carbon-isotope ratio, is a sensitive indicator of the effect of a sudden disequilibrium in the ecosystems on the carbon cycle.

6. CONCLUSIONS

Many questions remain unanswered about the mechanisms which promote the production of organic matter in lacustrine and marine environments. What is the source of the increased nutrient supply required to elevate the rate of surface-water productivity? What bottom-water conditions are best for significant preservation of the sedimented organic matter? What effects do changes in circulation, climate, and salinity have on productivity? Obviously, all of these factors are complexly interrelated and must be considered in any attempt to formulate workable models.

This paper has attempted to demonstrate that the carbon-isotope content of both lacustrine and marine carbonate sediments is a key indicator of productivity rates. The biogeochemical processes causing the fractionation of carbon isotopes during photosynthesis and respiration outwardly operate in a similar manner in all aquatic environments. Models developed for one environment may be transferable to another. Since lakes often are examples of basins experiencing extreme conditions or rapid changes—conditions which may have been operating in the geologic past but are no longer found in the modern marine environment—they can serve as models for the study of the effect of these conditions or changes on productivity. Lakes, then, assume a dictionary definition for the word model, being small copies or imitations of an existing object. In this sense, lake models are miniature representations of oceans. In conclusion, combining and synthesizing data from diverse chemical, biological, and sedimentological indicators is a rewarding approach to the study of modern environments and can provide actualistic knowledge to our understanding of the ancient record of paleoproductivity.

ACKNOWLEDGMENTS

I gratefully acknowledge the many members of the Zürich limnogeology group who have provided invaluable assistance to me since 1976. I thank especially K. J. Hsü and K. Kelts for their constant interest and many stimulating discussions on lacustrine carbonates. I acknowledge W. Stumm for his encouragement of limnogeology and for proposing the theme of this paper. I thank B. Das Gupta for her technical assistance and C. Lee for her constructive comments. This work was partially supported by the Swiss National Science Foundation (Limnogeology Project) and is Contribution No. 229 of the Laboratory for Experimental Geology, ETH, Zürich.

REFERENCES

1. G. Faure, *Principles of Isotope Geology*, Wiley, New York, 1977.
2. J. Hoefs, *Stable Isotope Geochemistry*, 2nd ed., Springer-Verlag, Berlin, 1980.
3. F. J. Pearson, Jr. and T. B. Coplen, "Stable Isotope Studies of Lakes." In A. Lerman (Ed.), *Lakes: Chemistry, Geology, Physics*, Springer-Verlag, Berlin, 1978, p. 325.
4. J. A. McKenzie, "Carbon-13 Cycle in Lake Greifen: A Model for Restricted Ocean Basins." In S. O. Schlanger and M. B. Cita (Eds.), *Nature and Origin of Cretaceous Carbon-Rich Facies*, Academic Press, London, 1982, p. 197.
5. S. Oana and E. S. Deevey, "Carbon 13 in Lake Waters, and Its Possible Bearing on Paleolimnology," *Am. J. Sci.* **258A,** 253–272 (1960).
6. H. Weber, *Sedimentologisches und geochemische Untersuchungen im Greifensee (Kanton Zürich, Schweiz)*, Unpublished Doctoral Thesis, Swiss Federal Institute of Technology (ETH), 1981.
7. W. S. Broecker and T.-H. Peng, *Tracers in the Sea*, Eldigio Press, New York, 1983.
8. P. M. Kroopnick, S. V. Margolis, and C. S. Wong, "$\delta^{13}C$ Variations in Marine Carbonate Sediments as Indicators of the CO_2 Balance between the Atmosphere and Oceans." In N. R. Andersen and A. Malahoff, *The Fate of Fossil Fuel CO_2 in the Oceans*, Plenum Press, New York, 1977, p. 295.
9. P. M. Kroopnick, "The Distribution of ^{13}C in the Atlantic Ocean," *Earth Planet. Sci. Lett.* **49,** 469 (1980).
10. K. Kelts and K. J. Hsü, "Freshwater Carbonate Sedimentation." In A. Lerman (Ed.), *Lakes: Chemistry, Geology, Physics*, Springer-Verlag, Berlin, 1978, p. 295.
11. K. Emrich, D. H. Ehhalt and J. C. Vogel, "Carbon Isotope Fractionation during the Precipitation of Calcium Carbonate," *Earth Planet. Sci. Lett.* **8,** 363 (1970).
12. H. Ambühl, "Der Anteil der Waschmittel-Phosphate bei der Eutrophierung der Seen," Special Publication of *Jahrbuch der Schweizerischen Naturforschenden Gesellschaft, wissenschaftlicher Teil*, 1978, p. 279.
13. B. Kunz, "Heterogene Nukleierung und Kristallwachstum von $CaCO_3$ in Natuerlichen Gewaessern," unpublished doctoral thesis, Swiss Federal Institute of Technology (ETH), 1983.
14. G. S. Lister, "Lacustrine Sediments in Lake Zürich and the Lower Limmat Valley," unpublished doctoral thesis, Swiss Federal Institute of Technology (ETH), 1984.
15. W. Finger, A. Frey, K. Kelts, and K. J. Hsü, "Carbonate Mineralogy and Organic Carbon Content of Zübo Sediments." In K. J. Hsü and K. Kelts (Eds.), *Quaternary Geology of Lake Zürich: An Interdisciplinary Investigation by*

Deep-Lake Drilling, Contributions to Sedimentology, E. Schweizerbart'sche Verlagsbuchhandlung, Stuttgart, 1984, Chapter 6.

16. M. Biolzi, "Stable Isotopic Study of Oligocene–Miocene Sediments from DSDP Site 354, Equatorial Atlantic, *Marine Micropaleo.* **8**,(2), 121–139, (1983).

17. W. Giger, C. Schaffner, and S. G. Wakeham, "Aliphatic and Olefinic Hydrocarbons in Recent Sediments of Greifensee, Switzerland, *Geochim. Cosmochim. Acta* **44**, 119 (1980).

18. J. A. McKenzie, G. Eberli, K. Kelts, and J. Pika, *Abstr. Prog. Geolog. Soc. Am.* **14**, 562 (1982).

19. L. H. Austin, "Lake Level Predictions of the Great Salt Lake." In J. W. Gwynn (Ed.), *Great Salt Lake: A Scientific, Historical and Economic Overview*, Utah Department of Natural Resources Bulletin 116, 1980, p. 273.

20. A. Katz, Y. Kolodny, and A. Nissenbaum, "The Geochemical Evolution of the Pleistocene Lake Lisan–Dead Sea System," *Geochim. Cosmochim. Acta* **41**, 1609 (1977).

21. P. A. Scholle and M. A. Arthur, "Carbon Isotope Fluctuations in Cretaceous Pelagic Limestones: Potential Stratigraphic and Petroleum Exploration Tool," American Association of Petroleum Geologists Bulletin **64**, 67 (1980).

22. S. E. Calvert, "Oceanographic Controls on the Accumulation of Organic Matter, in Marine Sediments." In J. Brooks and A. Fleet (Eds.), *Marine Petroleum Source Rocks*, Blackwell Scientific Publications Ltd, Oxford, 1984.

23. K. Perch-Nielsen, J. McKenzie, and Q. He, "Bio- and Isotope-Stratigraphy and the Catastrophic Extinction of Calcareous Nannoplankton at the Cretaceous/Tertiary Boundary." In L. T. Silver and P. H. Schultz (Eds.), *Geological Implications of Large Asteroids and Comets on the Earth*, Geological Society of America Special Paper No. 190, 1982, p. 353.

24. K. J. Hsü, Q. He, J. A. McKenzie, H. Weissert, K. Perch-Nielsen, H. Oberhänsli, K. Kelts, J. LaBrecque, L. Tauxe, U. Krähenbühl, S. F. Percival, Jr., R. Wright, A. M. Karpoff, N. Petersen, P. Tucker, R. Z. Poore, A. M. Gombos, K. Pisciotto, M. F. Carman, Jr., and E. Schreiber, "Mass Mortality and Its Environmental and Evolutionary Consequences," *Science* **216** 249 (1982).

25. K. J. Hsü and J. A. McKenzie, "A Strangelove Ocean in the Earliest Tertiary," In E. Sundquist and W. Broecker (Eds.), Natural Variations in Carbon Dioxide and the Carbon Cycle, American Geophysical Union Monograph, 1985.

6

REDOX-RELATED GEOCHEMISTRY IN LAKES: ALKALI METALS, ALKALINE-EARTH ELEMENTS, AND ^{137}Cs

Edward R. Sholkovitz

Woods Hole Oceanographic Institution, Woods Hole, MA 02543, USA

Abstract

Redox-driven reactions are important components of the biogeochemical cycles in many lake waters and sediments. The possible coupling between redox conditions and the aquatic chemistry of alkali metals (Na, K, Cs), alkaline-earth elements (Mg, Ca, Ba), and radioactive ^{137}Cs is discussed. Water column (dissolved and suspended particulate phases) and pore water data are used to show that anoxic conditions indirectly result in enhanced fluxes of alkali and alkaline-earth elements across the sediment–water interface. An important reaction appears to be the release of these elements from an adsorbed phase as the reduction of sedimentary Fe(III) oxides yields the more soluble Fe(II) species under anoxic conditions in lakes. As such, the alkali and alkaline-earth elements undergo a seasonal cycle in the hypolimnion of seasonally anoxic lakes.

Recent studies in lakes are used to hypothesize that anoxic conditions in certain types of lake sediments may result in enhanced mobilities of ^{137}Cs across the sediment–water interface and within the sediment column. The indirect coupling between ammonia production, which accompanies anoxic conditions, and the exchange of ammonium cations for ^{137}Cs may be an important process in sediments of certain mineralogical compositions. Be-

fore any generalities about ^{137}Cs *mobility can be made, much more data on* ^{137}Cs *in anoxic sediments and water columns are required.*

1. INTRODUCTION

The coupling between redox variations and the aquatic chemistry of inorganic elements is an important process in many lake systems (Morgan and Stumm (1). This coupling with respect to alkalis (Na, K, Cs) and alkaline-earths (Mg, Ca, Sr, Ba) will be the focus of this paper. The major objective is to ascertain if there are links between anoxia and the chemical behavior of selected alkali elements, alkaline-earth elements, and the fission product radionuclide ^{137}Cs. The first part of the paper will develop the concept that anoxia results in enhanced fluxes of alkali and alkaline-earth elements within the sedimentary column and across the sediment–water interface. Hence, the mobility and diagenetic chemistry of these elements may be greatly influenced by redox-related reactions.

The second part of this paper will discuss the aquatic chemistry of ^{137}Cs in lake systems where anoxia develops. The question of ^{137}Cs mobility under different redox conditions will be explored. The hypothesis that ^{137}Cs can be significantly mobilized under anoxic conditions in certain lake environments will be developed. The indirect coupling of ^{137}Cs mobility with redox reactions, if proven to be a general phenomenon, has important implications with respect to ^{137}Cs-based geochronology and the aquatic chemistry of this radionuclide. It is evident that a study of radionuclides cannot be undertaken in isolation from the general biogeochemical processes operating in lakes. For example, the studies of alkali and alkaline-earth elements should provide important anologs for the chemical behavior of ^{137}Cs and ^{90}Sr.

This paper is not intended to be a comprehensive review, but rather an overview of chemical changes occurring under anoxic conditions in lake waters and sediments. Specific examples for various elements will be chosen from the literature to illustrate distribution and flux properties.

2. BIOGEOCHEMICAL CYCLES

The biogeochemical cycle refers to movement of elements between various phases in the natural environment as a consequence of their involvement in biological and chemical reactions. A combination of these reactions will control the temporal and spatial distribution, concentration, and speciation of substances in water, suspended particles, and sediments. In contrast to the oceans where much research has been and is being undertaken into the biogeochemistry of elements (2), lake waters have received relatively little study. Lakes are dynamic environments (3) which contrast the open ocean where steady-state distributions exist (2). Although the transient nature of lake processes adds extra complexities, time-dependent variations (e.g., oxic/anoxic cycles) can be taken advantage of to establish the dynamics of biogeochemical cycles.

Oxidation/reduction (redox) cycles are important components of the biogeochemistry of many lake waters and sediments. Lakes with seasonally anoxic hypolimnia represent good natural laboratories in which to study redox-related geochemistry. Thermal stratification of the water column in many temperate lakes can lead to the development of oxygen-poor and/or anoxic conditions in hypolimnia and to a seasonally oxic/anoxic cycle (4–11). The development of a dynamic Fe and Mn redox cycle is one of the major biogeochemical processes in seasonally anoxic lakes (1,4–27).

3. FE AND MN REDOX CHEMISTRIES AND CYCLES

It has long been recognized that the Fe and Mn oxides could play an important role in controlling the concentration of many trace elements [e.g., Jenne (28)]. The redox chemistries of Fe and Mn could have pronounced effects on the adsorption of trace elements onto oxide surfaces and trace-element fluxes under different redox conditions. The redox cycles of Fe and Mn in lakes are well documented (1,4–27) as are the basic mechanisms and kinetics of the Fe(III)–Fe(II) and Mn(IV)–Mn(II) redox couples (1,29–35). Chapter 2 in this book by W. Davison discusses the Fe and Mn redox system in detail. I would like to illustrate the main features of the Fe/Mn redox cycles.

In many lakes there is a depletion of dissolved oxygen during summer stratification (and in some instances during winter ice formation), which drives the Fe and Mn redox cycles. With oxygen depletion the redox potential decreases causing the reduction of Mn(IV) and Fe(III) oxides to form the more soluble Mn(II) and Fe(II). This process begins in the surface sediments which become reducing earlier than the overlying water column. As such, large concentrations of Mn(II) and Fe(II) appear in the pore waters (23,27) and set up diffusional fluxes into the bottom water of lakes. With time the oxic/anoxic boundary moves up through the surface sediments, and in certain lakes the water column then becomes anoxic. Depending on the lake in question, the hypolimnion may become anoxic over a large proportion of its depth. Figure 6.1 illustrates the situation for Esthwaite Water, one of the most thoroughly studied seasonally anoxic lakes in the world (6,7,10–12,19). There is a moving oxic/anoxic boundary as Esthwaite Water goes through seasonal cycles in temperature, stratification, and dissolved-oxygen concentrations. These cycles result in dynamic Fe and Mn redox cycles as shown in Figures 6.1 and 6.2. See Davison, Chapter 2 (this volume), for more details of Fe and Mn in Esthwaite Water.

Two important differences in the redox chemistries of Mn and Fe must be emphasized: (1) the reduction of Mn(IV) oxides to dissolved Mn(II) occurs

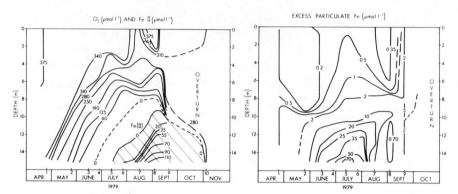

Figure 6.1. Distribution of dissolved O_2 and dissolved Fe(II) in Esthwaite Water for 1979. [Data from Sholkovitz and Copland (12).]

Figure 6.2. Distribution of the excess concentration of suspended particulate Fe in Esthwaite Water for 1979. [Figure from Sholkovitz and Copland (12).]

at higher redox potential (i.e., higher O_2 concentrations) than does the Fe(III) to Fe(II) reduction (29); and (2) the oxidation of dissolved Fe(II) to particulate Fe(III) oxides occurs much more rapidly than does that of dissolved Mn(II) to Mn(IV) oxide (29). These thermodynamic and kinetic differences have important implications. For example, Mn(IV) to Mn(II) reduction has been shown in laboratory experiments (23) with lake sediments to begin at a dissolved-oxygen concentration of about 2–3 mL liter^{-1} (or approximately one-third to one-fourth saturation levels). Fe(III) reduction, in contrast, occurs at essentially anoxic conditions. The redox chemistries of Mn and Fe should result in their having different temporal and spatial distributions in a seasonally anoxic lake with moving redox boundaries.

In lakes undergoing oxygen depletion one would expect that Mn(II) fluxes to the hypolimnion from sediments and pore waters would precede fluxes of Fe(II). Similarly, one would predict that dissolved Fe(II), fluxing from anoxic sediments into the hypolimnion, would rapidly precipitate at the oxic/anoxic boundary, while Mn(II) would be transported into the oxygenated part of the lake where it would undergo slow oxidation.

The Mn and Fe data in Sholkovitz and Copland (12) for Esthwaite Water and in Kjensmo (9) for Norwegian lakes show different redox behaviors. For example, Figures 6.1 and 6.2 show the distribution of dissolved Fe(II) and suspended particulate Fe(III) in Esthwaite Water, respectively. The increase in particulate Fe concentration (corrected for any contribution of detrital Fe) in the hypolimnion during the spring coincides with the onset of low-oxygen conditions in the water column (Fig. 6.1) and presumably anoxic conditions in the surface sediments. There is a continual buildup of suspended particulate Fe between May and mid-July resulting from the rapid oxidation of dissolved Fe(II) diffusing into the bottom water from the sediments (Fig. 6.1). As conditions become more reducing later in the summer, dissolved Fe(II) accumulates in the hypolimnion (Fig. 6.1) as the particulate Fe(III) particles are reduced and solubilized (Fig. 6.2).

Fe may also enter into important reactions with sulfides in certain lakes and under very reducing conditions. The Fe(II)–S system comes into operation in the water column of Esthwaite Water during the early fall. This subject is well documented for Esthwaite Water (16–18).

Manganese exhibits very different time and depth distributions than that of Fe in Esthwaite Water. Mn does not accumulate in the hypolimnion as suspended particles of Mn(IV) oxides because conditions are too reducing. Also, the rates of Mn(II) oxidation are slow enough that Mn(II) can be transported across the thermocline and into the epilimnion without an appreciable amount of Mn(IV) oxides precipitating [see Fig. 17 in Sholkovitz and Copland (12)]. For example, periods of several weeks were required for the total oxidation of dissolved Mn(II) in Lake Mendota bottom water under experimental conditions at pH 8.5 (21).

In summary, this section illustrates the dynamic and complex nature of the Fe and Mn redox cycle in seasonally anoxic lakes Although the main thermodynamic and kinetic features of the redox cycles will hold for all lakes, it should be appreciated that significant differences exist between lakes (9). The redox cycles will depend on such factors as bathemetry, climatic conditions, biological productivity, and sediment type. Hutchinson (4) and Wetzel (5) should be consulted for a broader view of chemical limnology.

4. FLUXES AT THE SEDIMENT–WATER INTERFACE UNDER ANOXIC CONDITIONS

The classic research of Mortimer (6,7) elegantly describes the major theme of this paper—that the development of anoxic conditions causes enhanced fluxes of many elements across the sediment–water interface of lakes. Using both field and laboratory studies and measurements of both the water column and sediments of Esthwaite Water, Mortimer showed that oxygen depletion and anoxicity cause a large increase in the conductivity of the hypolimnion and surface sediments. This conductivity increase was due, as Mortimer showed, to enhanced concentrations of dissolved ammonia, silica, phosphorus, alkalinity and, of course, Fe and Mn. Research, following on the direction of Mortimer, confirms that the development of reducing or anoxic conditions in lakes drive large fluxes of many inorganic elements across the sediment–water interface. It appears that the release of these elements is "triggered" by the dissolution of the Fe and/or Mn oxides in the oxic surface sediments during oxygen depletion. In an oversimplification of a complicated biogeochemical subject, it appears that the Fe(III) and/or Mn(IV) oxides at sediment–water interface act as traps or adsorbers for many different types of elements and substances. The phosphorus cycle is a classic example; the lake chemistry of P has been described in numerous publications including a recent review article by Böstrom et al. (37).

Fluxes across the sediment–water interface and migration within sediments must be controlled by pore-water gradients, diffusion, adsorption, sedimentation, and ground-water movement; this assumes insignificant amounts of bioturbation during periods of anoxicity. Studies have shown that Fe, Mn, and P undergo large seasonal cycles in their pore-water concentrations (23,27,38). Under reducing conditions present in the warmer months, pore-water concentrations of these elements increase and lead to large fluxes across the sediment–water interface to the bottom water.

5. ALKALI AND ALKALINE-EARTH ELEMENTS—DISTRIBUTION IN DISSOLVED, SUSPENDED, AND PORE-WATER PHASES

Chemical data from a variety of lakes indicate that there are higher concentrations of dissolved Na, K, Mg, Ca, and Ba in anoxic hypolimnia than in oxygenated epilimnia. Figures 6.3 to 6.6 illustrate this observation for four different lakes.

Figure 6.3 uses data from the classic study of Kjensmo (9) in which the chemical features of four meromictic lakes in Norway are described in detail. Figure 6.3 shows a time series of vertical profiles for total Fe, Mn, Ca, and Mg from Bjordammen, a seasonally anoxic lake. With the development of anoxic conditions, there are large increases in the hypolimnetic concentrations of total Fe and Mn, which result in large vertical gradients in the bottom three meters. Accompanying the increases in total Fe and Mn [which should exist as predominantly dissolved Fe(II) and Mn(II)], there are large increases in the concentrations of Ca and Mg. This relationship was also observed in the other lakes studied by Kjensmo (9). His data show a well-developed seasonal cycle in the hypolimnion for these alkaline-earth elements. This is particularly the case for Ca for which there is more detailed time-series data. Anoxicity in Bjordammen's hypolimnion also resulted in increases in conductivity, alkalinity, and phosphate. In interpreting his data, Kjensmo (9) eliminated dissolution of biogenic carbonates as the main source of Ca and Mg; he speculated that ion exchange with humic substances might be releasing these alkaline-earth elements to the hypolimnia.

In a study of Par Pond, a seasonally anoxic reservoir in South Carolina (U.S.A.), Evans et al. (39) have shown that anoxic conditions lead to enhanced hypolimnetic concentrations of Ca, Mg, Na, K, and [137]Cs. These increases, shown in the profiles of Figure 6.4, occur in association with anoxia and large increases in dissolved Fe, Mn, and ammonia. The ^{137}Cs data will be discussed in Section 6.

The author's own studies of Esthwaite Water have demonstrated that Ba is also indirectly involved in the redox cycle (12). Four profiles of dissolved Fe and dissolved Ba, taken from a larger 1979 time series, are shown in Figure 6.5. The main feature is the enhanced concentrations of Ba which accompany the increased concentrations of Fe during the anoxic conditions

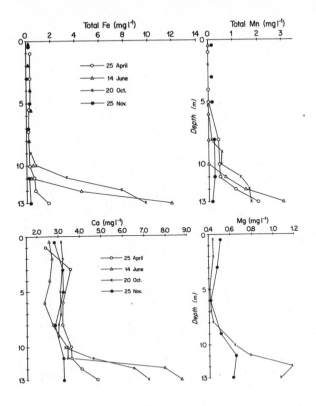

Figure 6.3.   A time series of vertical distributions of total Fe, Mn, Ca, and Mg concentrations from Bjordammen, a lake in Norway, for 1961. [Data from Kjensmo (9).]

in the hypolimnion. This seasonal cycle in Ba can be seen in more detail in Sholkovitz and Copland [see Fig. 23 in Ref. 12].

A recent study of the chemical composition of suspended particles in a seasonally anoxic lake further demonstrates how alkali and alkaline-earth elements participate in the Fe redox cycle. As illustrated in Figure 6.6 for Mg and K, time-series studies by Sholkovitz and Copland (12) of Esthwaite Water show the existence of pronounced seasonal cycles in which large "excess" (nondetrital) concentrations of particulate C, Fe, Mn, P, S, Mg, K, Ca, and Ba are being generated and lost in the water column. In the epilimnion, "excess" concentrations of P, S, Mg, Ba, and K follow that of particulate organic carbon. The data indicate that biochemical activity of growing phytoplankton in the spring and summer results in the *in situ* generation of excess particulate P, S, Mg, Ba, and K. As the hypolimnion becomes anoxic in the summer, the release of dissolved ferrous iron from the sediments results in oxidative precipitation of the Fe(III) oxide suspended

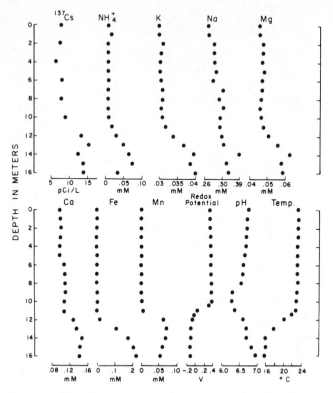

Figure 6.4. Vertical profiles of temperature, pH, redox potential, and dissolved Na, K, Mg, Ca, Fe, Mn, and ^{137}Cs concentrations from Par Pond, South Carolina, for September 29, 1979. [Figure reproduced with permission of authors, Evans et al. (39).]

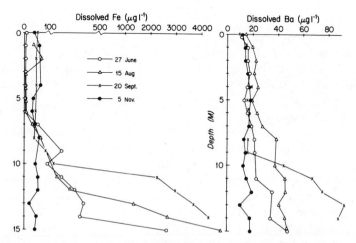

Figure 6.5. A time series of vertical profiles of dissolved Fe and Ba concentrations from Esthwaite Water for 1979.

particles (Fig. 6.2). Adsorption and/or coprecipitation onto these suspended oxides is responsible for the *in situ* accumulation of "excess" particulate S, P, C, Ca, Mg, Ba, and K, all of whose concentrations closely follow that of excess Fe (Figs. 6.2 and 6.6). As reducing conditions become more intense after July, the Fe(III) oxides are solubilized [i.e., reduced to Fe(II)] along with excess concentrations of S, P, Mg, Ca, Ba, K, and POC (particulate organic carbon). This indirect coupling of redox reactions and adsorption of alkali and alkaline-earth elements can be simplified as

$$(Fe(III)O_n-Mg)_p \underset{\text{reduced}}{\overset{\text{oxidized}}{\rightleftharpoons}} (Fe(II))_d + Mg_d$$

and/or

$$(Fe(III)O_n-\text{organics}-Mg)_p \underset{\text{reduced}}{\overset{\text{oxidized}}{\rightleftharpoons}} (Fe(II))_d + \text{organics}_d + Mg_d$$

where p is the particulate phase and d is the dissolved phase. The second equation implies that adsorbed organic matter might form a chemical link between excess Fe(III) oxides and inorganic elements (40).

Two points should be noted with respect to the above study (12). First, the affinity of K for particles of Fe(III) oxides is much less than the divalent Ca, Mg, and Ba (12). Second, the percentage of total K, Mg, and Ba associated with the Fe(III) particle phase is very small. Hence, the adsorption of these cations onto the oxidized Fe(III) particles should only have a small effect on their dissolved concentrations (less than 0.4, 3, and 2%, respectively).

There have been several studies of porewater concentrations which indicate the existence of elevated concentrations of Na, K, Ca, and Mg over those of lake bottom water (41–44). All these investigations conclude that mineral-dissolution reactions with pore waters are primarily responsible for

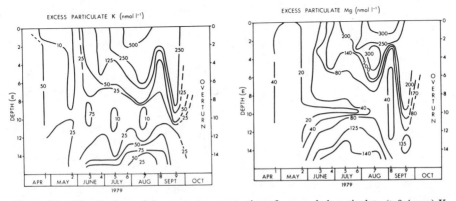

Figure 6.6. Distributions of the excess concentration of suspended particulate (>0.4 mm) K and Mg in Esthwaite Water for 1979. [Figure from Sholkovitz and Copland (12).]

the elevated concentrations and pore-water gradients of alkali and alkaline-earth elements. Sasseville and Norton (44), for example, report the pore-water composition of four Maine lakes. Their profiles show that the concentrations of K, Na, Ca, and Mg increase with depth in the sediment. These increases are usually accompanied by increases in Fe, Mn, and P. In Fish River Lake, where pronounced subsurface maxima in Fe and Mn exist, there is also a slight indication of subsurface maxima in Ca, Mg, and K. As the authors point out, there may be some handling artifacts which hamper interpretations. Nevertheless, the pore-water data of Sasseville and Norton provide evidence that alkali and alkaline-earth elements are enriched in pore waters In the survey of Brunskill et al. (42) of lakes in the Experimental Lake Area of Ontario, they point out that pore waters (from 0–20 cm sections) are enriched with respect to lake waters in Na, K, Ca, Mg, Mn, and Fe. Pore water/lake water ratios of these elements greatly exceed one, except for the two meromictic lakes where the ratios are near unity. Both the anoxic bottom waters and pore waters of the two meromictic lakes have much greater concentrations of alkali and alkaline-earth elements than do their counterparts in the holomictic lakes. Brunskill et al. (42) conclude that mineral dissolution in sediments is responsible for their observations in the holomictic lakes. In the two meromictic lakes, they suggest that the stagnant anoxic bottom waters have been in contact with the sediments for longer periods of time and hence show no gradients across the sediment–water interface.

The data just described on the distribution of particulate and dissolved Na, K, Mg, Ca, and Ba in lakes indicate that these elements are indirectly involved in the redox cycles operating in lakes. These cycles represent interesting and important examples of the coupling effect between anoxicity, redox reactions, and the behavior of elements not having redox chemistries of their own. The concentration, fluxes, and speciation of alkali and alkaline-earth elements appear to be indirectly influenced by the development of anoxic conditions and the Fe (and perhaps Mn) redox cycles. With respect to the pore-water studies, it is not clear to what extent the concentrations and fluxes of Na, K, Mg, Ca, and Ba are controlled by mineral dissolution and/or redox-related reactions (e.g., desorption or ion exchange).

The enhanced concentrations of dissolved alkali and alkaline-earth elements in anoxic hypolimnion may result from several processes. Five such processes are discussed below; these processes may be operating either individually or, more likely, in some combination.

One possibility is that higher concentrations may just be an "artifact" of stratification. In this explanation, pore-water gradients and upward fluxes are assumed to exist all year round. Thermal stratification then simply allows a summertime buildup of concentrations in the hypolimnion. This mechanism does not invoke redox-related geochemistry. The studies of Mortimer (6,7) and J. Hamilton-Taylor (36) suggest that the above mechanism is not the major one responsible for elevated hypolimnetic concentrations. Mortimer (6) set up two water–sediment systems in which the overlying lake water was aerated in one case and sealed from the air (anaerobic) in the second case. The anaerobic experiment showed large increases in concentrations of Fe, Mn, Si, P, and HCO_3 (and conductivity) in the overlying water, while the aerobic tank showed only small fluxes of these elements across the sediment–water interface. Although the major cations (e.g., K, Mg, Ca) were not measured, the results indicate that a redox-related reaction is necessary to cause fluxes of many inorganic elements from the sediment to the bottom water. From his time-series study of pore waters from Esthwaite Water sediments, Hamilton-Taylor (36) confirms that Ca and Mg, at least, have redox-related cycles.

A second possibility is that alkali and alkaline-earth elements, associated with the organic matter of phytoplankton in the epilimnion (12), are transported to the sediments and released to the hypolimnion. Although this cycle is probably occurring in lakes, studies of particle composition in Esthwaite Water (12) and Windermere (45) indicate that alkali and alkaline-earth elements are rapidly recycled off the organic matter in the epilimnion and/or thermocline. Nevertheless, biological incorporation must be considered in the overall cycling of these elements—see the recent study of Ba association (46).

A third mechanism may be ion-exchange reactions whereby certain alkali and alkaline-earth elements might be released. In this regard, ammonia produced in large amounts under anoxic conditions might be an effective cation exchanger. This particular mechanism will be discussed in detail in Section 6 with respect to ^{137}Cs geochemistry. The overall importance of cation-exchange reactions in causing the higher concentrations of K, Mg, Ca, and Ba in anoxic hypolimnia remains unresolved.

Adsorption/desorption of alkali and alkaline-earth elements is a fourth process which might explain the seasonal cycle of these elements in anoxic lake waters. In this case, iron oxides could be the reactive adsorbers. The adsorption properties of alkali and alkaline-earth elements have been reviewed by Kinniburgh and Jackson (47). In addition, Kinniburgh et al. (48) have measured the adsorption of Mg, Ca, Sr, and Ba onto a Fe gel (in 1 M $NaNO_3$ solution); they found the relative affinities to be Ba > Ca > Sr > Mg at equimolar solution concentrations. The extent of adsorption is highly pH dependent with the percentage absorption less than 5% at pH <6 and 30–60% at pH8.

Cation selectivity during absorption onto various substrates have been discussed in Kinniburgh and Jackson (47). Their compilation of data is given in Table 6.1. The alkali elements usually exhibit little specific adsorption to oxide surface and can show both a "normal" sequence of selectivity (Cs > Rb > K > Na > Li) and/or "reverse" sequence (Li > Na > K > Rb > Cs). Although the reverse sequence has been observed for iron oxides, the results pertain to conditions in laboratory experiments. The alkaline-earth elements show no consistent pattern in their relative selectivity of adsorption onto metal oxides (Table 6.1). However, Ba usually shows the greatest affinity followed by either Ca or Sr. Adsorption properties under natural conditions, such as those associated with the precipitation and dissolution of iron oxides in Esthwaite Water (Figs. 6.2 and 6.6), may be quite different.

The adsorption experiments are consistent with the suspended particle study of Sholkovitz and Copland (12) which shows that alkali and alkaline-earth elements can be adsorbed onto suspended particles of ferric oxide and, as such, indirectly participate in the Fe redox cycle (Fig. 6). Because of their greater adsorption affinity for iron oxides, the alkaline-earth elements are probably more affected by the dissolution–precipitation chemistry associated with the Fe(II)–Fe(III) redox cycle (see earlier part of this section). The mechanisms and magnitude of the interactions are not well known. Although this author has not found studies of Sr or Cs in seasonally anoxic lake water, one would expect (Table 6.1) Sr adsorption onto oxide particles to be occurring along with a much smaller amount of Cs adsorption. Chem-

Table 6.1. Selectivity Sequences for the Alkali-Metal Cations on Various Hydrous Metal Oxides[a]

Sequence	Oxide
A. Alkali Metals	
Cs > Rb > K > Na > Li	Si gel
K > Na > Li	Si gel
Cs > K > Li	SiO_2
Cs > K > Na > Li	SiO_2
Li > Na > K	SiO_2
K > Na > Li	Al_2O_3
Li > K ~ Cs	Fe_2O_3
Li > Na > K ~ Cs	Fe_2O_3
Cs ~ K > Na > Li	Fe_3O_4
Li > Na > Cs	TiO_2
Li > Na > K	$Zr(OH)_4$
B. Alkaline Earths	
Ba > Ca	Fe gel
Ba > Ca > Sr > Mg	Fe gel
Mg > Ca > Sr > Ba	α–Fe_2O_3
Mg > Ca > Sr > Ba	Al gel
Mg > Ca > Sr > Ba	Al_2O_3
Ba > Sr > Ca	α–Al_2O_3
Ba > Sr > Ca > Mg	MnO_2
Ba > Sr > Ca > Mg	δ–MnO_2
Ba > Ca > Sr > Mg	SiO_2
Ba > Sr > Ca	SiO_2

[a] Data from Kinniburgh and Jackson (47), pp. 111 and 113.

ical extractions of ^{90}Sr-contaminated sediments indicate that oxides of Fe and Mn can specifically adsorb this radionuclide (49,50). Therefore, the concentrations and fluxes (e.g., cycling) of Sr and ^{90}Sr may be significantly influenced by Fe redox reactions in lake water and sediments. As will be discussed in Section 6, Evans et al. (39) indicate that ion exchange rather than desorption off iron oxides, is the most important reaction for ^{137}Cs under anoxic conditions.

The fifth possibility that mineral dissolution in surface sediments might be responsible for enhanced fluxes of alkali and alkaline-earth elements in anoxic lake water has been previously discussed with respect to pore-water studies. Similarly, Matisoff et al. (51) reach the same conclusion for Ca, Mg, Fe, and manganese. These authors used anaerobic laboratory incubation experiments in which lake sediments, sealed in jars, had their pore-water compositions followed as a function of time and temperature. Their results showed that there are similar release patterns of NH_4^+, Ca, Mg, Fe, and alkalinity with time. The interpretation of Matisoff et al. (51) is that dissolution of carbonate minerals (e.g., calcite, aragonite, and dolomite), driven by decomposition of organic matter, is responsible for the release of Ca, Mg, Fe, and Mn. In both this and the pore-water studies (41–44), a complex set of biogeochemical reactions are occurring. It is not clear which specific reactions are most responsible for the release of alkali and alkaline-earth elements to the pore waters.

In summary, this section has shown that the development of anoxic conditions in lake hypolimnia can result in enhanced concentrations of alkali (Na, K) and alkaline-earth (Mg, Ca, Ba) elements. Although there is little information on their pore-water and diagenetic chemistries, large time-variable fluxes of these elements across the sediment–water interface must be operating in seasonally anoxic lakes. The exact mechanisms responsible for the fluxes are not known. This author favors a combination of fluxes driven by mineral dissolution and adsorption/desorption associated with iron oxides. The former process is not redox related and results in high pore-water concentrations of alkali and alkaline-earth elements. The latter process is redox related and can lead to an increased flux of K, Mg, Ca, and Ba to lake hypolimnia during anoxic conditions as iron oxides on the surface of lake sediments undergo reduction to soluble Fe(II). More systematic study of lake geochemistry is required before these processes can be generally applied to lakes of different types.

6. ^{137}Cs IN LAKES

6.1.   General Information

Radioactive Cs, ^{137}Cs, is a major fission product of nuclear reactions and has been introduced to the earth through global fallout associated with weapons testing, and planned and accidental releases from nuclear facilities (52). The chemical behavior of ^{137}Cs is important to understand because Cs is involved in biological processes and ^{137}Cs has a relatively long half-life (30 years). As such, the radioecology of ^{137}Cs has received a tremendous amount of study (53).

In this paper the focus will be on the chemical reactivity and mobility of ^{137}Cs in lake waters and sediments. It will be argued that the chemical behavior of ^{137}Cs in certain lakes can be indirectly controlled by redox reactions; there is evidence that anoxic conditions can result in increased mobility within the sediment columns and increased fluxes across the sediment–water interface. The existence of an active ^{137}Cs diagenetic chemistry in certain lake environments has important long- and short-range consequences in relation to ^{137}Cs-based geochronology and the environmental chemistry of this artificial radionuclide. The evidence, however, is small in mass and not yet general enough to apply across the board to lakes. Nevertheless, the hypothesis of redox-enhanced mobility of ^{137}Cs is still worth developing at this stage of our knowledge.

6.1.1. Water-Column Studies of Par Pond

The research of Alberts et al. (54) and Evans et al. (39) provides unequivocal evidence that the activity and mobility of ^{137}Cs can be strongly influenced by anoxic conditions. These papers deal with Par Pond, a manmade reservoir with a seasonally anoxic hypolimnion. Par Pond, located near two Savannah River Plant nuclear reactors, received an accidental release of ^{137}Cs (about 220 Ci) between 1958 and 1964 (39). Therefore, the sediments and waters of Par Pond contain considerably more ^{137}Cs than do lakes receiving global fallout. The chemical form of ^{137}Cs introduced to Par Pond might be different from that of global fallout. If the exchange of this reactor ^{137}Cs with stable Cs (i.e., ^{133}Cs) does not reach equilibrium with various sedimentary phases, then the distributions of ^{137}Cs in Par Pond may not reflect the geochemistry of stable Cs or fallout-derived ^{137}Cs in other lakes.

Alberts et al. (54) showed, with detailed time-series measurements of dissolved ^{137}Cs, that this radionuclide undergoes a redox-related seasonal cycle. Figure 6.7 illustrates this cycle by plotting the activities of dissolved ^{137}Cs in the epilimnion (5 m) and the hypolimnion (15 m) over the 2-year period of 1966 and 1967. The increased ^{137}Cs activity in the hypolimnion corresponds to the development of an anoxic hypolimnion in the summer months. The epilimnion also shows a cyclic pattern; but the activities are lower and the maxima appear 1 to 2 months later than in the hypolimnion. These differences between the cyclic patterns of the epilimnion and hypolimnion are consistent with Alberts et al. (54) interpretation that ^{137}Cs is being released to the hypolimnion from sediments under anoxic conditions and then transported across the thermocline to the epilimnion. The seasonal cycle is then completed during the fall and winter when mixing homogenizes the water column and removal of ^{137}Cs to the sediments occurs.

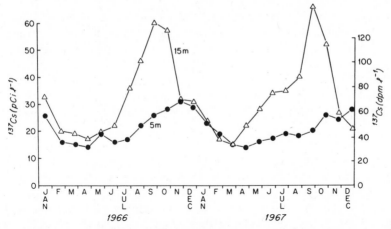

Figure 6.7. Seasonal variations in activity of dissolved ¹³⁷Cs for Par Pond, South Carolina, during 1966 and 1967. Variations at 5 and 15 m are shown. [Data taken from Alberts et al. (54).]

The geochemical mechanisms responsible for the seasonal cycle of ¹³⁷Cs in Par Pond were studied in more detail and recently reported in Evans et al. (39). They combined field observations with laboratory experiments to conclude that ion-exchange reactions are primarily responsible for the release of ¹³⁷Cs from Par Pond sediments. Their water-column profiles, previously discussed in Section 3 and shown in Figure 6.4, indicate that enhanced activities of dissolved ¹³⁷Cs occur in the anoxic hypolimnetic waters. This increase is accompanied by large increases in the concentration of Fe, Mn, NH_4^+, K, Na, Mg, and Ca.

Based on a variety of chemical extraction techniques using Par Pond sediments, Evans et al. (39) conclude that (1) the role of iron and/or manganese oxides in the redox-related release of ¹³⁷Cs is a minor one and (2) ion-exchange with NH_4^+ is probably the major process leading to ¹³⁷Cs release from sediments. With respect to the latter process, ion-exchange experiments showed the displacement selectivity of ¹³⁷Cs from Par Pond sediments to be $NH_4^+ > H^+ > Cs^+ > K^+ > Ca^{2+} > Na^+ = Mg^{2+} > Mn^{2+} = Fe^{2+}$. This order of selectivity does not follow the normal or reverse order discussed in Section 3. Further ion-exchange experiments with NH_4^+ showed the percentage of ¹³⁷Cs released as a function of dissolved NH_4^+ concentration; Figure 6.8 indicates the resulting nonlinear relationship.

The study of Evans et al. (39) provides good evidence that ion-exchange reactions, in Par Pond, can be responsible for significant amounts of ¹³⁷Cs mobility. They suggest an intriguing indirect coupling between the development of anoxic conditions, and the aquatic chemistry of ¹³⁷Cs. High ammonia concentrations are found in anoxic environments where sedimentary organic matter contents are high and where there is bacterially mediated degradation of organic N to NH_4^+ (6,7). Ammonia concentrations in pore waters can be orders of magnitude higher than those in bottom waters.

Hence, Evans et al. (39) suggested that high pore-water NH_4^+ concentrations, associated with anoxic conditions, are responsible for the release of ^{137}Cs from sediments. Whether this process is a general one which can be applied to all lake systems is a subject in need of more research. A recent measurement of ^{137}Cs in the pore water of Pond B sediments (this pond is connected to Par Pond) confirm that elevated activities of ^{137}Cs can exist. Systematic pore water studies would prove valuable in determining the geochemistry of ^{137}Cs.

The mobility of ^{137}Cs in lake sediments will greatly depend on the mineral assemblages present. Evans et al. (39) show that in Par Pond ^{137}Cs is held in two exchangeable sites and one nonexchangeable site (interlayer). They propose that ^{137}Cs in held on (1) surface sites, (2) edge or "wedge" sites, and (3) interlattice sites where ^{137}Cs is not readily exchangeable with any cation. Evans et al. (39) suggest that NH_4^+, because of its size and high concentration, is the key cation in releasing ^{137}Cs from the surface and exchangeable wedge sites.

Estimates based on the ^{137}Cs activities in Par Pond (Fig. 6.5) indicate that about 0.3–0.7% of the sediment inventory is released to the water column each year (39). At the ammonia concentrations in the hypolimnion and with the experimental release information in Figure 6.8, these authors calculate that less than 20% of bottom sediments need to be exchanged with NH_4^+ to account for the water-column increase. On the good assumption that NH_4^+ pore-water concentrations are much higher, a smaller fraction of the sediments will be required to supply the ^{137}Cs.

As emphasized in Evans et al. (39), the extent of chemical diagenesis and mobility of ^{137}Cs in lake sediments will be greatly influenced by the clay

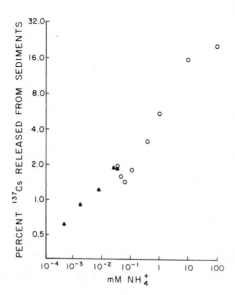

Figure 6.8. Percentage displacement of ^{137}Cs from Par Pond sediments at various concentrations of ammonia. Triangles are results for anaerobic laboratory incubation experiments in which sediments, deionized water, and dextrose were mixed and sealed for 2 weeks. Results for open circles come from extractions in which sediments were mixed with solutions of NH_4NO_3. [Figure from Evans et al. (39).]

minerals present. The clay minerals of Par Pond sediment are predominantely kaolinite (85%) with smaller amounts of illite (7%) and mixed chlorite/vermiculite (8%). Kaolinite does not strongly bind ^{137}Cs, while there is much information to show that ^{137}Cs is strongly held to micaceous clay minerals, such as illite, where ^{137}Cs can substitute for K (55,56). Seasonally anoxic lakes in the colder climatic regions, where kaolinite is a minor clay mineral and micaceous clays are more abundant, may exhibit a different type of behavior with respect to ^{137}Cs. The translocation of ^{137}Cs from exchangeable sites to nonexchangeable sites is another important kinetic process which Evans et al. (39) discuss.

In marine sediments it is thought that high pore-water ammonia concentrations are responsible, through ion-exchange reactions, for enhanced concentrations of pore-water Rb (57).

In summary of the Par Pond studies, there is convincing data to indicate that the activity, distribution, and fluxes of ^{137}Cs in lake systems can be significantly altered by the development of anoxic conditions. As such, ^{137}Cs is indirectly involved in the redox-related processes whereby ^{137}Cs is postulated to be released via ion-exchange reactions with NH_4^+ produced in large concentrations under anoxic conditions. However, at present it would be premature to extrapolate this diagenetic process to lake systems other than Par Pond. Clearly more research in different lake systems is required to ascertain if there are general relationships between anoxicity, NH_4^+ production, and ^{137}Cs geochemistry.

6.1.2. ^{137}Cs in Lake Sediments

In this section studies of ^{137}Cs and ^{239,240}Pu in lake sediments will be presented to explore the possibility that ^{137}Cs can undergo postdepositional mobility under anoxic conditions. It should be emphasized that the distributions of solid-phase components of cores do not usually provide sensitive indicators of chemical diagenesis. There are exceptions such as Mn which has an active redox-driven mobilization. In general, small but significant (i.e., <10%) changes in solid-phase distributions are difficult to detect and interpret. In contrast, pore-water chemistry provides the most direct and sensitive means of studying and modeling diagenesis (58). However, no published measurements of ^{137}Cs in pore waters of lake sediments presently exist.

^{137}Cs and ^{239,240}Pu profiles from many sedimentary cores in Lake Michigan have been reported and discussed by Edgington and Robbins (59) and Edgington et al. (60). Lake Michigan is an example of a large oligotrophic lake whose water column remains oxygenated throughout the year. The near-surface sediments are oxic. Pu, in the form of ^{239,240}Pu, provides a good contrast to ^{137}Cs. ^{239,240}Pu is also predominantly delivered to lakes from global fallout associated with the weapons' testings in the 1950s and 1960s (61).

The main feature observed in Lake Michigan sediments is that the vertical profiles of ^{137}Cs and ^{239,240}Pu activity track each other well. There appears to be no separation of the two artificial radionuclides with depth. In addition, sedimentation rates based on ^{210}Pb geochronology of the same cores agree well with those derived from assuming a 1963 maximum in the fallout of ^{137}Cs and ^{239,240}Pu. The profiles of ^{137}Cs (and ^{239,240}Pu) have different shapes at different locations. These shapes depend on the rate of sedimentation, sediment mixing rates, and atmospheric fluxes of radionuclides, and have been modeled by Edgington and Robbins (59).

The studies of the oxic Lake Michigan sediments indicate that ^{137}Cs (and ^{239,240}Pu) do not undergo measurable amounts of postdepositional diagenesis and mobility [as distinguished from physical transport, Edgington and Robbins (59)]. One would not expect the anoxic-related ion-exchange processes of Evans et al. (39) to be operating in oxic Lake Michigan sediments.

Unfortunately, there have not been systematic studies of ^{137}Cs, ^{239,240}Pu, and ^{210}Pb in sediments of lakes where anoxic conditions exist. There are, however, several studies which "point to" a redox-related mobility of ^{137}Cs. This evidence is tenuous, intriguing, and gives the author some latitude for interpretation. Pennington and co-workers (62–64) report vertical profiles of ^{137}Cs from sediments in the central basins of Esthwaite Water, Blelham Tarn, Windermere, Ennerdale Water, and Wastwater. The five lakes in the English Lake District have very different trophic and redox conditions. Esthwaite Water and Blelham Tarn are small, shallow (15 m) eutrophic lakes which have seasonally anoxic hypolimnia. Windermere is a large, deep (40 m, South Basin) mesotrophic lake, where the hypolimnion remain oxygenated but the surface sediments are mildly reducing. Ennerdale Water and Wastwater are large, deep (42 and 76 m, respectively), oligotrophic lakes where the water column and surface sediments are oxic. Pennington et al. (62) also report sediment profiles of ^{137}Cs taken from five different locations of Blelham Tarn on the same date (1974).

The most obvious feature in comparing the five lakes is that the ^{137}Cs profiles in Windermere, Ennerdale Water, and Wastwater exhibit sharp and well-defined maxima in the top 3 cm of the cores, while in Esthwaite Water and Blelham Tarn ^{137}Cs exhibits broad maxima. These latter maxima extend for about 5–8 cm in the top 10 cm of each core and are followed by elongated tails which extend to 15–25 cm. One interpretation of the Lake District data is that the anoxic conditions of Esthwaite Water and Blelham Tarn sediments may be responsible for enhanced mobility of ^{137}Cs and hence the broad ^{137}Cs maxima and tails in the profiles. This interpretation is at odds with that of Pennington et al. (62) who consider the ^{137}Cs profile in Esthwaite Water to be a demonstration of the lack of significant vertical movement of ^{137}Cs. They do suggest that the elongated tails may have resulted from diffusion or bioturbation. Significant amounts of bioturbation or bioirrigation do not occur in the anoxic sediments of Esthwaite Water and Blelham Tarn (70). High concentrations of ammonia are produced in the anoxic hypolimnion of these lakes during the summer months; for Esthwaite Water, concentrations up to 2.5 mg liter^{-1} of Esthwaite Water NH$_4{}^+$ exist in the bottom waters (6,7). With the available ^{137}Cs information for the Lake District sediments, it is fair to state that one cannot rule out the existence of significant amounts of postdepositional mobilization in the seasonally anoxic Esthwaite Water and Blelham Tarn. The NH$_4{}^+$ cation-exchange mechanism previously described may be operating. It would prove informative to now analyze Lake District sediment cores in the mid-1980s, 10 years after the measurements of Pennington and co-workers.

The use of ^{137}Cs sediment profiles to decipher diagenetic chemistry and mobility becomes more tenuous when one compares the distribution of ^{137}Cs from five locations in Blelham Tarn [see Fig. 2 in Pennington et al. (64)]. Both the shape of the profiles (e.g., depth of the subsurface maximum associated with the 1963 maximum in global fallout) and the sedimentary inventory of ^{137}Cs differ significantly from location to location. There are still, however, a broad maxima and elongated tails in all cores as previously discussed in this subsection. As Pennington et al. (64) discuss, there must be different rates of sedimentation within Blelham Tarn. It is now recognized that sediment stratigraphy of small lake basins is complicated by sediment focusing, time-variable rates of deposition, and biological mixing (65–69). The study of ^{210}Pb and P by Evans and Rigler (68) is particularly relevant.

New results reported in this book by Martin (71) compare the sediment profiles of ^{210}Pb, ^{137}Cs, and ^{239,240}Pu in Pavin Crater Lake, France. This meromictic lake has diatomaceous sediments which contain only negligible amounts of detrital minerals. Both ^{137}Cs and ^{239,240}Pu are found at depths of about 15 cm in the sediments. Based on sedimentation rates derived from ^{210}Pb and ^{32}Si geochronologies, the fallout ^{137}Cs and ^{239,240}Pu should only be found in the upper 3–5 cm of the unbioturbated sediments of Pavin Crater Lake. Hence, it appears that enhanced mobilities of ^{137}Cs (and ^{239,240}Pu) occur under anoxic conditions. As proposed by Martin (see Chapter 8), the siliceous sediments of Pavin Crater Lake probably provide less of a barrier to mobility than do clay mineral sediments. Martin (71) also notes that the permanently anoxic bottom waters of Pavin Crater Lake have significantly higher concentrations of dissolved ^{137}Cs and ^{239,240}Pu than does the oxic upper waters (700 vs. 325 fCi/liter and 0.96 vs. 0.37 fCi/liter, respectively). Again it appears that ^{137}Cs (and ^{239,240}Pu) activities are influenced by redox conditions.

In summary, the extent to which ^{137}Cs undergoes chemical mobility in lake sediments remains an open question. I would argue from the studies of Par Pond (39,54) and the English Lake District (62–64) that redox-induced mobility of ^{137}Cs in certain lake sediments should not be ruled out as a potentially significant process. At present, however, the evidence is certainly not unequivocal. More detailed investigations of sediments, pore waters, and water columns of anoxic systems are required to understand the relationship between anoxia, redox reactions, and ^{137}Cs geochemistry. Comparative studies of seasonally anoxic and oxic lakes in similar drainage basins would provide very useful information about the redox-related geochemistries of radionuclides and stable inorganic elements. Regions such as the English Lake District and the Experimental Lake Area of Canada would be ideal ones for such research.

ACKNOWLEDGMENTS

The author would like to thank J. Alberts and D. W. Evans for the use of figures from their papers. M. Hess's help in organizing and typing this paper is greatly appreciated. The writing of this paper was supported, in part, by Contract DE-AC02-83ER60172 from the U.S. Department of Energy. This is Contribution No. 5536 from the Woods Hole Oceanographic Institution.

REFERENCES

1. J. J. Morgan and W. Stumm, "The Role of Multivalent Metal Oxides in Limnological Transformations as Exemplified by Iron and Manganese," in Proceedings of the Second International Water Pollution Conference, Tokyo, 1964, p. 103.
2. W. S. Broecker and T.-H. Peng, *Tracers in the Sea*, Lamont-Doherty Geological Observatory, Palisades, New York, 1982.
3. A. Lerman (Ed.), *Lakes: Chemistry, Geology and Physics*, Springer-Verlag, New York, 1978.
4. G. E. Hutchinson, *A Treatise on Limnology*, Vols. 1 and 2, Wiley, New York, 1975.
5. R. G. Wetzel, *Limnology*, Saunders College Publishing, Philadelphia, 1975.
6. C. H. Mortimer, "The Exchange of Dissolved Substances between Mud and Water in Lakes: I and II, *J. Ecol.* **29**, 280 (1941).
7. C. H. Mortimer, The Exchange of Dissolved Substances between Mud and Water in Lakes: III and IV, *J. Ecol.* **30**, 147, (1942).
8. E. Gorham, Observations on the Formation and Breakdown of the Oxidized Microzones at the Mud Surfaces of Lakes, *Limnol. Oceanogr.* **3**, 291 (1958).
9. J. Kjensmo, "The Development and Some Main Features of Iron-Meromictic Soft Water Lakes," *Arch. Hydrobiol. Suppl.* **32**,137 (1967).
10. R. Goulder, "The Seasonal and Spatial Distributions of Some Benthic Ciliated Protozoa in Esthwaite Water," *Freshwater Biol.* **9**, 127 (1974).
11. W. Davison, S. I. Heaney, J. F. Talling, and E. Rigg, "Seasonal Transformations and Movements of Iron in a Productive English Lake with Deep-Water Anoxia," *Schweiz. Z. Hydrol.* **42**, 196 (1981).
12. E. R. Sholkovitz and D. Copland, "The Chemistry of Suspended Matter in Esthwaite Water, a Biologically Productive Lake with Seasonally Anoxic Hypolimnion," *Geochim. Cosmochim. Acta* **46**, 393 (1982).
13. P. Campbell and T. Jorgensen, "Maintenance of Iron Meromixis by Iron Re-

deposition in a Rapidly Flushed Monimolimnion,'' *Can. J. Fish. Aquat. Sci.* **37,** 1303 (1980).

14. L. M. Mayer, F. P. Gotta, and S. A. Norton, ''Hypolimnetic Redox and Phosphorus Cycling in Hypereutophic Lake Sebasticok,'' *Maine Water Res.* **16,** 1196 (1982).

15. W. Davison, ''Transport of Iron and Manganese in Relation to the Shapes of Their Concentration–Depth Profiles,'' *Hydrobiologia* **62,** 463 (1982).

16. W. Davison and S. I. Heaney, ''Determination of the Solubility of Ferrous Sulphide in a Seasonally Anoxic Marine Basin,'' *Limnol. Oceanogr.* **25,** 153 (1980).

17. W. Davison, ''The Polographic Measurement of O_2, Fe^{2+}, Mn^{2+} and S^{2-} in Hypolimnetic Water,'' *Limnol. Oceanogr.* **22,** 746 (1977).

18. W. Davison and S. I. Heaney, ''Ferrous Iron-Sulfide Interactions in Anoxic Hypolimnetic Waters,'' *Limnol. Oceanogr.* **23,** 1194 (1978).

19. W. Davison, ''Supply of Iron and Manganese to an Anoxic Lake Basin,'' *Nature* **290,** 241 (1981).

20. M. R. Hoffmann and S. J. Eisenreich, ''Development of a Computer-Generated Equilibrium Model for the Variation of Iron and Manganese in the Hypolimnion of Lake Mendota,'' *Environ. Sci. Technol.* **15,** 339 (1981).

21. J. J. Delfino and G. F. Lee, ''Chemistry of Manganese in Lake Mendota,'' *Wisconsin Environ. Sci. Technol.* **2,** 1094 (1968).

22. J. A. Robbins and E. Callender, ''Diagenesis of Manganese in Lake Michigan Sediments,'' *Am. J. Sci.* **275,** 512 (1975).

23. V. W. Ostendorp and T. Frevert, ''Untersuchungen zur Manganfreisetzung und zum Mangangehalf der Sedimentoberschicht im Bodensee,'' *Arch. Hydrobiol. Suppl.* **55,** 255 (1979).

24. T. Frevert, ''Phosphorus and Iron Concentrations in the Interstitial Water and Dry Substance of Sediments of Lake Constance (Obersee),'' *Arch. Hydrobiol. Suppl.* **55,** 298 (1979).

25. V. U. Tessenow and Y. Baynes, ''Redoxchemische Einflusse von Isoetes lacustris L. in Litoralsediment des Feldsees (Hochschwarzwald),'' *Arch. Hydrobiol.* **82,** 20 (1978).

26. V. U. Tessenow, Losungs-, diffusionsund Sorptionsporozesse in der Oberschicht von Sessedimenten,'' *Arch. Hydrobiol. Suppl.* **47,** 1 (1974).

27. H. Verdouw and E. M. J. Dekkers, ''Iron and Manganese in Lake Vechten: Dynamics and Role in the Cycle of Reducing Power,'' *Arch. Hydrobiol.* **89,** 509 (1980).

28. E. A. Jenne, ''Controls on Mn, Fe, Co, Ni, Cu and Zn Concentrations in Soils and Water: The Significant Role of Hydrous Mn and Fe Oxides.'' In R. A. Baker (Ed.), *Trace Inorganics in Water* (Adv. Chem. Ser. No. 73), American Chemical Society, Washington, D.C., 1968, p. 337.

29. W. Stumm and J. J. Morgan, *Aquatic Chemistry*, Wiley–Interscience, New York, 1981.

30. J. J. Morgan, ''Chemical Equilibria and Kinetic Properties of Manganese in Natural Waters.'' In S. D. Faust and J. V. Hunter (Eds.), *Principles and Applications in Water Chemistry*, Wiley–Interscience, New York, 1967, p. 561.

31. W. Sung and J. J. Morgan, "Kinetics and Product of Ferrous Iron Oxygenation in Aqueous Systems," *Environ. Sci. Technol.* **14,** 561 (1980).

32. J. F. Pankow and J. J. Morgan, "Kinetics for the Aquatic Environment," *Environ. Sci. Technol.* **15,** 1306 (1981).

33. J. F. Pankow and J. J. Morgan, "Kinetics for the Aquatic Environment," *Environ. Sci. Technol.* **15,** 1155 (1981).

34. W. Stumm and C. F. Lee, "The Chemistry of Aqueous Iron," *Schweiz. Z. Hydrol.* **22,** 295 (1960).

35. W. Davison, "A Critical Comparison of the Measured Solubilities of Ferrous Sulphide in Natural Waters," *Geochim. Cosmochim. Acta* **44,** 803 (1980).

36. J. Hamilton-Taylor, Lancaster University, England, personal communication and to be published. Chapter 2 in this book.

37. B. Böstrom, M. Johnson, and C. Forsberg, "Phosphorus Release from Lake Sediments," *Arch. Hydrobiol. Beih.* **18,** 5 (1982).

38. G. C. Holdren, D. E. Armstrong, and R. F. Harris, "Interstitial Inorganic Phosphorus Concentrations in Lake Mendota and Wingra," *Water Res.* **11,** 1041 (1977).

39. D. W. Evans, J. J. Alberts, and R. A. Clark, III, "Reversible Ion-Exchange Fixation of Cesium-137 Leading to Mobilization from Reservoir Sediments," *Geochim. Cosmochim. Acta* **47,** 1041 (1983).

40. E. Tipping, "The adsorption of Aquatic Humic Substances by Iron Oxides," *Geochim. Cosmochim. Acta* **45,** 191 (1981).

41. G. Müller, "Diagenetic Changes in Interstitial Waters of Holocene Lake Constance," *Nature* **224,** 258 (1969).

42. G. J. Brunskill, D. Povoledo, B. W. Graham, and M. P. Stainton, "Chemistry of Surface Sediments of Sixteen Lakes in the Experimental Lakes Area, Northwestern Ontario," *J. Fish. Res. Bd. Can.* **28,** 277 (1971).

43. R. R. Weiler, "The Interstitial Water Composition in the Sediments of the Great Lakes. 1. Western Lake Ontario," *Limnol. Oceanogr.* **18,** 918 (1973).

44. D. R. Sasseville and S. A. Norton, "Present and Historic Geochemical Relationships in Four Maine Lakes," *Limnol. Oceanogr.* **20,** 699 (1975).

45. E. R. Sholkovitz and D. Copland, "The Major-Element Chemistry of Suspended Particles in the North Basin of Windermere," *Geochim. Cosmochim. Acta* **46,** 1921 (1982).

46. B. J. Finlay, N. B. Hetherington, and W. Davison, "Active Biological Participation in Lacustrine Barium Chemistry," *Geochim. Cosmochim. Acta* **47,** 1325 (1983).

47. D. G. Kinniburgh and M. L. Jackson, "Cation Adsorption by Hydrous Metal Oxides and Clays," In M. A. Anderson and A. J. Rubin (Eds.), *Adsorption of Inorganics at Solid-Liquid Interfaces*, Ann Arbor Science Publ., Ann Arbor, Michigan, 1981, p. 91.

48. D. G. Kinniburgh, M. L. Jackson, and J. K. Syers, "Adsorption of Alkaline Earth, Transition and Heavy Metal Cations by Hydrous Oxide Gels of Iron and Aluminum, *Soil Sci. Am. J.* **40,** 796 (1976).

49. T. E. Cerling and R. R. Turner, "Formation of Freshwater Fe–Mn Coatings on Gravels and the Behavior of ^{60}Co, ^{90}Sr and ^{137}Cs in a Small Water-Shed," *Geochim. Cosmochim. Acta* **46,** 1333 (1982).

50. R. E. Jackson and K. J. Inch, "Partitioning of Strontium-90 Among Aqueous and Mineral Species in a Contaminated Aquifer," *Environ. Sci. Technol.* **17,** 231 (1983).

51. G. Matisoff, J. B. Fisher, and P. L. McCall, "Kinetics of Nutrient and Metal Release from Decomposing Lake Sediments," *Geochim. Comochim. Acta* **45,** 2333 (1981).

52. G. Haury and W. Schikarski, "Radioactive Inputs into the Environment; Comparison of Natural and Man-Made Inventories." In W. Stumm (Ed.), *Global Chemical Cycles and Their Alterations by Man*, Dahlem Konferenzen, Berlin, 1977, p. 165.

53. F. W. Whicker and V. Schultz, *Radioecology: Nuclear Energy and the Environment*, Vol. 1, CRC Press, Boca Raton, Florida, 1982.

54. J. J. Alberts, L. J. Tilly, and T. J. Vigerstad, "Seasonal Cycling of Cesium-137 in a Reservoir," *Science* **203,** 649 (1979).

55. C. W. Francis and F. S. Brinkley, "Preferential Adsorption of ^{137}Cs to Micaceous Minerals in Contaminated Freshwater Sediments," *Nature* **260,** 511–513 (1976).

56. T. F. Lomenick and T. Tamura, "Naturally Occurring Fixation of Cesium-137 on Sediments of Lacustrine Origin," *Soil Sci. Soc. Proc. Am.* **29,** 383 (1965).

57. J. M. Gieskes, H. Elderfield, J. K. Lawarence, J. Johnson, B. Meyers, and A. Campbell, "Geochemistry of interstitial water and sediments, Leg 64, Gulf of California," in J. Curry and G. Moore, *Initial Reports of the Deep Sea Drilling Project*, U.S. Government Printing Office, Vol. 64, Part II, 1982, pp. 675–694.

58. R. A. Berner, *Early Diagenesis*, McGraw-Hill, New York, 1980.

59. D. N. Edgington and J. A. Robbins, "The behavior of Plutonium and Other Long-Lived Radionuclides in Lake Michigan." In *Impacts of Nuclear Releases into Aquatic Environment*, Proceedings of the International Atomic Energy Agency, Vienna, Vol. IAEA-SM-198140, 1975, p. 245.

60. D. N. Edgington, J. J. Alberts, M. A. Wahlgren, J. O. Karttunen, and C. A. Reeve, "Plutonium and Americium in Lake Michigan Sediments," In *Transuranium Nuclides in the Environment*, Proceedings of the International Atomic Energy Agency, Vienna, Vol. IAEA-SM-199147, 1976, p. 493.

61. E. R. Sholkovitz, "The Geochemistry of Plutonium in Fresh and Marine Water Environments," *Earth Sci. Rev.* **19,** 95 (1983).

62. W. Pennington, R. S. Cambray, and E. M. Fisher, "Observations on Lake Sediments Using Fallout ^{137}Cs as a Tracer," *Nature* **242,** 324, (1973).

63. W. Pennington, "The Recent Sediments of Windermere," *Freshwater Biol.* **3,** 363 (1973).

64. W. Pennington, R. S. Cambray, J. D. Eakins, and D. D. Harkness, "Radionuclide Dating of the Recent Sediments of Blelham Tarn,' *Freshwater Biol.* **6,** 317 (1976).

65. J. A. Robbins, J. R. Krezoski, and S. C. Mozley, "Radioactivity in Sediments of the Great Lakes: Post-Depositional Redistribution by Deposit Feeding Organisms," *Earth Planet. Sci. Lett.* **36,** 325 (1977).

66. J. A. Robbins and D. N. Edgington, "Determination of Recent Sedimentation Rates in Lake Michigan Using Pb-210 and Cs-137," *Geochim. Cosmochim. Acta* **39,** 285 (1975).

67. M. B. Davis and M. S. Ford, "Sediment Focusing in Mirror Lake, New Hampshire," *Limnol. Oceanogr.* **27,** 137 (1982).

68. R. D. Evans and F. H. Rigler, "A Test of Lead-210 Dating for the Measurement of Whole Lake Soft Sediment Accumulation," *Can. J. Fish. Aquat. Sci.* **40,** 506 (1983).

69. K. L. Von Damm, L. K. Benninger, and K. K. Turekian, "The ^{210}Pb Chronology of a Core from Mirror Lake, New Hampshire," *Limnol. Oceanogr.* **24,** 434 (1979).

70. W. Davison, personal communication.

7

MECHANISMS CONTROL-LING THE SEDIMENTATION SEQUENCE OF VARIOUS ELEMENTS IN PREALPINE LAKES

Hans-Henning Stabel

Limnological Institute, University of Constance, D-7750 Konstanz, Federal Republic of Germany

Abstract

Sedimentation rates in several prealpine lakes revealed considerable differences both in quantity and annual patterns as shown by sediment-trap technology. The phytoplankton seems to play a major role in the formation of particles in these lakes, while allochthonous minerals were of minor importance.

In Lake Constance the annual succession of different algal species supposedly controls the seasonal patterns of the settling fluxes of organic matter, phosphorus, nitrogen, and silicon. Three major types of settling material were evaluated: (1) particulate organic matter; (2) autochthonously precipitated calcite; and (3) allochthonous minerals.

The annual flux of particulate organic matter increased from oligotrophic to eutrophic lakes. Turnover of particulate organic carbon in the euphotic zone was shown to be enhanced in more productive lakes. In Lake Constance, flagellates, blue-greens, and dinoflagellates were shown to be nearly completely remineralized in the euphotic zone, while diatoms and their remains transferred the bulk of particulate organic matter, silicon, and phosphorus to the lake bottom. Pronounced maxima of sedimentation rates of calcium were not correlated with maxima of supersaturation, but presum-

*ably were triggered by several algal species. Settling fluxes of silicon de-
riving from lacustrine sources were closely related to biomass development
of diatoms throughout the year.*

1. INTRODUCTION

The cycling of many elements in freshwater is intensively coupled with the
process of sedimentation of particulate matter. The formation of particles,
adsorption and desorption reactions, and redissolution of solids contribute
significantly to changes in the chemical composition of lake waters. Regardless of whether they are mediated by biological or by chemical reactions,
settling processes followed by a permanent deposition form a major sink for
many elements. In contrast to the oceans, settling processes in lakes can be
studied easily to greater detail within short time intervals and at many depths.
The transfer of particles from the watershed and from the atmosphere in
most cases complicates analysis of lacustrine particle formation. Moreover,
resuspension of bottom sediments and inputs from the littoral zone often
contribute to particle concentration and settling fluxes in the water. Due to
seasonality of the phytoplankton and the hydraulic load chemically very
different substances constitute the pool of particulate matter.

However, for the establishment of whole-lake budgets of various elements, as well as for the understanding of the obliteration of nutrients from
the euphotic zone and subsequent enrichment of the deeper layers, knowledge on vertical transfer of masses is essential.

Three methodologically different approaches are used to evaluate the
masses of material settling in a given time:

1. Deposition rates have been calculated from differences between the an-
    nual income and outflow of suspended matters.
2. Krishnaswami and Lal (1) reviewed methods to determine deposition
    rates from chemically stratified sediment cores; certain layers of these
    cores can be dated by radionuclide limnochronology (i.e., measurements
    of ^{137}Cs, ^{210}Pb, ^{226}Ra, etc.).
3. In numerous studies the vertical flux of particles is assessed by the use
    of sediment traps.

One of the major disadvantages in the use of the "one-box model" is that
lacustrine processes of particle formation, of resuspension, and of chemical
reactions (dissolution, adsorption, etc.), are not recorded. This approach
was applied to Lake Constance by Wagner (2) and Golterman (3). It should
be stressed also that data from radionuclide measurements merely record a
final deposition. Moreover, Sturm (4) and Sturm and Matter (5) have shown
that a continuous formation of bottom sediments often is disturbed by interference of slumps and turbidities.

Bloesch and Burns (6) and Blomqvist and Hakanson (7) have reviewed different types of sediment traps and have suggested special approaches to how to best record the gross flux of particulate material. By the application of this technique it is possible to describe the settling process due to both seasonal variations and depth-dependent biological and chemical transformations.

The formation, transportation, and settling processes of particulate matter (PM) from several prealpine lakes investigated by the sediment-trap technique are the subject of this paper. Essentials for the understanding of sedimentary fluxes are concomitant measurements of thermal stratification and water movements, knowledge on the changes of the respective dissolved elements, and records of the phytoplankton biomass and primary productivity. Therefore, only a limited number of lakes are included in this comparison, which have been studied intensively for all these subjects. Siebeck (8) and Lehmann (9) reported on Königsee (Bavaria), while data from the Lake of Lucerne are available through the works of Stadelmann (10), Bloesch (11), and Bloesch et al. (12). Stabel and Tilzer (13), Stabel and Kleiner (14), and Tilzer (15) studied Überlinger See/Lake Constance, while data from Kelts and Hsü (16), Pelletier (17), and Jaquet et al. (18) provide information on Lake Zürich and Lake Geneva, respectively. All these lakes represent a similar chemical constitution concerning alkalinity and concentrations of dissolved earth metals, but vary widely in their trophic states.

2. ANALYSIS OF SEDIMENTATION PROCESSES

2.1. Sources, Fluxes, and Chemical Composition of Particulate Matter

Particles deriving from outside the lake include minerals transported by rivers, overland flow, shore erosion, and atmospheric transport, as well as organic debris from terrestrial vegetation. Within the lake particles are subjected to physical forces which move and sort the solids. Due to their temperature and density, waters from tributaries tend to further flow within the water column, thus forming subsurface sediment plumes. Several studies performed in prealpine lakes are reported by Wagner and Wagner (19), Lehn (20), Nydegger (21), Nümann (22), and others. According to their specific weight, grain sizes, and driving forces of the currents, particles are deposited along horizontal gradients. With increasing distances from the delta, the horizontal flux of allochthonous particulate matter (PM) will be reduced, and gradually gravity forces will regulate particle movements. Therefore, different spectra of particle grain size and mineralogical composition will occur in different parts of the lake. The total load of solids varies due to changing discharge of the tributaries as regulated by rainfall and snowmelt.

Phytoplankton is a predominant source of PM produced in lakes. Through photosynthesis many elements are transferred from the dissolved to the solid state. Since light is a prerequisite for primary production of algae, photo-synthetic particle formation is restricted to the euphotic zone of lakes. Due to their minimal excess density over the environment, algae tend to settle into the aphotic zone. Further losses of particulate organic matter occur through rapid utilization within the limnetic food chain, leaving living or-ganisms as well as their debris and dissolved remains. Therefore, significant changes in the patterns of dry-weight matter mostly are found in the up-permost 20-m layers as demonstrated for the Überlinger See (Fig. 7.1). In the deeper strata occasionally some regions of enhanced particle concen-trations can be recorded. These concentrations probably originate from hor-izontal subsurface fluxes of allochthonous material (Fig. 7.2).

Tilzer (15), Sommer (23,24), and Mohammed and Müller (25) have shown that phytoplankton biomass and primary production in Lake Constance ex-hibit a somewhat regular seasonal pattern.

Changes in the concentrations of particle dry-weight matter (i.e., dried for 24 hours at 110°C) as measured in the euphotic zone seem to be in tune with fluctuations of the phytoplankton biomass. Therefore, more or less typical variations in the mean concentrations of particulate matter are re-flecting changes of biomass (Fig. 7.3). Low and rather constant particle concentrations (ranging from 0.1 to 0.5 mg liter^{-1}) prevailed during winter, while maxima of nearly 5 mg liter^{-1} occurred in the stratified period. Prior to minima regularly occurring in June, PM concentrations peaked in May. Summer maxima were found in mid-July and in August, while increases of PM concentrations in the fall were less pronounced (Fig. 7.3). Unfortunately, comparable data on particle distribution are not available for other prealpine lakes.

Settling fluxes of PM exhibited rather similar patterns from 1980 through

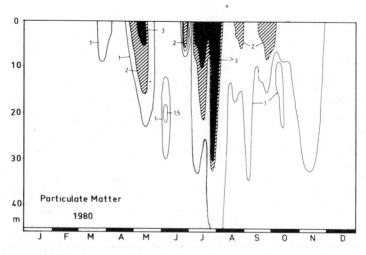

Figure 7.1. Isopleths of particulate matter concentrations in Lake Constance in 1980 (dry weight in mg liter^{-1}).

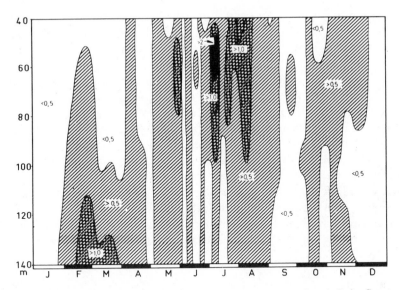

Figure 7.2. Hypolimnetic isopleths of particulate matter concentrations in Lake Constance in 1980, showing resuspension (February through March) and horizontal subsurface plumes of particulates (May through September).

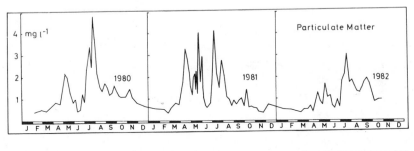

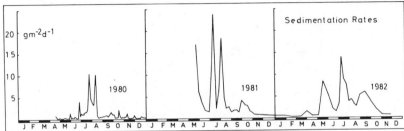

Figure 7.3. Comparison of particulate matter (PM) concentrations and settling fluxes of dry-weight matter in Lake Constance/Überlinger See from 1980 through 1982 (mean values of the upper 20-m layers).

1982, with minima in winter (0.01 g m^{-2} d^{-1}) and several distinct maxima in the stratified period (exceeding 25 g m^{-2} d^{-1}). Maxima of PM concentrations were not always reflected in enhanced sedimentation rates. In contrast to these enormous fluctuations found in Lake Constance/Überlinger See, sinking rates in the Königsee were rather low and constant as demonstrated by Lehmann (9), with only two small peaks in June and September and daily maximum fluxes barely exceeding 1.4 gm^{-2} (Fig. 7.4). Although sampling in Königsee revealed only fortnightly integrals of the PM flux, pronounced changes in the annual pattern are not to be expected for shorter time intervals. Settling rates in Lake of Lucerne were low during winter (0.1 g m^{-2} d^{-1}) and 100 times higher in May and July. A minimum was found in June (11). In Rotsee a more or less unimodal pattern was found (Fig. 7.4).

The annual mean composition of settling material from Königsee, Lake of Lucerne, Rotsee, and Lake Constance was rather similar (Table 7.1). The chemistry of the particulates collected just below the euphotic zone was

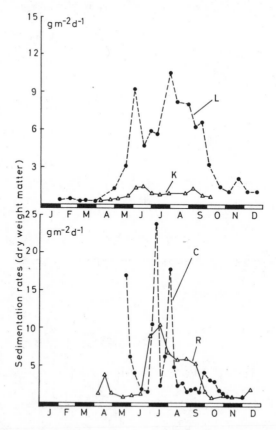

Figure 7.4.   Settling fluxes in four prealpine lakes (gm^{-2} d^{-1}); K = Königsee, L = Lake of Lucerne, R = Rotsee, and C = Lake Constance; redrawn from (9) and (11) and from author's data.

Table 7.1. Mean Chemical Composition of Settling Material (Percent of Dry Weight)

	Königsee 1980[a]	Lake of Lucerne 1969[b]		Rotsee 1969[b]		Lake Constance	
		Stagnation	Circulation	Stagnation	Circulation	1980	1981
POM[c]	21.5	9.5	15.7	17.45	46.67	14.4	11.31
POC[d]	9.67	4.47	8.3	10.54	23.07	6.2	5.07
PON[e]	—	0.96	1.24	0.75	3.06	0.96	0.83
CO₃	42.5	—	—	—	—	26.0	28.3
Ca	28.3	22.14	10.17	12.8	19.82	17.3	17.9
Si	—	—	—	—	—	—	11.31
Al	—	—	—	—	—	—	4.49
Fe	—	1.20	2.52	0.93	1.32	—	2.08
K	—	—	—	—	—	—	1.18
Mg	—	0.61	0.93	0.75	0.49	—	0.92
S	—	—	—	—	—	—	0.176
Ti	—	—	—	—	—	—	0.157
P	—	0.077	0.155	0.136	0.35	0.098	0.134
Mn	—	0.049	0.172	0.032	0.303	—	0.079
Cl	—	—	—	—	—	—	0.046

[a] Calculated from Lehmann (9).
[b] Calculated from Bloesch (11).
[c] POM = particulate organic matter
[d] POC = particulate organic carbon.
[e] PON = particulate organic nitrogen.

predominated by calcium carbonate, silicon-containing compounds, and organic matter (POM). Constituents of minor importance are Al, Fe, K, and Mg, while S, Ti, P, Mn, and Cl are below the 1% level. Sigg et al. (26) and Sigg (Chapter 13 in this book) measured trace metals in both the water column and the PM from the central part of Lake Constance, showing that the total amount of Cu, Zn, Pb, and Cd barely exceeded 0.002% of the particle dry weight.

2.2. Seasonality of Elemental Fluxes

Data from Lake Lucerne and from Rotsee (11) indicated that seasonally different compositions of sedimentary material will occur within the same lake. Due to chemically very different material dominating the respective suspended matter in Lake Constance, the composition of settling material changed considerably within short periods. The settling fluxes of the various elements fluctuated greatly throughout the year. This is demonstrated by the variations in the annual patterns of sedimentation rates of Mg, P, and Si (Fig. 7.5), showing discrepancies in the time courses of extremes.

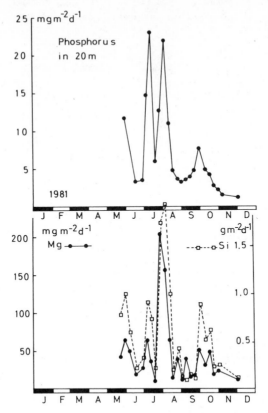

Figure 7.5. Settling fluxes of Mg, P, and Si in Lake Constance in 1981, showing differences in seasonal patterns.

In order to classify the elements, a statistical comparison of the annual patterns of sedimentation rates can be applied. Cluster analysis is a useful technique to gain understanding of the relations between many variables more or less regulated by a limited number of factors. A table of intercorrelations between time courses of the fluxes of the elements under study (Table 7.2) was tested for any defined clusters of geochemicals indicating their common origin. Data from Lake Constance revealed three chemically different types of material (Fig. 7.6):

Very close correlations existed between particulate organic matter (POM), organic nitrogen (PON), organic carbon‡ (POC), and phosphorus, suggesting that phosphorus is transported to the lake bottom mainly by organic material.

‡ Obtained from the difference between total carbon content and carbon content after combustion at 520°C (4 hours) by use of an elemental analyzer.

Table 7.2. Correlation Matrix for Daily Sedimentation Rates of Various Constituents of Particulate Matter Trapped in the Überlinger See (1981) (Limits of confidence = 95%)

	POM[a]	POC[b]	PIC[c]	PN[d]	P	Mg	Al	Si	S	Cl	K	Ca	Ti	Mn	Fe
POM	1.00														
POC	0.99	1.00													
PIC	−0.70	−0.68	1.00												
PN	0.92	0.91	−0.71	1.00											
P	0.93	0.95	−0.72	0.90	1.00										
Mg	0.39	0.37	−0.85	0.53	0.50	1.00									
Al	0.28	0.23	−0.77	0.36	0.28	0.92	1.00								
Si	0.12	0.05	−0.33	0.10	−0.06	0.41	0.70	1.00							
S	0.0	−0.03	−0.07	−0.04	−0.13	0.10	0.27	0.70	1.00						
Cl	−0.05	−0.06	−0.15	−0.08	−0.10	0.20	0.32	0.61	0.96	1.00					
K	0.22	0.16	−0.70	0.28	0.17	0.84	0.98	0.80	0.37	0.39	1.00				
Ca	−0.45	−0.41	0.31	−0.30	−0.20	0.01	−0.19	−0.63	−0.49	−0.39	−0.29	1.00			
Ti	0.39	0.36	−0.83	0.53	0.46	0.97	0.91	0.40	−0.03	0.03	0.84	0.04	1.00		
Mn	0.47	0.43	−0.80	0.62	0.45	0.84	0.86	0.66	0.47	0.47	0.83	−0.37	0.80	1.00	
Fe	−0.02	−0.07	−0.36	0.09	−0.11	0.54	0.75	0.88	0.70	0.69	0.81	−0.45	0.46	0.74	1.00

[a] POM = particulate organic matter.

[b] POC = particulate organic carbon.

[c] PIC = particulate inorganic carbon.

[d] PN = particulate nitrogen.

Correlations between calcium and inorganic carbon were rather weak, because the calcium content of the samples originated from autochthonously precipitated calcite as well as from allochthonous minerals. No interrelations to any other constituent could be found.

Sedimentation rates of elements such as Al, K, Mg, Mn, Fe, Si, Ti, S, and Cl formed a third cluster, due to their occurrence in allochthonous minerals such as clay, quartz, and silicates.

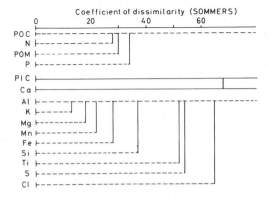

Figure 7.6. Cluster dendrogram depicting relations between patterns of sedimentation fluxes of elements in Lake Constance. Decreasing coefficients indicate increasing similarity.

2.2.1. Sedimentation of Particulate Organic Matter

Except for Rotsee, the lakes discussed in this chapter comprise large and deep water bodies. They probably accept little input of particulate organic matter (POM) from their surroundings and from their narrow littoral zones. Therefore, POM settling within these lakes was suspected to originate almost entirely from phytoplankton production.

It is commonplace in limnology that the flux of organic carbon into the aphotic zone increases with enhanced primary production. Several important theories and conceptual models are based on this fundamental concept. Therefore it is most surprising that quantitative experimental data for a functional description of the relation between primary production (PPR) and POM flux settling out of the epilimnion are scarce. For prealpine lakes, only Königsee, Lake of Lucerne, Rotsee, and Lake Constance were investigated for both PPR and sedimentation rates of solid organic matter on an annual scale. Annual data on PPR (17) and from some monthly integral values of the POM flux (18) from Lake Geneva suggest limnological conditions and sedimentary processes close to those found in Lake Constance.

Except for Rotsee, a small and relatively shallow lake which is heavily fertilized, these prealpine lakes exhibited an inverse relation between the annual primary production rate and the percentage of this material that settles out of the euphotic zone (Table 7.3). It seems that carbon turnover is much faster and more effective in fertilized lakes (Fig. 7.7). There is still uncertainty in the interpretation of measurements by the ^{14}C method, whether it determines net or gross primary production (27). Moreover, assessments of gross sedimentary fluxes are amenable to methodological imprecision (6,7). Therefore, a valid correlation cannot be deduced from these few investigations, but a definite trend seems to exist.

Table 7.3. Primary Production Rates (PPR, g C m^{-2} yr^{-1}), Fluxes of Particulate Matter [PM, g (Dry Weight) m^{-2} yr^{-1}], and Sedimentation Rates of Organic Carbon (POC, g C m^{-2} yr^{-1}, and Percentage of PPR) of Several Prealpine Lakes

	Depth (m)	Year	Status	PPR	PM	POC	Percent of PPR
Königsee	20	1980	Oligotrophic	77	238.5	35.7	46.4
	170	1980	Oligotrophic	—	130.6	15.1	19.6
Überlinger See	20	1980	Mesotrophic/ Eutrophic	250	452.0	53.8	21.5
	20	1981	Mesotrophic/ Eutrophic	330	1137.0	57.7	17.5
Lake of Lucerne	15	1969	Mesotrophic	415	1153.6	58.0	14.0
Rotsee	5	1969	Eutrophic	511	992.2	153.7	30.1

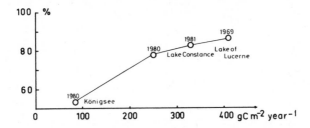

Figure 7.7. Relation between primary production rate (PPR as carbon in $gm^{-2} yr^{-1}$) and annual turnover of organic carbon (as percentage of the PPR) in the euphotic zones of prealpine lakes.

A more detailed study of primary production rates (PPR) and of the distribution and settling of organic matter in Lake Constance elucidated several mechanisms governing the algal biomass development. Although primary production varied from year to year (15) and exhibited distinct seasonal patterns, fluctuations in the pool of POM (biomass and its debris) were shown to be more regular (Fig. 7.8). Biomass development was coupled to production and losses as follows: After the onset of phytoplankton development in early spring, POM concentrations peaked in May ("spring bloom"). Deep mixing of the algae by wind-induced turbulences sometimes interrupted their continuous growth. A well-defined minimum of particulate organic carbon appeared in the beginning of June ("clearwater phase"). Regularly, during this period, a mass development of the crustacean zooplankton occurs and the grazing activity scavenges the algal biomass almost completely as shown by Lampert and Schober (28) and Geller (29). The water becomes considerably more transparent and the pool of dissolved nutrients is regenerated to some extent. After the starvation of the zooplankton there are again favorable preconditions for the development of the summer maximum of phytoplankton. Towards the end of the summer, algal biomass declines gradually until a more or less pronounced autumnal maximum of diatoms appears. Destratification and deep mixing of algae then contribute to the further decline of POM concentrations, which are maintained low through the whole period of destratification.

Maximum sedimentation rates of organic matter, however, only occa-

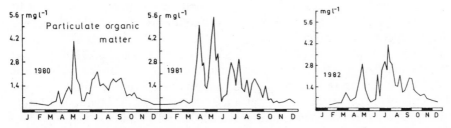

Figure 7.8. Annual distribution patterns of mean POM concentrations in the euphotic zone of Lake Constance (1980 through 1982).

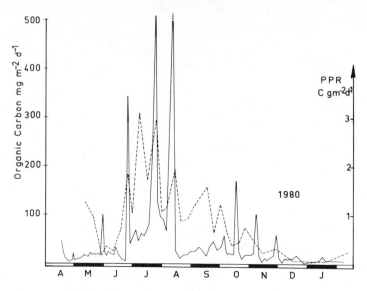

Figure 7.9.   Daily primary production rates (PPR, dotted line) and sedimentation rates of organic carbon (POC, solid line) in Lake Constance (1980).

sionally corresponded to peaks of primary production rates (Fig. 7.9). Whenever mass developments of the flagellates *Rhodomonas sp.* and *Cryptomonas sp.* occurred (23), sedimentation rates of particulate organic matter (POM) remained low, due to a nearly quantitative grazing, especially during the clearwater phase. In late summer primary production was maintained mainly by *Ceratium hirundinella*, *Cryptomonas sp.*, and to a lesser extent by blue-green algae. These species also did not contribute significantly to POM sedimentation. The maximum settling fluxes of POM in May, July, August, and the beginning of October (Fig. 7.9) were associated with "blooms" of diatoms and green algae.

In the ocean, phosphorus is known to exclusively settle as a component of the organic residues (30). During the sinking process POM is decomposed by zooplankton and microheterotrophes releasing phosphorus into the aphotic zone. Only a very small proportion of particulate phosphorus leaving the euphotic zone finally is buried in the bottom sediments.

Although the external load of particulate phosphorus to Lake Constance varies according to precipitation and weathering (31), allochthonous mineral phosphorus will exceed particulate organic phosphorus in water bodies close to the main inlets. With growing distances mineral particulate phosphorus will decline. The settling flux of phosphorus measured in Überlinger See and in a central part of Lake Constance (Obersee) was shown to be closely correlated to the sedimentation rates of POM (cf. Table 7.2 and Fig. 7.6). Therefore, as in the oceans, phosphorus in Lake Constance is suggested to settle as a component of the organic phase. In the bottom sediments, however, only one-third of the total phosphorus content was fixed in organic material, as determined by Frevert (32).

2.2.2. Sedimentation of Calcium Carbonate

During the vegetation period of the phytoplankton, prealpine lakes tend to precipitate calcite, provided supersaturation conditions for $CaCO_3$ are maintained by high-temperature and high-pH values. A further prerequisite is the existence of a low Mg/Ca ratio. Most of the prealpine lakes are hard-water lakes and calcite formation has been reported for many of them (8,16,33–39). Although "summer whitings" as observed for Lake Michigan by Strong and Eadie (40) were not demonstrated for central European lakes, a distinct seasonality in the deposition of calcareous sediments was shown for Lake Zürich (34,16).

In Lake Constance calcite precipitation was restricted to very short periods only. In 1980 sedimentation rates of calcium exhibited peaks in July, August, and in late September, while in 1981 maxima occurred in May, July, and August (Figs. 7.10 and 7.12). The cluster analysis (see Section 2.2) revealed no correlations to any other component of the settling PM, except for inorganic carbon (Table 7.2). In order to separate allochthonous from autochthonous calcium carbonate sedimentation, seasonal variations in the concentrations of Ca in the solids were compared with those of Ti.

Titanium was used as a tracer for allochthonous clastic material since this element is known for its widespread distribution and constant concentration in every geological formation thus far investigated (41). Moreover, titanium is reported not to be involved in biological reactions (42).

Particles in Lake Constance showed parallel fluctuations in their Ca and Ti contents from April through June 1980 and from mid-November through Mid-February 1981 (Fig. 7.10). The period of autochthonous calcite formation was interposed. The correlation can be expressed by a linear equation:

$$[Ti] = 0.0409 + 0.01342 \times [Ca]$$

In this equation concentrations are given in percent of dry weight; $r = 0.706$, $n = 31$, and confidence limit $= 95\%$. To some extent this linear regression reflects stoichiometric relations of Ca and Ti within allochthonous material in Lake Constance. It allows us to calculate the allochthonous portion of the total Ca content from measurements of Ti.

In order to estimate the "production" of lacustrine calcite, the annual settling flux of allochthonous calcium was calculated from the mean Ti content (cf. Table 7.1) by use of the equation and was subtracted from the total flux. In 1980 the total flux of calcium amounted to 76.9 gm^{-2}. Autochthonous calcite formation contributed 51.6% (39.7 gm^{-2}) of the annual rate. It must be emphasized that these data were obtained from sediment traps suspended just below the maximum extension of the euphotic zone (i.e., 20-m depth).

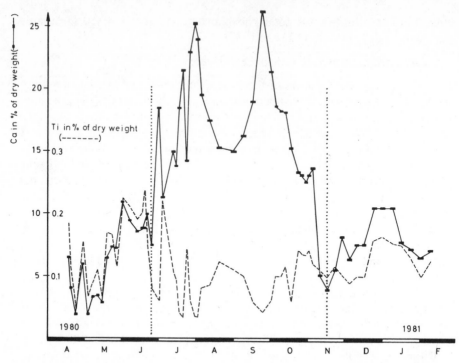

Figure 7.10. Ca (in % of dry weight) and Ti (in % of dry weight) as constituents of settling material in Lake Constance in 1980 showing parallel fluctuations in spring and autumn. Interposed is the period of lacustrine calcite formation. Horizontal bars indicate the extension of sampling periods.

In order to find mechanisms triggering events of high calcite precipitation, saturation conditions for $CaCO_3$ were calculated for the upper layers of Lake Constance/Überlinger See using a set of equations as outlined by Stumm and Morgan (43). Concentrations of dissolved Ca and $HCO_3{}^-$ were measured weekly in vertical profiles, along with pH, temperature, and conductivity. As indicated by the graph of calcite saturation ratios (Fig. 7.11), significant supersaturation was restricted only to the uppermost 20-m layers. Periods of high supersaturation were succeeded by intervals of low indices, supposedly generated by the combined effects of cooling, deep mixing, and phytoplankton breakdown (declining pH values). Generally, in the deeper layers (i.e., below 30-m depth), saturation prevailed but redissolution of calcite crystals may occur in deeper strata.

In the euphotic zone high saturation ratios were calculated (Fig. 7.12), exceeding those estimated by Kelts and Hsü (16) for Lake Zürich. Patterns of high supersaturation ratios in Lake Constance were not in phase with extremes of the calcium sedimentation rates. Sometimes conditions of high supersaturation did not lead to significant calcite precipitation. For example, this was the case during the clearwater phase in June.

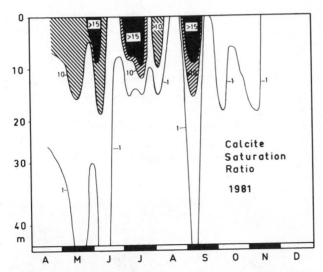

Figure 7.11. Isolines of calcite saturation ratios (Ionic Activity Product/Solution Constant IAP/K_{SO}) in Lake Constance (1981).

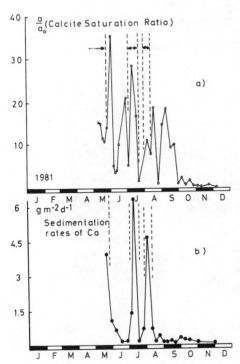

Figure 7.12. (*a*) Calcite saturation ratios in 1-m depth and (*b*) sedimentation rates of calcium measured below the euphotic zone (Lake Constance).

157

While Brunskill (44) concluded that temperature was the direct factor initiating calcite precipitation in Fayetteville Green Lake (New York) and that phytoplankton activities play a secondary role, in Lake Constance events of sudden calcite formation may be triggered by several algal species. Several authors reported on seasonal blooms of pelagic algae depleting dissolved CO_2 and thus inducing calcite precipitation (38,45,16). Kelts and Hsü (16) reviewed the literature on biologically induced supersaturation in freshwaters. SEM (Scanning electron microscope) studies of the particles trapped in Lake Constance revealed polyhedral crystals of 40-μm diameter (Fig. 7.13), similar to those occurring in Lake Zürich (16). Remains of the diatom

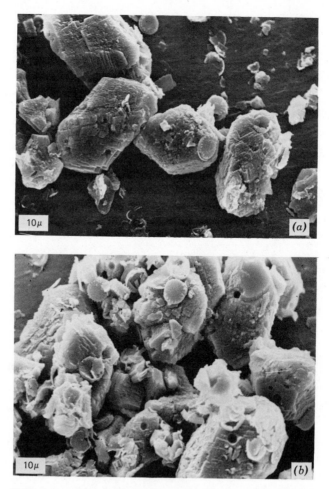

Figure 7.13. SEM photographs showing calcite crystals trapped in the water column of Lake Constance: (*a*) freshly precipitated crystals; (*b*) calcite crystals with half-spherical holes of 2–3 μm diameter; (*c*) deteriorated calcite crystals with remains of the diatom *Stephanodiscus hantzschii*; (*d*) calcite with remains of the green alga *Pandorina sp.*

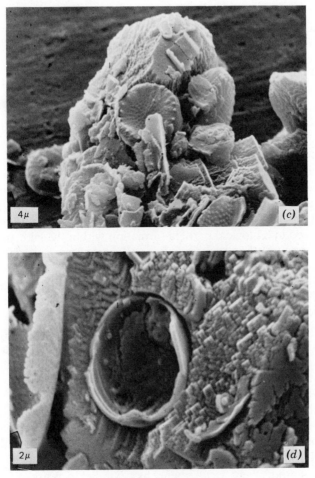

Figure 7.13. (*continued*)

Stephanodiscus hantzschii were occluded in most of the calcite crystals, and many of them contained half-spherical holes of nearly 2-μm diameter, probably due to the presence of small round algae like *Chlorella*. In addition, sometimes the remains of *Pandorina sp.* and of *Phacotus lenticularis* could be detected (Fig. 7.13). From the ratios of supersaturation calculated for the epilimnetic zone of Lake Constance, it can be concluded that homogeneous nucleation is not the initiating step in calcite formation [cf. (43)]. It seems that several algal species were able to induce heterogeneous nucleation, while during some periods with high saturation indices nucleation was hindered.

Recently, Kunz (45) showed through laboratory experiments that algae were able to serve as reactants in heterogeneous nucleation of calcite. He also found that phosphate, even in micromolar concentrations, hindered calcite precipitation in supersaturated solutions. During the clearwater phase in Lake Constance a recovery of the phosphate concentrations might prevent calcite formation in this period.

2.2.3. Sedimentation of Silicon

The silicon content of settling particles in lake waters is derived either from biogenically synthesized diatom frustules or from allochthonous minerals. Therefore, sedimentation rates of silicon exhibited only weak correlations to the fluxes of elements constituting the cluster of clastic minerals (Table 7.2). In order to estimate the autochthonous contribution of silicon in the settling material, sedimentation rates were corrected for those supported by external mineral fluxes. Varying mineral input was indicated by changes in the aluminum content of the settling particles. Aluminum was assumed not to be involved in biological particle formation within the lake and to exhibit linear relations to other elements of the allochthonous material. Whenever diatom biomass and growth is low, changes in the silicon content of the particles will correlate with fluctuations of the aluminum content. Breakdown of diatom populations, however, will alter the Si/Al ratio in the settling material. Sholkovitz and Copland (46) calculated "excess" concentrations of various elements in particles using a similar approach. These "excess" concentrations were indicative for lacustrine PM in Esthwaite Water (U.K.).

Settling rates of silicon in Lake Constance were related to those of aluminum and the resultant "excess" rates were plotted (Fig. 7.14). The annual patterns were in phase with fluctuations of the dissolved silicon content and with distribution of diatom population maxima and minima, while the uncorrected silicon fluxes were not (see Fig. 7.5). As pointed out earlier, POM was transported into the hypolimnion mainly as a constituent of diatoms and their debris. Whenever high autochthonous sedimentation of silicon occurred, the diatoms were accompanied by a parallel increase of POM settling to the lake bottom. *Asterionella formosa*, *Fragilaria crotonensis*, *Stephanodiscus hantzschii*, and *Stephanodiscus binderanus* were the main species regulating the POM and silicon sedimentation.

2.2.4. Sedimentation of Allochthonous Matter

Allochthonous mineral particles were of minor importance in the settling material from the Überlinger See as compared to autochthonous contributions of POM and calcite. The amount of minerals supported by external fluxes changed due to rainfall in the drainage area and water-level changes. For prealpine lakes, snowmelt and precipitation maxima provide the highest load of particulate matter in midsummer. Although the forces driving erosion vary stochastically from one season to another, settling rates of clastic minerals peaked in July 1980 and 1981 (Überlinger See). Major minerals were quartz, calcite, dolomite, feldspar, chlorite, and illite.

Conservative elements like Al, Mg, and Ti have proven to be suitable

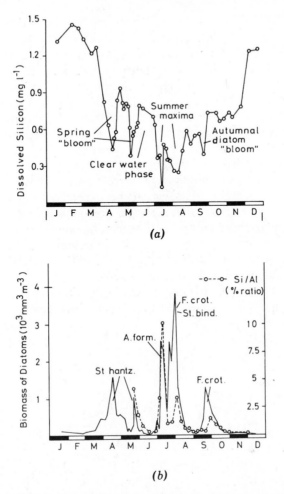

Figure 7.14. (a) Annual patterns of dissolved silicate (mean concentration upper 20-m layer). (b) Diatom biomass development in Lake Constance in 1981 and Si/Al ratios (% of dry weight) in particles indicating "excess silicon"; St. hantz. = *Stephanodiscus hantzschii*; A. form. = *Asterionella formosa*; F. crot. = *Fragilaria crotonensis*; and St. bind. = *Stephanodiscus binderanus*. (Data on cell volumes were kindly provided by U. Sommer.)

indicators of allochthonous minerals. Maxima of their settling fluxes, therefore, can be used as tracers for horizontal subsurface plumes of external minerals and for resuspension of bottom sediments.

2.3. Stoichiometric Relations in the Organic Material

Stoichiometric relations for micromolar amounts of nitrate, phosphate, and dissolved inorganic carbon taken up during photosynthesis in the sea were reported by Redfield et al. (47). A simple relation was established to characterize algal protoplasm chemically with respect to the nutrients in the proportion C/N/P = 106:16:1. This ratio may vary from one habitat to another, but was found to exist for nutrient uptake in Lake Constance (26). However, the molar ratio of organic carbon and phosphorus trapped just below the euphotic zone fluctuated widely throughout the year (Fig. 7.15). The annual mean was estimated for material trapped in Überlinger See (1981) deviating from the stoichiometric value given by Redfield (i.e., 89 instead of 106). On the other hand, C/N ratios revealed an annual mean of 7.1 — not much deviating from Redfield's stoichiometry.

The phosphorus concentrations in the particulates, however, were found to be significantly higher than was to be expected from Redfield's ratio established for organic material. A simple explanation for the stoichiometric "surplus" of phosphorus in the settling material might be the occurrence of inorganic phosphorus either bound in allochthonous minerals, such as apatite, or in coprecipitates with calcite. Rossknecht (48) reported on coprecipitation of calcite and phosphorus in Lake Constance. Formation of coprecipitates with calcite also might explain the occasionally low C/P ratios of settling material in summer. Whenever calcium sedimentation rates were

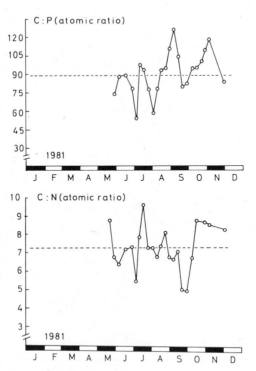

Figure 7.15. Atomic ratios in settling particles in Lake Constance (1981). Dotted line indicate calculated annual means.

high (cf. Fig. 7.12), the stoichiometric relation between phosphorus and carbon was shifted to low values (Fig. 7.15).

Dissolved soluble reactive phosphorus was found to adsorb on clay minerals carried into Lake Constance by the river Rhine (49). This effect might also contribute to the relatively enhanced phosphorus concentrations of the particles. Keeping these abiotic reactions in mind, it must be emphasized that in Lake Constance the bulk of phosphorus is sedimented with organic residues.

3. CONCLUSIONS

Except for allochthonous input due to runoff from the drainage area, formation of particles in Lake Constance was mediated by biological as well as by chemical processes. It seems that primary production of the phytoplankton and the annual succession of the dominant algal species play an important role in the sedimentation process.

Photosynthesis mediates high-pH values which in turn accelerate supersaturation with respect to calcite. As a consequence high amounts of calcium were found in the particles of prealpine hard-water lakes in summer. Due to the rather low primary production in the oligotrophic Königsee, resulting in rather unchanged biomass concentrations throughout the year (8), sedimentation rates of the dry-weight matter exhibited only weak fluctuations. In more eutrophic lakes the annual biomass concentration was at least bimodal, that is, the "spring bloom" and the "summer maximum." For Lake Constance the minimum regularly occurring in June was shown to correspond to high grazing pressure of the crustacean zooplankton (28,29).

Many algal species were shown not to settle into deeper strata of Lake Constance. These species either were flagellates with cell walls rather sensitive to physical damages, or were buoyant blue-green algae. Still others were dinoflagellates like *Ceratium hirundinella* (23). Their cells rapidly deteriorate and the respective elements are recycled. Therefore, not only biomass, but also cell structure of the various species, affect the settling of particulate organic matter (POM). As phosphorus precipitates simultaneously with POM, biomass development of the respective algae also impacts the phosphorus flux.

The succession of mass developments of predominant algal species was very similar throughout the years as outlined by Sommer (23,24). Competition for nutrients, the capability not to settle out of the euphotic zone, edibility by herbivorous zooplankton, and regulation of the spatial distribution of populations were the most important factors governing the sequence of different algal species. Their populations generated a regular pattern of settling fluxes of POM, phosphorus, and silicon and probably mediated the precipitation of calcite.

Using conservative elements as tracers it was possible to separate autochthonous from allochthonous calcium and silicon sedimentation rates. Diatom biomass development proved closely related to the autochthonous settling flux of silicon (Fig. 7.14). Autochthonous precipitation of calcium occurred in prealpine lakes whenever PPR and pH were high. It is supposed that due to these regulating mechanisms the proportion of autochthonously precipitated calcium is higher in more eutrophic habitats.

Therefore, stoichiometric relations in the settling particles depend on the production and on the inventory of algal species as far as calcium and silicon is concerned. In order to characterize biogenic seston by elemental relations, at least annual mean values corrected for allochthonous contributions should be used for prealpine lakes. In Lake Constance, due to the allochthonous mineral input of phosphorus, stoichiometric relations of carbon, nitrogen, and phosphorus were shifted from the so-called "Redfield ratio" in the year's course, while annual means of carbon-to-nitrogen ratios in Lake Constance were close to the value required by Redfield's stoichiometry.

The referred mechanisms play an important role in the transportation of the various elements from the euphotic to the aphotic zone. Degradation of organic matter and redissolution processes of calcite and silica frustules within the hypolimnion will change the chemical constitution of particles settling towards the bottom. Diagenetic processes are known to occur within the uppermost layers of the sediments. Therefore, the chemical composition of the particles does not allow the interpretation of the dynamics of the sedimentation process.

REFERENCES

1. S. Krishnaswami and D. Lal, "Radionuclide Limnochronology." In A. Lerman (Ed.), *Lakes—Chemistry, Geology, Physics*, Springer, Berlin, 1978, p. 153.

2. G. Wagner, "Simultationsmodelle der Seeneutrophierung, dargestellt am Beispiel des Bodensee-Obersees, Part II" *Arch. Hydrobiol.* **78,** 1 (1976).

3. H. L. Golterman, "Phosphate Models, a Gap to Bridge," *Hydrobiologia* **72,** 61 (1980).

4. M. Sturm, "Die Oberflächensedimente des Brienzersees", *Eclogae Geol. Helv.* **69,** 111 (1976).

5. M. Sturm and A. Matter, "Turbidites and varves in Lake Brienz (Switzerland): Deposition of Clastic Detritus by Density Currents." In A. Matter and M. E. Tucker (Eds.), *Modern and Ancient Lake Sediments*, Blackwell Scientific Publ., Oxford, 1978, p. 147.

6. J. Bloesch and N. M. Burns, "A Critical Review of Sedimentation Trap Technique," *Schweiz. Z. Hydrol.* **42,** 15 (1980).

7. S. Blomqvist and L. Hakanson, "A Review on Sediment Traps in Aquatic Environments," *Arch. Hydrobiol.* **91,** 101 (1981).

8. O. Siebeck, "Der Königsee—Eine limnologische Projektstudie," *Nationalpark Berchtesgaden Forschungsber.* **5,** 131 (1982).

9. R. Lehmann, "Untersuchungen zur Sedimentation in einem oligotrophen Alpensee (Königsee) während der sommerlichen Schichtung", *Arch. Hydrobiol.* **96**, 486 (1983).

10. P. Stadelmann, "Stickstoffkreislauf und Primärproduktion im mesotrophen Vierwaldstättersee (Horwer Bucht) und im eutrophen Rotsee, mit besonderer Berücksichtigung des Nitrats als limitierender Faktor," *Schweiz. Z. Hydrol.* **33**, 1 (1971).

11. J. Bloesch, "Sedimentation und Phosphorhaushalt im Vierwaldstättersee (Horwer Bucht) und im Rotsee", *Schweiz. Z. Hydrol.* **36**, 71 (1974).

12. J. Bloesch, P. Stadelmann and H. Bührer, "Primary Production, Mineralization and Sedimentation in the Euphotic Zone of Two Swiss Lakes," *Limnol. Oceanogr.* **22**, 511 (1977).

13. H.-H. Stabel and M. M. Tilzer, "Nährstoffkreisläufe im Überlinger See und ihre Beziehungen zu den biologischen Umsetzungen", *Verh. Ges. Ökol.* **9**, 23 (1981).

14. H.-H. Stabel and J. Kleiner, "Endogenic Precipitation of Mn in Lake Constance," *Arch. Hydrobiol.* **98**, 307–316 (1983).

15. M. M. Tilzer, "The Importance of Fractional Light Absorption by Photosynthetic Pigments for Phytoplankton Productivity in Lake Constance," *Limnol. Oceanogr.* **28**, 833 (1983).

16. K. Kelts and J. Hsü, "Freshwater Carbonate Sedimentation." In A. Lerman (Ed.), *Lakes—Chemistry, Geology, Physics*, Springer, Berlin, 1978, p. 295.

17. J. P. Pelletier, "Production primaire du Léman et de quelques lacs francais voisins", *Verh. Internat. Verein. Limnol.* **20**, 921 (1978).

18. J. M. Jaquet, G. Nembrini, J. Garcia, and J. P. Vernet, "The Manganese Cycle in Lac Léman, Switzerland: The Role of *Metallogenium*", *Hydrobiologia* **91**, 323 (1982).

19. G. Wagner and B. Wagner, "Zur Einschichtung von Flußwasser in den Bodensee-Obersee," *Schweiz. Z. Hydrol.* **40**, 231 (1978).

20. H. Lehn, "Vom Abfluß des Bodensee-Obersees," *Verh. Ges. Ökol.* **6**, 163 (1978).

21. P. Nydegger, "Untersuchungen über den Feinstofftransport in Flüssen und Seen, über Entstehung von Trübungshorizonten und zuflussbedingten Strömungen im Brienzersee und einigen Vergleichsseen," *Beitr. Geol. Schweiz. Hydrol.* **16**, 1 (1967).

22. W. Nümann, "Die Verbreitung des Rheinwassers im Bodensee," *Int. Revue. Ges. Hydrobiol.* **36**, 501 (1938).

23. U. Sommer, "The Role of *r*- and *K*-Selection in the Succession of Phytoplankton in Lake Constance," *Acta Oecolog./Oecolog. General.* **2**, 327 (1981).

24. U. Sommer, "Phytoplanktonbiocoenosen und -sukzessionen im Bodensee/Überlinger See," *Verh. Ges. Ökol.* **9**, 33 (1981).

25. A. A. A Mohammed and H. Müller, "Zur Nährstofflimitierung des Phytoplanktons im Bodensee," *Arch. Hydrobiol.* (*Suppl.*) **59**, 151 (1981).

26. L. Sigg, M. Sturm, J. Davis, and W. Stumm, "Metal Transfer Mechanisms in Lakes," *Thalassia. Jugoslav.* **18**, 293 (1983).

27. P. J. le B. Williams, K. R. Heinemann, J. Marra, and D. A. Purdie, "Com-

parison of ^{14}C and O_2 Measurements of Phytoplankton in Oligotrophic Waters," *Nature* **305**, 49 (1983).

28. W. Lampert and U. Schober, "Das Regelmäßige Auftreten von Frühjahrsalgenmaximum und Klarwasserstadium im Bodensee als Folge von Klimatischen Bedingungen und Wechselwirkungen zwischen Phyto- und Zooplankton," *Arch. Hydrobiol.* **82**, 364 (1978).

29. W. Geller, "Stabile Zeitmuster in der Planktonsukzession des Bodensees (Überlinger See)," *Verh. Ges. Ökol.* **8**, 373 (1980).

30. W. S. Broecker, "The Ocean," *Sci. Am.* **249**, 100 (1983).

31. G. Wagner, "Die Untersuchung von Sinkstoffen aus Bodenseezuflüssen," *Schweiz. Z. Hydrol.* **38**, 191 (1976).

32. T. Frevert, "Phosphorus and Iron Concentrations in the Interstitial Water and Dry Substance of Sediments in Lake Constance (Obersee)," *Arch. Hydrobiol. (Suppl.)* **55**, 289 (1979).

33. G. Müller, G. Irion, and M. Förstner, "Formation and Diagenesis of Inorganic Ca–Mg Carbonates in the Lacustrine Environment," *Naturwissenschaften* **59**, 158 (1972).

34. H. Züllig, "Sedimente als Ausdruck des Zustandes eines Gewässers," *Schweiz. Z. Hydrol.* **18**, 487 (1956).

35. R. Thompson and K. Kelts "Holocene Sediments and Magnetic Stratigraphy from Lakes Zug and Zurich, Switzerland," *Sedimentology* **21**, 577 (1974).

36. C. Serruya, "Problems of Sedimentation in the Lake of Geneva," *Verh. Int. Verein. Limnol.* **17**, 208 (1969).

37. H. Rossknecht, "Zur autochthonen Calcitfällung im Bodensee-Obersee," *Arch. Hydrobiol.* **81**, 35 (1977).

38. L. Minder, "Über biogene Entkalkung im Zürichsee," *Verh. Int. Verein. Limnol.* **1**, 20 (1922).

39. M. Sturm, U. Zeh, J. Müller, L. Sigg, and H.-H. Stabel, "Schwebstoffuntersuchungen im Bodensee mit Intervall-Sedimentationsfallen," *Eclogae Geol. Helv.* **75**, 579 (1982).

40. A. Strong and B. J. Eadie, "Satellite Observations of Calcium Carbonate Precipitation in the Great Lakes," *Limnol. Oceanogr.* **23**, 877 (1978).

41. K. K. Turekian and K. H. Wedepohl, "Distribution of the Elements in Some Major Units of the Earth's Crust," *Bull. Geol. Soc. Am.* **72**, 175 (1961).

42. V. M. Goldschmidt, *Geochemistry*, Clarendon Press, Oxford, 1954.

43. W. Stumm and J. J. Morgan, *Aquatic Chemistry*, 2nd ed., Wiley, New York, 1981, pp. 780.

44. G. J. Brunskill, "Fayetteville Green Lake New York. II: Precipitation and Sedimentation of Calcite in a Meromictic Lake with Laminated Sediments," *Limnol. Oceanogr.* **14**, 858 (1969).

45. B. Kunz, "Heterogene Nucleierung und Kristallwachstum von $CaCO_3$ (Calcit) in natürlichen Gewässern," Ph.D. thesis, ETH-Zürich (1983).

46. E. R. Sholkovitz and D. Copland, "The Chemistry of Suspended Matter in Esthwaite Water, A Biologically Productive Lake with Seasonally Anoxic Hypolimnion," *Geochim. Cosmochim. Acta* **46**, 393 (1982).

47. A. C. Redfield, B. H. Ketchum, and F. A. Richards, "The Influence of Organisms on the Composition of Sea Water." In M. N. Hill (Ed.), *The Sea*, Vol. 2, Wiley, New York, 1963, p. 26.

48. H. Rossknecht, "Phosphatelimination durch autochthone Calcitfällung im Bodensee-Obersee," *Arch. Hydrobiol.* **88,** 328 (1980).

49. M. Geiger, "Abiotische Adsorption und Desorption von o-Phosphat an Sinkstoffen des Bodensees," Diploma-Thesis, Freiburg, 1982, 68 pp.

8

THE PAVIN CRATER LAKE

Jean-Marie Martin

Laboratoire de Géologie, Ecole Normale Supérieure, 46, rue d'Ulm—75230 Paris Cedex 05, France

Abstract

The Pavin Crater Lake (Massif Central, France) is a small and well-defined ecosystem characterized by the occurrence of two stratified oxic and anoxic layers. It provides a promising site to register atmospheric fallout of trace metals and artificial radionuclides and to understand the basic processes which govern the fate of these elements in the aquatic environment.

1. INTRODUCTION

In view of the complexity of the marine environment, small and well-defined ecosystems, such as crater lakes, can be used as useful "test tubes" to elucidate trace-metal and radionuclides transfer processes in aquatic systems (1,2). Besides, lacustrine deposits represent an ideal site to register atmospheric fallout of pollutants over a short-term period.

Due to its situation and geochemical and morphological characteristics, the Pavin Lake in the Massif Central (France) appeared to be most suitable for such a purpose (Fig. 8.1).

The Pavin Lake (1197 m) is located at 45°55' N–2°54' E in a remote area far from important industrial activities. The crater, which is part of the Mont–Dore range, a major feature of Massif Central, has been formed 3500 years ago by the explosion of a trachytic maar. It represents the most recent volcanic activity in the Massif Central (3).

Its shape is circular with a mean diameter of 750 m, corresponding to an area of 0.44 km². Despite this small dimension, its depth reaches 92 m. Its

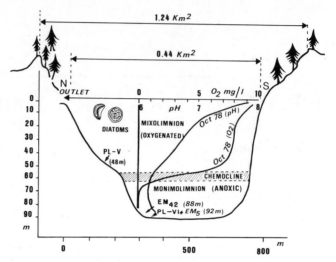

Figure 8.1. Schematic representation of Pavin Crater Lake characteristics. PL-V, PL-VI, EM_5, and EM_{42} are the locations of the cores mentioned in this chapter.

drainage basin (0.8 km²) is totally forested with beeches and conifers—a limiting factor to mechanical erosion.

Phytoplanktonic communities have been studied by Devaux (4) and Amblard and Devaux (5). Ecological succession described by Margalef (6,7) for marine phytoplankton was observed but the ecosystem structure remains relatively simple with the predominance of diatoms species leading to the formation of an almost pure diatomite deposit at the bottom of the lake, with negligible detrital material.

From a chemical point of view the lake is characterized by the presence of two stratified layers. The upper layer of the hypolimnion (or mixolimnion) is affected by winter mixing and is oxygenated, while the deeper layer (or monimolimnion) is totally anoxic and a permanent feature (8).

The goals of this paper are

1. To describe the overall hydrological and chemical functioning of the lake.
2. To study the effect of atmospheric fallout of some representative trace metals and radionuclides on the sediment composition.
3. To understand the major factors which affect the fate of these elements in the lake water and sediments.

2. THE ENVIRONMENTAL FRAMEWORK

2.1. The Water Balance

The water balance of Pavin Lake has long been studied by Glangeaud (9), but is still poorly known. Meybeck et al. (8) estimated the average annual input to the lake via rainfall, runoff, and surficial springs to be 58 liter s⁻¹

from which rainfall amounted to 19 liter s^{-1}. The output from the lake is in two forms: (1) via evaporation; and (2) discharge/outflow from the lake. We have estimated the average annual evaporation to be about 6 liter s^{-1}; so the average annual outflow from the lake would be 52 liter s^{-1}. But four measurements of this outflow during different seasons yielded 65, 69, 87, and 360 liter s^{-1}, whereas the maximum reported discharge is as high as 1400 liter s^{-1}. Then it seems reasonable to balance the water budget of the lake in order to hypothesize an additional input to the lake by underlacustrine freshwater springs (8,10,11).

Unfortunately, we have no means to measure this input by the classical hydrological method. Mathematical modeling based on tritium data has allowed us to quantify the underlacustrine discharge and to highlight the functioning of the Pavin Lake system, such as the water residence time in each compartment (44).

In Figure 8.2(a) is shown a schematic model of the lake. We assume that the input Q_R to the mixolimnion has a tritium concentration C_r (t) which is that of rainwater. A short local history of these concentrations is available and has been extended by linear regression to the long history of these concentrations at Thonon-les-Bains (Haute-Savoie, France). In addition, we assume that the underground-freshwater-springs discharge Q_2 has negligible tritium activity. The time functions C_1 (1) and C_2 (t), which correspond to the tritium concentration of mixolimnion ($V_1 = 18 \times 10^6$ m^3) and of monimolimnion ($V_2 = 5 \times 10^6$ m^3), respectively, are solutions of a system of linear differential equations, which are the tritium conservation equations for the two compartments of the lake [Fig. 8.2(b)].

We have estimated the unknown values of the parameters Q_R, Q_2, Q_0, and q (three of which are linked by the equation $Q_0 = Q_2 + Q_R$) by minimizing the difference (in the least-squares way) between the actual and the computed values of C_1 and C_2. The best fit is achieved for the values $Q_R = 53$ liter s^{-1}, $Q_2 = 40$ liter s^{-1}, and $Q_0 = 93$ liter s^{-1} with an exchange rate between compartments $q = 5$ liter s^{-1} [Fig. 8.2(b)]. The accuracy of the results, which is governed by that of the tritium analyses, cannot be better than 10 or 20%.

This water balance emphasizes the major importance of underlacustrine springs especially in the anoxic compartment, which makes the Pavin Lake a more complex example than previously thought. Carbon-14 concentration in sediments and chemical composition of the anoxic water provide some additional evidence of the underlacustrine springs.

As previously mentioned (3), the crater has been formed 3500 B.P. However, all deposition rates of sediments on the lake bottom systematically provide older ages ranging from 4700 to 8600 B.P. This anomaly can only be explained by a diluting effect of carbon-14 by nonactive carbon originating from the dissolution of the parent rocks by underground waters flowing through the watershed.

Additional information is provided by the comparison of highly mineralized bottom water and nearby springs such as Fontaine Goyon (Fig. 8.3),

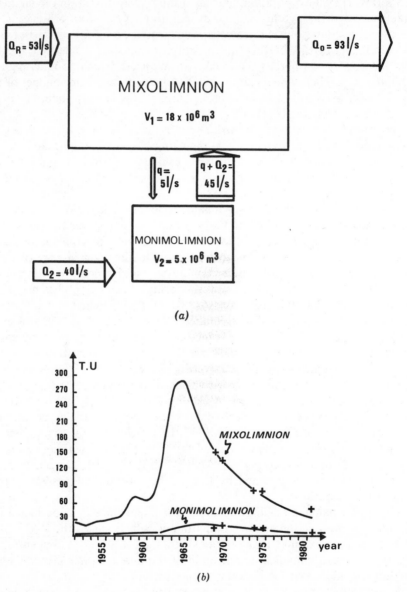

Figure 8.2. (*a*) Schematic model of Pavin Lake used for hydrological budget (V_1 = volume of mixolimnion; V_2 = volume of monimolimnion; Q_0 = evaporation + outlet; Q_R = rainfall, runoff, and superficial springs; Q_2 = underlacustrine springs; and q = exchange between 1 and 2. (*b*) Experimental and model-based tritium concentration in surface (mixolimnion) and anoxic (monimolimnion) waters.

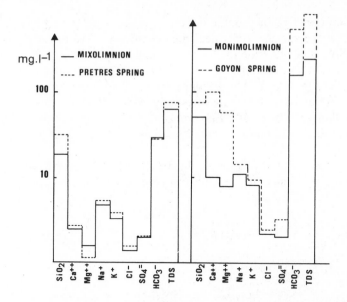

Figure 8.3.   Dissolved major-elements concentrations in lake water and neighboring springs.

which presents very similar silica but also minor-elements concentrations such as iron (46 mg liter^{-1} in Goyon Spring). The use of major elements as geothermometers, such as Na–K–Ca, and minor elements, such as Li, provides some useful indication on the original temperature of the underlacustrine springs, which would average 100°C (45).

The water budget allow us to compute the mean residence times of water in the two compartments (Fig. 8.2) of Pavin Lake without a knowledge of the whole hydrological system. If we replace the system by a number of hypothetical reservoirs of volume V_i ($i = 1, 2, 3, \ldots$) linked together, the total residence time in the whole system will then be the sum of the partial residence times in each reservoir (12).

The mean residence times in the two compartments, mixolimnion and monimolimnion, are calculated to be 6.1 and 1.7 years, respectively. The slower renewal time for deep rather than surface water is surprising but reflects the major significance of underlacustrine springs flowing to the deep compartment, which has a rather small volume as compared to the upper compartment. Such an interesting paradox has also been reported by Campbell and Torgersen (13) in Lake 120 (Canada), where they extensively discussed the maintenance of meromixis in a similar rapidly flushed monimolimnion.

2.2.   The Chemical Functioning

From a chemical point of view the lake structure is characterized by the occurrence of two stratified layers separated by a chemocline at a depth of

60 m. The upper layer of mixolimnion is oxygenated with a pH decreasing from 7.5 at the surface to 6.3 near the chemocline. The lower layer or monimolimnion is a permanently anoxic feature which is not affected by the winter mixing as the upper one is. Its pH is almost constant ranging from 6.2 to 6.4 at the bottom. The volume of the mixolimnion and monimolimnion are 18×10^6 and 5×10^6 m^3, respectively. Typical pH and dissolved-oxygen concentration are given as a function of depth on Fig. 8.1. According to the limnological classification, Pavin Lake can be considered as strictly meromictic.

The chemical stratification is illustrated by the vertical distribution of dissolved silica, manganese, and iron (Fig. 8.4). The density stratification

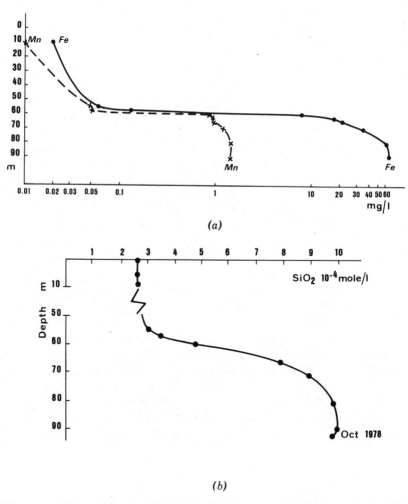

(a)

(b)

Figure 8.4. Vertical distribution of dissolved Fe, Mn, and Si concentration in Pavin Lake waters.

originates from the fact that a major portion of these elements is supplied to the lake by the highly mineralized underlacustrine springs flowing to the monimolimnion and by their efficient chemical recycling. For instance, the mechanisms which explain the transport of iron and manganese in water bodies have received a great deal of attention because of the central role they can play in the geochemical cycling of other elements; their chemical properties, cycling, and modeling are discussed at length by W. Davison (Chapter 2 in this book), and will not be further detailed in this paper. In short, one can consider that the degradation of organic matter associated to diatom tests is primarily responsible for the low redox potential occurring in the monimolimnion. The recycling of iron at the chemocline by an $Fe^{2+} \rightarrow Fe^{3+} + e^-$ reaction greatly favors the delineating of the chemocline. Iron may be transported from the monimolimnion to the mixolimnion as dissolved ferrous ion; once it reaches the mixolimnion iron it is oxidized to Fe^{3+} and forms a precipitate which is efficiently returned to the monimolimnion. Such a mechanism has been described as an "iron wheel" by Campbell and Torgersen (13).

2.3. Sedimentation Rate and Pavin Lake History

2.3.1. Sedimentation Rate

Geochronology of lake sediments has been determined with two naturally occurring radionuclides: ^{210}Pb ($t_{1/2} = 21$ yr) and ^{32}Si ($t_{1/2} = 108$ yr) (14). Both nuclides are produced in the atmosphere–^{210}Pb by the decay of ^{222}Rn, a member of the ^{238}U decay series ($t_{1/2} = 3.8$ days) escaped from the earth surface, and ^{32}Si by interactions of cosmic-ray particles with argon nuclei. These methods which assume a constant rate of deposition per unit area have been first applied to sediment chronology in Pavin Crater Lake by Krishnaswami et al. (15).

The excess ^{210}Pb concentration versus depth is plotted in (Fig. 8.5). For the anoxic layer the data obtained for three cores are given. The data show an average accumulation rate of 1.5 mm yr^{-1} (11.5 mg cm^{-2} yr^{-1}) at 92 m. With regard to the oxic compartment, only the PL-V core has been dated. The data also give a very close sedimentation rate of $\simeq 1$ mm yr^{-1} (7.7 mg cm^{-2} yr^{-1}) up to about 10 cm, below which the sedimentation ratio is approximately four times faster. In order to investigate the sedimentation rates for the older layers, ^{32}Si appeared to be the most suitable nuclide owing to its half-life and the diatomous nature of the sediment. Preliminary data indicate a sedimentation rate of 0.5 mm yr^{-1} from the surface to 10–15 cm, a value very close to that deduced from ^{210}Pb profiles. Below this depth, there is a straightforward increase by a factor of 5 to 6, resulting in a rate of 3 mm yr^{-1}. Several explanations have been discussed to explain this

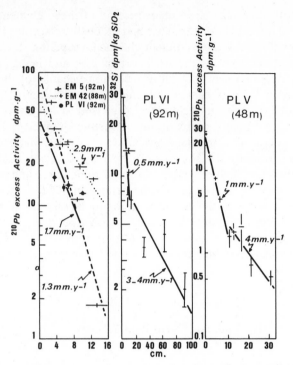

Figure 8.5. ^{210}Pb excess and ^{32}Si activities (in disintegrations per minute) vs depth (in cm) plots for sediment cores.

reduction of the accumulation rate observed in the most recent deposits (48):

1.   A sediment slump could have occurred at about 100–200 B.P.

2.   The decrease in the deposition rate is due to the tree plantation that took place over the catchment area of the lake during the middle of the last century (based on historical records). If aluminum is considered (as a first approximation) as a tracer of detrital erosion, this hypothesis is unlikely due to the slight increase of this element in the more recent periods.

3.   There has been a change in the chemistry of the lake related to a progressive shift from a well-mixed to meromictic type (formation of two layers—the upper oxygenated one and the bottom anoxic one). The new (present) situation would prevent winter mixing and the subsequent regeneration of nutrients (C, N, P, Si) from the bottom sediment to the productive upper layer. The sedimentological consequences would be a clear decrease in the diatom productivity and accumulation rate from a few to 1 mm yr^{-1} observed for the top sections. To check on this, we have measured the Fe, Mn, and As concentrations; indeed there is a progressive increase of these elements down core. Manganese and iron could be mobilized from the sediment to the overlying water when anoxic conditions appeared; however,

the observed gradient does not correspond exactly to the sedimentation rate decrease (Fig. 8.6).

An attempt has been made to find two metals, the ratio of which could serve as an indication for the redox state of ancient bottom waters. Following Hallberg (16), we selected Cu and Zn, because of the different solubility products of their sulfides and because copper is more favored than zinc for precipitation in a reduced environment. In an oxidized environment, Cu and Zn have about the same solubility products, resulting in an increase of the Cu/Zn ratio in an anoxic environment. However, because of a straightforward increase of these two metals due to atmospheric pollution, it has not been possible to use them as a redox index. Indeed, both metals show higher concentrations in the upper layers of lake sediments, indicating a noticeable atmospheric contamination which is efficiently scavenged by biogenic sedimentation, as demonstrated by Sigg et al. (2) in Lake Constance.

4.   Finally, instead of a sedimentation rate decrease one can put forward a progressive increase of diatom test dissolution related to a water circulation through the sediment.

2.3.2.   Short-Term Paleoclimatology

The oxygen-isotopic ratio of diatom silica depends upon the isotopic ratio and the temperature of the water in which the diatoms grow (17). In lakes, both the water temperature and isotopic ratio are functions of the climatic regime. The study of the diatom isotopic changes along cores should there-

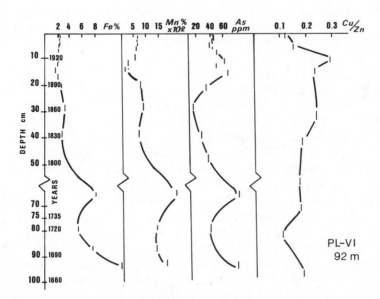

Figure 8.6. Fe, Mn, and As concentrations and Cu/Zn ratios vs depth in the PL-VI sediment core.

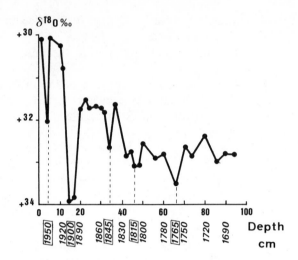

Figure 8.7. δ^{18} vs SMOW of the silica of diatom valves extracted from the core PL-VI as a function of depth in the core.

fore produce a continuous stratigraphy and paleoclimatic record. The first application of such study, on Lake Pavin core PL-VI, has been carried out by Labeyrie and Juillet (46).

As part of the silica, oxygen is isotopically exchangeable (18). A standard treatment has been devised which gives reliable measurements. Fossil diatom valves were sampled, purified from contaminants, and the oxygen extracted following the procedure of Labeyrie (19). Results are expressed versus SMOW (Standard Mean Ocean Water), as $\delta^{18} = [(^{18}O/^{16}O) \text{ sample} - (^{18}O/^{16}O)_{st.} - 1] \times 1000$.

The results are presented in Figure 8.7 as the δ^{18} function of depth in the core. The amplitude of the variations, 4‰, corresponds to a 12–15°C change in the mean lake-water temperature during spring—the maximum growth period for the diatoms—if the water isotopic changes are neglected. Because the long-term dependence upon climate of the lake-water isotopic ratio is not known presently, a detailed paleoclimatic interpretation is not possible. However, the general trend of the isotopic curve is in agreement with what is known of the European climate in the last few hundred years; using the sedimentation rate given by ^{210}Pb and ^{32}Si for PL-VI, we get an age of A.D. 1750 for the cold maximum (δ max) at 65-cm depth, which is in very good agreement with the time when the Little Ice Age cooling was stronger in Europe. Some unexpected minima are observed for three points corresponding to 1900 and 1950 sediment layers; no clear explanation can be presently given for such deviations. A possibility would be a diatom bloom occurring before the normal periods as indicated by oxygen-isotopic data in carbonate reported by McKenzie (Chapter 5 in this book).

3. ATMOSPHERIC DEPOSITION OF POLLUTANTS

3.1. Trace Metals

The atmospheric deposition for Zn, Cu, and Pb, which shows a gradual increase since the beginning of industrialization, is given in Figure 8.8 and Table 8.1, where it is compared to Lake Constance and the North Atlantic. It is noteworthy that the atmospheric deposition on Lake Pavin is one order of magnitude lower than on Lake Constance and very close to that observed on the North Atlantic.

More insight has been given on lead contamination by studying lead-isotopic composition (49). Contrary to other elements, the variations of lead-isotopic composition may significantly depend on its origin. Basically, there are four stable lead isotopes: ^{204}Pb or common lead and ^{208}Pb, ^{207}Pb and ^{206}Pb, which represent the end products of the ^{238}U, ^{235}U, and ^{232}Th decay series. It is then likely that their quantity will vary according to the uranium and thorium concentration and the age of the parent ores.

It is well known that most of the lead present in the atmosphere during the past decade is derived from lead alkyls, which are added to gasoline because of their antiknock properties. The isotopic composition of lead is thus restricted to the range of the major lead ores, which are limited to a small number due to economical and political reasons (21).

Several studies have already demonstrated the feasibility of using lead-isotopic composition as a powerful tracer of sources of pollution in different environments (glaciers, sediments, etc.). In particular, the lead-isotopic composition has previously been used to differentiate the origin of lead in a few ponds (22,23).

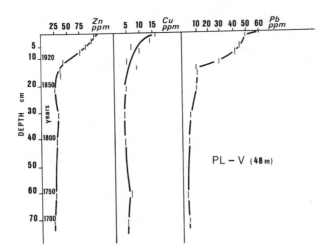

Figure 8.8. Vertical distribution of Zn, Cu, and Pb in the PL-V core.

Table 8.1.   Atmospheric Deposition of Cu, Pb, and Zn in Lake Pavin as Compared to Lake Constance and the North Atlantic Ocean ($ng\ cm^{-2}\ yr^{-1}$)

Water Bodies	Elements		
	Cu	Pb	Zn
Lake Pavin	100	($\approx$500)	1000
Lake Constance[a]	714	11,000	8400
North Atlantic[b]	25	310	130

[a] Sigg et al. (2).

[b] Buat-Menard and Chesselet (20).

The $^{206}Pb/^{207}Pb$ ratio is plotted as a function of depth in Figure 8.9. The $^{206}Pb/^{207}Pb$ ratio (to avoid confusion, only this ratio will be discussed) increases from 1.149 at the surface (0–2 mm layer which represents the 1978 horizon) to the more radiogenic (enriched in ^{206}Pb) value (1.167), which remains more or less constant with time up to about 60 B.P. These isotopic ratios are very similar to those measured in leads originating from metallurgy and, to a lesser extent, from fuel and coal burning as observed by Petit (22) in Belgium.

Beyond this time, the ratio gradually increases to 1.192 (at about 40-cm

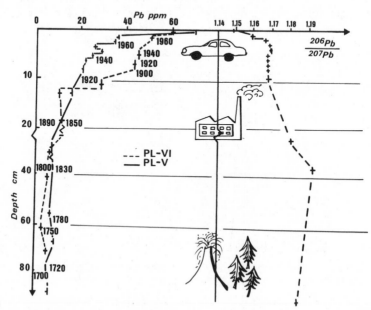

Figure 8.9.   Pb concentrations and $^{206}Pb/^{207}Pb$ isotopic ratio variations vs depth in core PL-VI.

depth) representing more radiogenic lead sources, which have a ratio comparable to those found in the upper part of the earth's crust (24). The $^{206}Pb/^{207}Pb$ ratio found in the surface sediments is primarily due to anthropogenic lead (the total Pb concentrations in Fig. 8.9, which are about an order of magnitude higher than the background levels, support this deduction) and is thus representative of atmospheric lead. The slightly low radiogenic ratio, 1.148, in the surface layer is consistent with 1.128 (total Pb = 150 ppm) measured in a grass sample collected near a highway in Paris which indicates Pb exhausts from automobiles.

3.2. Artificial Radionuclides

This study has focused on plutonium and ^{137}Cs isotopes (50). As far as plutonium is concerned, most of the works which have already been published deal with the Great Lakes. It has been shown that due to its high partition coefficient, plutonium is readily removed from the water column, (25,26). However, in Mono Lake, Simpson et al. (27) have shown that the dissolved concentration is two orders of magnitude higher than in the Great Lakes and that at least 50% of the plutonium fallout is still in solution, most likely due to its complexation by carbonates. Moreover, Alberts et al. (28) and Edgington et al. (29) consider the potentiality of plutonium migration under anaerobic conditions, but do not provide any field evidence. This hypothesis has also been evoked by Bowen et al. (30). Conversely, the data of Carpenter and Beasley (31) in the oxic and anoxic sediments off the Washington and Oregon coasts, as well as the laboratory and field studies of Alberts and Orlandini (32), showed evidence against remobilization of plutonium in oxic, as well as anoxic, environments.

Indeed many trace metals and radionuclides such as plutonium are known to be adsorbed onto the oxyhydroxides of iron and manganese (29,33); therefore the redox characteristics of Pavin Lake appeared to be very interesting to study their changes in concentration and potential mobilization.

With regard to ^{137}Cs, there are few examples of mobilization under anaerobic conditions as reported by Alberts et al. (34) and Baturin (35). This mobilization could be explained by an exchange reaction of ^{137}Cs with NH_4^+ as suggested by Evans (47).

3.2.1. Dissolved Radionuclides

The $^{239+240}Pu$ concentrations of Pavin waters are 0.37 ± 0.06 and 0.96 ± 0.15 fCi liter^{-1} in the oxic and anoxic compartments, respectively, and these values are similar to those found in the waters of the Laurentian Great Lakes (36). However, the concentrations are three times higher in anoxic water than in surface water, suggesting a possible influence of redox conditions upon plutonium concentrations, an observation which has also been made by Sholkovitz et al. (37) in a seasonally anoxic lake. The inventory of

$^{239+240}$Pu in the lake water is calculated to be 11.5×10^{-6} mCi, which represents only 0.5% of the integrated plutonium fallout. The calculated residence time of plutonium isotopes in the water column is one 1 year; again these values are in good agreement with those obtained by Edgington et al. (29) and Wahlgren and Nelson (38), who attributed the rapid downward migration of plutonium isotopes in the water column to intense biological activity. The dissolved ^{137}Cs concentrations are 700 ± 140 and 325 ± 65 fCi liter^{-1} in the surface and bottom layers, respectively, and these values are about an order of magnitude higher than those measured in the Laurentian Great Lakes (36). This is probably due to the siliceous nature of Pavin Lake sediments which are probably characterized by a low exchange capacity for dissolved ^{137}Cs as compared to clay minerals. The ^{137}Cs inventory in the Pavin waters is 17.6 mCi from which only 5 mCi is the contribution from the anoxic layer. The laker water ^{137}Cs inventory represents 20% of the total fallout activity.

3.2.2.   The Sedimentary Column

The $^{239+240}$Pu and ^{137}Cs concentrations of the Pavin cores PL-V and PL-VI are plotted as a function of depth (time) in Figures 8.10 and 8.11. In both

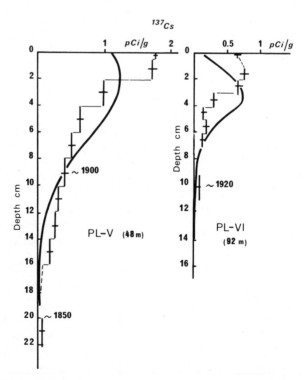

Figure 8.10. ^{137}Cs vertical profiles in the cores PL-V and PL-VI (experimental and computed values).

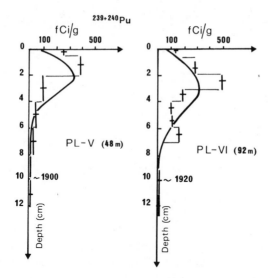

Figure 8.11. $^{239+240}$Pu vertical profiles in cores PL-V and PL-VI (experimental and computed values).

cores, the $^{239+240}$Pu concentrations increase from 0.11 to 0.25 pCi g^{-1} in the surface layers to 0.4 to 0.5 pCi g^{-1} at 2–3 cm depth and then decrease with depth to reach the detection limit below 15 cm. The ^{238}Pu/$^{239+240}$Pu ratios range from 0.03 to 0.04, which is typical of atmospheric fallout from nuclear weapons' testing (39). The ^{137}Cs concentration, in PL-V, decreases exponentially with depth from 1.8 ± 0.2 pCi g^{-1} at the surface to the detection limit of 0.08 pCi g^{-1} at about 22 cm, whereas in PL-VI the decrease is from 0.6 pCi g^{-1} to the detection limit at 11 cm. The $^{239+240}$Pu and ^{137}Cs inventories of the two Pavin cores are 0.077 and 7.4 pCi cm^{-2}, respectively, for PL-V and 0.12 and 2.3 pCi cm^{-2}, respectively, for PL-VI.

Though the production of artificial radionuclides through bomb activities was confined to the post-1950s (this production can be associated with elements found in the top 3-cm deposition for PL-V and 5-cm deposition in the case of PL-VI), both $^{239+240}$Pu and ^{137}Cs are found in measurable concentrations up to about 10–15 cm in both cores (Figs. 8.10 and 8.11).

In order to explain the observed ^{137}Cs profiles, it appears necessary to have a diffusional process combined with adsorption–desorption equilibrium. We attempted to simulate the ^{137}Cs concentrations using a model adopted by Aller (40) and Olsen et al. (41):

$$w_a \frac{\partial C}{\partial t} = D \frac{\partial^2 c}{\partial x^2} - V w_a \frac{\partial c}{\partial x}$$

where w_a = apparent porosity = $w + (1 - w) \rho_s K_D$, with w = porosity = 0.93 and ρ_s = density ≈ 1, $D = wd$ with d = molecular diffusion, and V = sedimentation rate cm yr^{-1}.

Table 8.2. ^{137}Cs and $^{239+240}$Pu Diffusion Coefficient in Pavin Lake Sediment.

Isotopes	Core	$D(cm^2\ yr^{-1})$	w_a	$D_a(cm^2\ yr^{-1})$
^{137}Cs	PL-V	10^2	176	5.7×10^{-1}
	PL-VI	10	127	7.9×10^{-2}
$^{239+240}$Pu	PL-V	2.10^3	4.55×10^4	4.4×10^{-2}
	PL-VI	2.10^3	2.1×10^4	9.5×10^{-2}

$D = wd$ is almost equal to the molecular diffusion d because of the very high porosity ($w \approx 0.93$).

D_a is the apparent diffusion coefficient, that is, D/w_a.

We assume that the radionuclide is transported to the sediment along with the settling particles and that the diffusion coefficient D and the distribution coefficient K_D are depth invariant (due to lack of radionuclide data in the pore waters and equal to the value computed from the water concentration and the surface sediment activity, i.e., $K_D = 2500$ for PL-V and 1800 cm^3 g^{-1} for PL-VI).

With regard to plutonium, we used a distribution coefficient of 6.7×10^5 and 3×10^5 cm^3 g^{-1}. The best-fit curves are obtained using the diffusion coefficients given in Table 8.2.

The noticeable diffusion which is obtained for plutonium in Pavin Lake sediments involves a release from diatom particles. This could be explained by the formation of more soluble complexes with carbonate or dissolved organic matter which is almost four times higher in pore waters than in the water column.

For both isotopes the integrated fallout for lake water and sediments appeared to be about half the average fallout under these latitudes (42,43). Although an outflow from the lake is possible for ^{137}Cs, in the case of plutonium this hypothesis is unlikely, due to its fast removal from the water column along with biogenic particles. Indeed less than 1% of the plutonium integrated fallout still remains in solution (versus 20% for ^{137}Cs). However, there is no clear evidence of plutonium remobilization, except for the slightly higher concentration found in the anoxic waters.

4. CONCLUSIONS

The characteristics of Pavin Lake have been reassessed by using the lake as a "test tube" to investigate aquatic processes. The lake's location in a remote area where detrital input is insignificant greatly favors its use as a giant "rain gauge" to register atmospheric fallout. Redox processes can be easily studied because of lake stratification and meromixis. The simplicity of the ecosystem, which is basically constituted of diatoms, might facilitate the study of biological cycling of trace metals and artificial radionuclides. The major difficulties arise from the occurrence of underlacustrine springs, which could possibly disturb sediment stratification and complicate the setup of chemical mass balance. Finally, it must be noted that these underlacustrine springs can be considered as a potential hydrothermal activity, making Pavin Lake an actual "microocean"!

ACKNOWLEDGMENTS

This paper represents a summary of several studies carried out during the last decade. The author is especially indebted to F. Elbaz-Poulichet, G. Figueres, C. Jeandel, A. Juillet, P. Hubert, L. Labeyrie, M. Meybeck, D. Petit, G. Olivier, M. Poulin, and B. L. K. Somayajulu who participated in this study. Research was partly supported by the CNRS Action Thématique Programmée "Océanographie Chimique."

REFERENCES

1. H. L. Volchok, M. Feiner, M. J. Simpson, W. S. Broecker, V. Noshkin, V. T. Bowen, and E. Willis, Ocean Fall-Out. The Crater Lake Experiment," *J. Geophys. Res.* **75,** 1084 (1970).

2. L. Sigg, M. Sturm, J. Davis, and W. Stumm. "Metal Transfer Mechanisms in Lakes," *Thalassia Jugoslav.* **18,** 1–4 (1982).

3. J. M. Peterlongo, *Le Massif Central. Guides Géologiques Régionaux*, 2nd ed., Masson, Paris, 1978, 224 pp.

4. J. Devaux, "Structure des Populations Phytoplanctoniques dans Trois lacs du Massif-Central: Succession Écologique et Diversité," *Oecol. Gener.* **1,** (1), 11 (1980).

5. C. Amblard and J. Devaux, "Structure et Orientation Énergétique d'un Peuplement Phytoplanctonique (Lac Pavin, France)," *Acta Oecolog., Oecol. Gener.* **2,**(2), 101 (1981).

6. R. Margalef, "Corrélations entre Certains Caractères Synthétiques des Populations de Phytoplancton," *Hydrobiologia* **18,** 155 (1961).

7. R. Margalef, *Perspectives in Ecological Theory*, Chicago series in Biology, Univ. of Chicago Press, Chicago, 1968, 111 pp.

8. M. Meybeck, J. M. Martin, and P. Olive, "Géochimie des Eaux et des Sédiments de Quelques lacs Volcaniques du Massif-Central," *Verh. Int. Verein. Limnol.* **19,** 1150 (1975).

9. P. Glangeaud, "Le Cratère lac du Pavin et le Volcan de Montchalm," *C. R. Acad. Sci.* (Paris) **162,** 428 (1916).

10. J. Alvinerie, B. Degot, P. Leveque, and M. Vigneaux, "Activité en tritium et Caractéristiques Chimiques des Eaux du lac Pavin," *C. R. Acad. Sci. (Paris) Series D*, **262,** 846 (1966).

11. J. P. Pelletier, "Un lac méromictique, le Pavin (Auvergne)," *Ann. St. Biol. Besse en Chandesse* **3**, 147 (1968).

12. L. G. Gibilaro, "Mean Residence Times in Continuous Flow Systems," *Nature* **273**, 77 (1977).

13. P. Campbell and T. Torgersen, "Maintenance of Iron Meromixis by Iron Redeposition in a Rapidly Flushed Monimolimnion," *Can. J. Fish. Aquat. Sci.* **37**, 1303 (1980).

14. D. Elmore, N. Anataram, H. W. Fullbright, H. E. Gove, H. S. Hans, K. Nishizumi, M. T. Murrell, and H. Honda, "Half-Life of ^{32}Si Using Tandem Accelerator Mass Spectrometry," *Phys. Rev. Lett.* **45** 589 (1980).

15. S. Krishnaswami, D. Lal, J. M. Martin, and M. Meybeck, "Geochronology of Lake Sediments," *Earth Planet Sci. Lett.* **11**, 407 (1971).

16. R. O. Hallberg, "The Microbiological C.N.S. Cycles in Sediments and Their Effect on the Ecology of Sediment–Water Interface," *Oikos Supplementum* **15**, 51 (1973).

17. L. D. Labeyrie, "New Approach to Surface Sea Water Temperatures Using ^{18}O/^{16}O Ratio in Silica of Diatoms Frustules," *Nature* **248**, 40 (1974).

18. L. D. Labeyrie and A. Juillet, "Oxygen Isotopic Exchangeability of Diatom Valve Silica," *Geochim. Cosmochim. Acta* **46**, 967 (1982).

19. L. D. Labeyrie, *Composition Isotopique de l'Oxygène de la Silice des Diatomées*, Rapp. C.E.A. 2086, Saclay, 1979.

20. P. Buat-Menard and R. Chesselet, "Variable Influence of the Atmospheric Flux on Trace Metals Chemistry of Oceanic Suspended Matter," *Earth Planet. Sci. Lett.* **42**, 344 (1979).

21. J. J. Chow, "Isotopic Identification of Industrial Pollutant Lead." In *Second International Clean Air Congress*, H. M. England (Ed.), Academic Press, New York, 1971, 348 pp.

22. D. Petit, "^{210}Pb et Isotopes Stables du Plomb des Sédiments Lacustres," *Earth Planet. Sci. Lett.* **23**, 199 (1974).

23. M. Shirahata, R. W. Ellis, and C. C. Patterson, "Chronological Variations in Concentrations and Isotopic Compositions of Anthropogenic Atmospheric Lead in Sediments of a Remote Sub-Alpine Pond," *Geochim. Cosmochim. Acta* **44**, 149 (1980).

24. J. S. Stacey and J. D. Kramers, "Approximation of Terrestrial Lead Isotope Evolution by a Two Stages Model," *Earth Planet. Sci. Lett.* **26**, 207 (1975).

25. M. A. Wahlgren and J. S. Marshall, "The Behaviour of Plutonium and Other Long-Lived Radionuclides in Lake Michigan: I—Biological Transport, Seasonal Cycling and Residence Times in the Water Column." In *Impacts of Nuclear Releases into the Aquatic Environment*, Proceedings of the IAEA, Vienna, 1975, p. 227.

26. M. A. Wahlgren, J. J. Alberts, D. M. Nelson, and K. A. Orlandini, "Study of the Behaviour of Transuranics and Possible Chemical Homologues in Lake Michigan Waters and Biota." In *Transuranium Nuclides in the Environment*, I A E A, 1976, p. 9.

27. H. J. Simpson, R. M. Trier, C. R. Olsen, D. E. Hammond, A. Ege, L. Miller,

and J. M. Melack, "Fall-Out Plutonium in an Alkaline-Saline Lake," *Science* **207,** 1071 (1980).

28. J. J. Alberts, M. A. Wahlgren, D. M. Nelson, and P. J. Jehn, "Submicron Particle Size and Charge Characteristics of $^{239+240}$Pu in Natural Waters," *J. Environ. Sci. Technol.* **11,** 676 (1977).

29. D. N. Edgington, J. J. Alberts, M. A. Wahlgren, J. O. Kartunen, and C. A. Reeves, "Plutonium and Americium in Lake Michigan Sediments." In *Transuranium nuclides in the environment*, Proceedings of the IAEA, Vienna, 1976, p. 493.

30. W. T. Bowen, H. D. Livingston, and J. C. Burke, "Distribution of Transuranium Nuclides in Sediments and Biota of the North Atlantic Ocean." In *Transuranium Nuclides in the Environment*, Proceedings of the IAEA, Vienna, 1976, p. 107.

31. R. Carpenter and J. M. Beasley, "Plutonium and Americium in Anoxic Marine Environments: Evidence Against Remobilization," *Geochim. Cosmochim. Acta* **45,** 1917 (1981).

32. J. J. Alberts and K. A. Orlandini, "Laboratory and Field Studies of the Relative Mobility of $^{239+240}$Pu and ^{241}Am from Lake Sediments under Oxic and Anoxic Conditions," *Geochim. Cosmochim. Acta* **45,** 1931 (1981).

33. M. P. Bacon, P. B. Brewer, D. W. Spencer, J. W. Murray, and J. Goddard, "Lead 210, Polonium-210, Manganese and Iron in the Cariaco Trench," *J. Deep Sea Res.* **27A,** 119 (1980).

34. J. J. Alberts, L. T. Tilly, and T. J. Vigerstad, "Seasonal Cycling of Cesium-137 in a Reservoir," *Science* **203,** 649 (1979).

35. G. N. Baturin, "Uranium in the Modern Marine Sedimentary Cycle," *Geochim. Int.* 1031 (1973).

36. J. J. Alberts and M. A. Wahlgren, "Concentrations of $^{239+240}$Pu, ^{137}Cs and ^{90}Zr in Waters of Laurentian Great Lakes," *Environ. Sci. Technol.* **15,** 94 (1981).

37. E. R. Sholkovitz, A. E. Carey, and J. K. Cochran, "Aquatic Chemistry of Plutonium in Seasonally Anoxic Lake Waters," *Nature* **300** (5888), 159 (1982).

38. M. A. Wahlgren and O. M. Nelson, "Plutonium in the Laurentian Great Lakes: Comparison of Surface Waters," *Verh. Int. Verein. Limnol.* **19,** 317 (1975).

39. M. Koide and E. D. Goldberg, "Transuranic Nuclides in Two Coastal Marine Sediments Off Peru," *Earth Planet. Sci. Lett.* **57,** 263 (1982).

40. R. C. Aller, "The Influence of Macrobenthos on Chemical Diagenesis of Marine Sediments," Ph.D. thesis, Yale University, New Haven, Connecticut 1977, 600 pp.

41. C. R. Olsen, H. J. Simpson, H. H. Peng, R. F. Bopp, and R. M. Trier, "Sediment Mixing and Accumulation Rate Effects on Radionuclide Depth Profiles in Hudson Estuary Sediments," *J. Geophys. Res.* **86,** 11020 (1981).

42. A. J. Jakubick, "Migration of Plutonium in Natural Soils." In *Transuranium Nuclides in the Environment*, Proceedings of the IAEA, Vienna, 1976, p. 47.

43. M. Thein, S. Ballestra, A. Yamato, and R. Fukai, "Delivery of Transuranic Elements by Rain to the Mediterranean Sea," *Geochim. Cosmochim. Acta* **44,** 1091 (1980).

44. P. Hubert, J. M. Martin, and M. Meybeck, work in preparation.

45. J. Boulègue, Paris, personal communication.

46. L. D. Labeyrie, A. Juillet, work in preparation.

47. D. W. Evans, J. J. Alberts, and R. A. Clark, III, "Reversible Ion-Exchange Fixation of Cesium-137 Leading to Mobilization from Reservoir Sediments," *Geochim. Cosmochim. Acta* **47**, 1041 (1983).

48. J. M. Martin, M. Meybeck, V. Nijampurkar, and B. L. K. Somayajulu, "^{210}Pb and ^{32}Si in Pavin Lake", submitted to *Limnol. Oceanogr.*

49. F. Elbaz-Poulichet, J. M. Martin, and D. Petit, "Atmospheric deposition of Zn, Cu, and Pb in Pavin Crater Lake", submitted to *Limnol. Oceanogr.*

50. C. Jeandel, J. M. Martin, and M. Poulin, work in preparation.

9

PHOSPHATE INTERACTIONS AT THE SEDIMENT–WATER INTERFACE

Peter Baccini*

Institute for Water Resources and Water Pollution Control (EAWAG), Zürich, and University of Neuchâtel, Neuchâtel, Switzerland

Abstract

Lake sediments of different chemical composition and different trophic levels show strong seasonal variations of their phosphorus fluxes. These phenomena cannot be explained solely on the basis of the redox conditions and the iron/phosphorus chemistry at the sediment–water interface. A more detailed physical, chemical, and biological analysis of the system "boundary layer" elucidates the processes which are mainly responsible for the dynamics of phosphorus transport.

In an aerobic zone (called OXY box) the microbial decomposition of incoming particulate organic matter utilizes the electron acceptors O_2 and NO_3^-. Freshly imported phosphorus is kept in the solid phase. Bioturbation increases the flux of dissolved species to the hypolimnion. The manganese (II) flux can be used to quantify its contribution. The observed net flux of phosphorus stems from the anaerobic zone (RED box), where there is no bioturbation. At the interface of the two zones a chemical barrier for phosphate is formed offering newly formed iron oxides as adsorbing surfaces. The capacity of this phosphorus sink depends on the stoichiometric Fe/P ratio in the dissolved phase of the anaerobic zone. A dynamic physical-chemical model uti-

---

* This work was realized in collaboration with Endre Laczkó, Jürg Ruchti, and Michael Sturm, Institute for Water Resources and Water Pollution Control (EAWAG), Zürich, Switzerland.

lizing a set of eight reactions with the state variables organic carbon, oxygen, iron (II), iron (III), and phosphate is introduced.

1. INTRODUCTION

The sediment–water interface may act as either a source or sink for phosphorus. In the overall phosphorus dynamics of lakes the subsystem "sediment boundary layer" can thus play the role of a buffer to maintain a certain phosphorus level in the overlying water. More sophisticated physical–chemical lake models for restoration programs ask for a quantitative description of such a buffer capacity which could be applied generally.

Since the classical works of Einsele (1) and Mortimer (2) almost half a century ago, limnological research has refined the phenomenology of phosphorus transfer at the sediment–water interface and confirmed by this the pioneers' hypothesis, namely that iron must be a master variable determining the phosphorus flux in early diagenesis. However, a quantitative description based solely on the dynamics of iron chemistry and its interactions with phosphorus cannot predict the phosphorus fluxes with the necessary accuracy for any lake.

During the last decade several field and laboratory studies have lead to the conclusion that the phosphorus release from aerobic lake sediments can be as important as the release under anaerobic conditions (e.g., 3–5) if one considers the phosphorus cycle of the whole lake. It became clear that the sediment–water interface had to be investigated as a biological subsystem of the lake. This subsystem's capacity to regulate the phosphorus flux is also a result of the lake's trophic evolution.

In a recent review of the related phenomenology published since 1973, Boström et al. (6) summarized the factors governing the phosphorus release in well-aerated lakes. Two groups of phosphorus transfer mechanisms have to be distinguished:

1. On a *molecular level* phosphorus mobilization can be described with *physicochemical processes*, namely, desorption, dissolution, and ligand exchange, and *biochemical processes* due to enzymatic decomposition of organic substances.

2. On a *compartment level* the phosphorus transfer from the compartment sediment to the compartment hypolimnion is to be characterized with *hydrodynamic mechanisms*, namely, diffusion, gas ebullition, bioturbation, and wind-induced turbulence.

For the first group of processes the "environmental factors" pH, pε, and temperature are most important. For the second group the lake's orographic, hydrodynamic, and trophic properties have to be considered in addition.

With these mechanisms and factors known, the phosphate interactions at the sediment–water interface could be described in a conceptual model of relatively high complexity.

At present the shortcomings are not on the level of adequate conceptual models. Berner (7) has presented good mathematical tools to treat the chemical and physical dynamics of early diagenesis. The difficulties lay mainly in the methodology to:

Elucidate the factors which govern the rates for organic phosphorus decomposition, authigenic mineral precipitation, and phosphate adsorption.

Develop experimental boundary conditions very close to natural ones for a determination of the above-mentioned factors.

Refine chemical speciation techniques for phosphorus in this subsystem.

This chapter presents the methodology applied in the SEDIPHOS project (8), which was designed to elaborate a physical–chemical model for the subsystem "sediment boundary layer". The main goal is to predict the phosphorus release as a function of the input of organic carbon (OC), phosphorus, and molecular oxygen.

2.   CONCEPTS AND METHODOLOGY

The sediment boundary layer is treated as a vertically stratified biological system with three main compartments and two main reactions in each sublayer (Fig. 9.1). The chemical speciation analysis of phosphorus in the dissolved phase is restricted to the determination of soluble reactive phosphorus (SRP = operationally defined orthophosphate concentration with the molybdenum blue method) and the total content (DP = SRP + DOP).

The following strategy of procedure is applied:

A.   Sediment samples in cores from three different lakes were chosen to study the seasonal variation in the chemical composition of the solid phase and the interstitial water. The purpose of this investigation was to find

    1.   The approximate thickness of the boundary layer where early diagenesis is quantitatively relevant for the phosphorus dynamics.

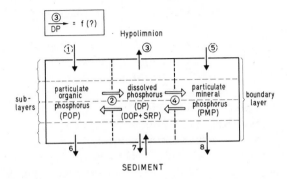

Figure 9.1.   Flux scheme for the phosphorus transport at the sediment water interface. DOP: Dissolved organic phosphorus; SRP: Soluble reactive phosphorus ($= DP - DOP$).

2. The necessary number of subdivisions to distinguish different types of chemical processes. (At first the choice of sublayer thickness was arbitrary.)

B. Simultaneously, separate cores from the same sampling stations were taken to measure phosphorus fluxes at the boundary layer in the laboratory. The fluxes are influenced by varying the input of molecular oxygen and defined particulate organic carbon.

C. The microbiological processes governing reaction 2 (Fig. 9.1) are studied with microorganisms from natural sediments and defined organic substances.

D. The abiotic processes controlling reaction 4 (Fig. 9.1) are studied with model substances based on the phenomenology and boundary conditions found in subprojects A and B.

E. The rate and equilibrium constants determined in subprojects C and D are applied in a conceptual model to simulate the phenomenology observed in A and B. The main goal is to predict correctly the flux 3 (Fig. 9.1), knowing the fluxes 1 and 5, and the input of the electron acceptors (e.g., O_2 and NO_3^-).

F. The thereby developed "organic carbon/phosphorus/oxygen model" is to be tested by "*in situ* measurements" in natural lakes.

The present discussion is limited to results obtained in subprojects A and B. Further information on sampling locations, separation techniques, chemical analysis, and flux measurement are given in the Appendix.

3. PHENOMENOLOGY OF PHOSPHORUS TRANSPORT PROCESSES

3.1. Chemical Characteristics of the Solid Phase

The sediments of the three chosen lakes show different chemical compositions. Lake Zug, an eutrophic lake, has the smallest allochthonous contribution of particles. The total iron and total phosphorus concentration in the solid phase is relatively low, whereas the organic carbon content is the highest (Fig. 9.2). On the contrary the eutrophic Lake Alpnach sediments are formed mainly by allochthonous particles (9). Their iron contents are the highest and their organic content is lowest (Fig. 9.2). The mesotrophic sediments from the Kreuztrichter occupy an intermediate position. The different composition due to different allochthonous contribution is illustrated best by comparing the aluminum to calcium ratio [Fig. 9.3(*c*)]. It is highest for Lake Alpnach and lowest for Lake Zug. For Lake Zug and Kreuztrichter there is a pronounced increase of phosphorus concentration in the first two centimeters [Fig. 9.2(*b*)]. In the same layer the organic carbon to phosphorus ratio is increasing from 0 to 2 cm [Fig. 9.3(*a*)]. This is also true for the iron to phosphorus ratio [Fig. 9.3(*b*)].

From other investigations we know that the thicknesses of the annual sed-

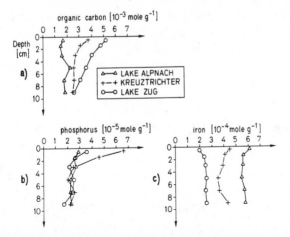

Figure 9.2. Concentration of (*a*) organic carbon, (*b*) phosphorus, and (*c*) iron as a function of depth in three different lake sediments (mean values from three to five cores).

iment layers are also different. For Lake Alpnach a layer of about 10 mm yr^{-1} is formed, whereas for Lake Zug the sedimentation does not exceed 2 to 3 mm. By choosing arbitrarily slices of the same thickness for all sediments, one does not compare necessarily synchronous steps of diagenesis. However, if the uppermost layers are a biological system demanding a certain depth for the activity of the biocenose at the sediment–water interface, the differences in sedimentation rates would only change the time of residence of incoming matter in this zone, whereas the type of reactions in and the structure of this system should be similar. The fact that the OC/P and Fe/P ratios are increasing in the uppermost layer of Lake Zug and the Kreuztrichter leads to the following

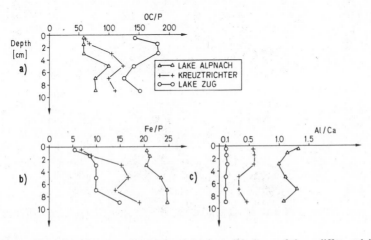

Figure 9.3. Element ratios (atomic members) in the solid phase of three different lake sediments: (*a*) organic carbon to phosphorus, (*b*) iron to phosphorus, and (*c*) aluminum to calcium.

hypothesis: The decomposition of organic carbon and the corresponding mobilization of phosphorus is not according to the stoichiometry in the educts. The residence time of phosphorus in the boundary layer is longer than the residence times of organic carbon and iron. In Lake Alpnach this effect is strongly diluted by the large flux of allochthonous material.

3.2.   Seasonal and Spatial Variation of the Chemical Composition of Pore Waters

Isopleths of soluble reactive phosphorus (SRP) and dissolved iron are presented for the upper 10 cm of each type of sediment (Fig. 9.4). For the three different sediments the following similarities are observed:

The highest concentration gradients for SRP and iron are in the first 2 cm.

In the second half of the stagnation period (July–October) the concentration of the two parameters is increasing most strongly in the first 2 cm.

The concentration peaks of SRP at the sediment boundary layer (0–2 cm) are at a value of $50–70 \times 10^{-6}$ M.

The differences between the three sediments are most pronounced in the iron concentration and their variation of concentration gradients as a function of time and depth. The same is true for dissolved manganese (Fig. 9.5). Contrary to the pore waters of Lake Alpnach, the dissolved phases of Lake Zug and

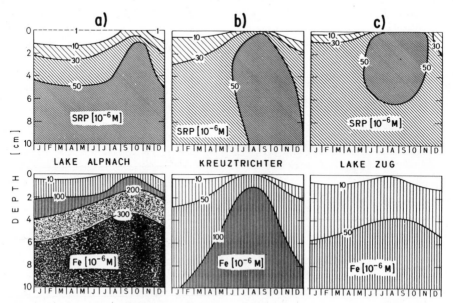

Figure 9.4.   Seasonal and spatial variation of soluble reactive phosphorus (SRP) and dissolved iron in pore waters of three different lake sediments.

MANGANESE CONCENTRATION [10^{-6}M] IN PORE WATERS

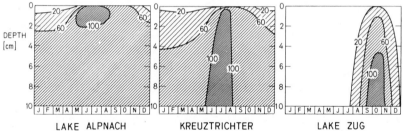

Figure 9.5. Seasonal and spatial variation of dissolved manganese in pore waters of three different lake sediments.

Kreuztrichter sediments show also concentration changes at depths below 4 cm.

In general the pore-water data support the above-mentioned hypothesis, namely, that the most dynamic zone of the sediment–water layer is to be found in the first 2 cm of the sediment, independent of the chemical composition of the solid phase and the sedimentation rate.

3.3. Phosphate and Manganese Fluxes at the Sediment Boundary Layer

The manganese fluxes from the sediment to the hypolimnion vary strongly in time and from lake to lake. If one assumes a pure molecular diffusion process for manganese (II) ions according to Fick's first law [Fig. 9.6(*a*)], one would expect a value for the diffusion coefficient D_{Mn} of approximately 0.3 cm^{2}d^{-1} ($= D_o$). Such a value can be observed for sediments without significant bioturbation [Fig. 9.6(*b*)]. Lake Zug sediments show very few benthic organisms all year, although its surface is still aerobic in spring. The species Beggiatoa is a good indicator for the interface separating oxic and anoxic conditions in the boundary layer (11). If the zone of highest population density of Beggiatoa is at a depth $\leq$ 1 mm, bioturbation drops practically to zero [Fig. 9.6(*b*)]. The higher the bioturbation the higher the value of the diffusion coefficient of manganese, D_{Mn} [Fig. 9.6(*a*)]. The parameter D_{Mn} is thus an empirical parameter to quantify the physical influence of bioturbation on the transport process. From a chemical point of view it has to be assumed simultaneously that reduced manganese passes the oxic layer without being oxidized significantly and thereby precipitated partially. Such an assumption is supported by two observations:

1. Abiotic oxygenation of manganese (II) at pH $\leq$ 8 is extremely slow (12).
2. Biotic oxygenation is taking place preferably at an O_2 concentration between 1 to 2 mg liter^{-1}, a condition which is fulfilled just above the boundary layer (13).

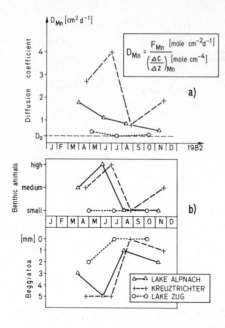

Figure 9.6. (*a*) Diffusion coefficient of manganese as a function of time in the sediment boundary layer of three different lakes; $D_0 = 0{,}3$ cm^2 d^{-1}; and value of the concentration factor for Mn^{2+} assuming pure diffusion and considering the following relations: $D_0 = \Phi^2 D_e$, where $\Phi = 0{,}9$ (porosity) and $D_e = 4 \times 10^{-6}$ cm^2 sec^{-1} (at 10°C)(F_{Mn} = manganese flux from the sediment phase to the hypolimnion) (10). (*b*) Relative abundance of benthic animals (larvae, worms, and protozoa) and depth of Beggiatoa layer (maximum density) as a function of time in three different lake sediments.

It is postulated that the manganese cycle passes through the hypolimnic water even if the uppermost sediment layer is aerobic.

The corresponding phosphorus fluxes are presented in Figure 9.7. The seasonal variations vary in a range of one to two orders of magnitude. The most pronounced increase is observed in Lake Alpnach during the second half of the stagnation period when bottom waters reduce the O$_2$ concentrations to ≤ 1 mg liter^{-1}. However, such an increase could not be expected solely by the increase of the SRP concentration gradient at the interface [Fig. 9.4(*a*)]. In addition one would expect a relative reduction of the flux due to a decrease in bioturbation (Fig. 9.6). A comparison of the diffusion coefficients of SRP

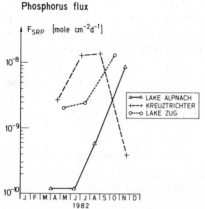

Figure 9.7.   Phosphorus fluxes F_{SRP} as a function of time at three different sediment boundary layers.

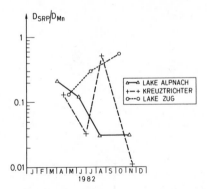

Figure 9.8. Relative diffusion factors of soluble reactive phosphorus, defined as D_{SRP}/D_{Mn} [see Fig. 9.6(a)], as a function of time, for three different sediment boundary layers.

and manganese (Fig. 9.8) makes clear that the SRP fluxes are not only controlled by diffusion and bioturbation. Otherwise, the D_{SRP}/D_{Mn} ratio should be approximately one. This is only observable once in Lake Zug and in Kreuztrichter, respectively. In all other cases D_{SRP} is much smaller than D_{Mn} indicating a potent sink function for phosphorus in the boundary layer. This sink function can be attributed to the adsorption of SRP on freshly precipitated iron (III) hydroxides (14). In the same process orthosilicate and dissolved organic carbon are adsorbed (8). In a first approximation the capacity of such a chemical barrier for the SRP transport can be estimated by the Fe/SRP ratio in the pore water of the reduced zone (Fig. 9.9). From this one would expect that Lake Zug has the smallest, and Lake Alpnach the highest, capacity to reduce the SRP flux. Therefore it is intelligible why the boundary layer of Lake Alpnach disposes of a second negative feedback mechanism when turning more anoxic. By increasing simultaneously the Fe/SRP ratio, the SRP adsorption capacity is increased. The boundary layer of Lake Zug does not dispose of such an iron reservoir. It is therefore explicable why a boundary layer under oxic hypolimnic water (e.g., Lake Zug with 0_2 concentration approximately 8–6 mg liter^{-1}) shows the same phosphorus flux

IRON TO PHOSPHORUS RATIO IN PORE WATERS

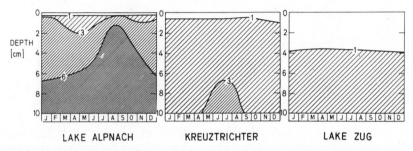

Figure 9.9. Dissolved iron to soluble reactive phosphorus ratio DFe/SRP (atomic ratios) in pore waters of three different lakes. Comparison of seasonal and spatial variations.

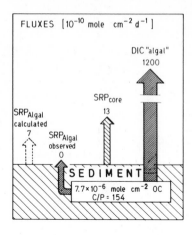

Figure 9.10. Dissolved inorganic carbon (DIC) and soluble reactive phosphorus (SRP) fluxes from a sediment core with newly added and labeled algal matter (*Fragilaria crotonensis* cultivated on nutrients with ^{14}C and ^{32}P).

as a boundary layer under almost anoxic conditions (e.g., Lake Alpnach with O_2 concentration 1–0.5 mg liter^{-1}).

The phosphorus flux could be influenced strongly by an additional source from freshly precipitated organic matter which is mineralized directly on the surface. There is experimental evidence that this contribution is negligible. Newly added organic phosphorus ($^{14}C/^{32}P$ labeled algal matter) is not transferred to the dissolved phase although there is mineralization of organic carbon (Fig. 9.10). The biota of the oxic part of the boundary layer is therefore capable to store phosphorus and decrease the C/P ratio (Fig. 9.3). In addition a fraction of the SRP diffusing upward from the reduced zone is deposited in the same zone. Two internal mechanisms are thus responsible for the observed P enrichment in the uppermost layers (15). It follows from this that the observable net flux of phosphorus has its source in the reduced zone of the boundary layer.

4. A TWO-BOX MODEL FOR THE SEDIMENT BOUNDARY LAYER

4.1. Structure and Biochemical Properties of the Boundary Layer

The sediment boundary layer determines the phosphorus exchange between hypolimnion and sediment. It can be divided into two boxes having the following properties (Fig. 9.11):

OXY Box. The biological processes decomposing incoming particulate organic carbon (POC) utilize the electron acceptors O_2 and NO_3^-. If the thickness of the OXY layer exceeds several millimeters, bioturbation determines the flux of dissolved species between the sediment and the hypolimnion. Manganese (II) passes the OXY box without being oxidized significantly. Its flux can be used to quantify the contribution of bioturbation to the overall flux of other chemical parameters. The incoming POP is transformed within the biotic particulate phase.

RED Box.   The microbial biocenose is anaerobic. There is no bioturbation. The decomposition of POC is controlled by fermentative processes. The production of sulfide due to desulfurization is not detectable ($\leq 10^{-6}$ M S^{2-}). Methanogenesis as the last step in the decomposition of POC can be observed (16). The dissolved phosphate concentration is controlled by the solubility of vivianite (17).

OXY–RED Interface.   The OXY–RED interface is indicated by the organism Beggiatoa sp. At the same level a chemical barrier for phosphate, silicate, and (dissolved organic carbon) DOC is formed offering newly formed iron oxides as adsorbing surfaces. Thereby the flux of phosphorus into the OXY box is reduced. The extent of this reduction depends on the stoichiometric ratio Fe/P in the dissolved phase of the RED box. In the investigated lakes this ratio varies from 0.5 to 6.

The OXY-RED interface moves upward if the incoming POC flux leads to an oxygen consumption which exceeds the incoming O_2 flux to maintain a constant size of the OXY box. Eventually, the interface is transferred into the hypolimnion leaving only a RED box in the sediment boundary layer.

This model does not include the cases in which sulfide concentration starts to control the reduced iron concentration. In addition it must be emphasized that in real sediments the two subsystems OXY and RED are not found only in a vertical separation. They exist in niches on the same level forming a "patchwork" of aerobic and anaerobic zones.

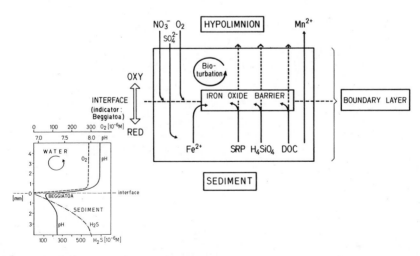

Figure 9.11.   Schematic presentation of the two-box model for the boundary layer of lake sediments.

Table 9.1. Schematic Presentation of Chemical Reactions and Corresponding Kinetics Applied in the Two-Box Model of the Sediment Boundary Layer

Reactions	R_{O_2}	R_{DOC}	$R_{Fe^{3+}}$	$R_{Fe^{2+}}$	R_{SRP}
OXY box					
(1) $\{POC\}^c + O_2 \rightarrow DOC + CO_2$	$\ominus^a$	$\oplus^b$			
(2) $DOC + O_2 \rightarrow CO_2$	$\ominus$	$\ominus$			
(3) $Fe^{2+} + \frac{1}{4}O_2 + 2OH^- + \frac{1}{2}H_2O \rightarrow \{Fe(OH)_3\}$	$\ominus$		$\oplus$	$\ominus$	
(4) $\{Fe(OH)_3\} + SRP \rightarrow \{Fe(OH)_3SRP\}$					$\ominus$
RED-box					
(5) $\{POC\} + A^d \rightarrow DOC + CO_2 + SRP + D^e$		$\oplus$			$\oplus$
(6) $DOC + A \rightarrow CO_2 + SRP + D$		$\ominus$			$\oplus$
(7) $Fe^{3+} + D \rightarrow Fe^{2+} + A$			$\ominus$	$\oplus$	
(8) $Fe^{2+} + SRP \rightarrow \{FeSRP\}$				$\ominus$	$\ominus$

^{a} $\ominus$ Consumption of.
^{b} $\oplus$ Formation of.
^{c} $\{\}$ Indicates a solid phase.
^{d} $A = e^-$ acceptor.
^{e} $D = e^-$ donor.

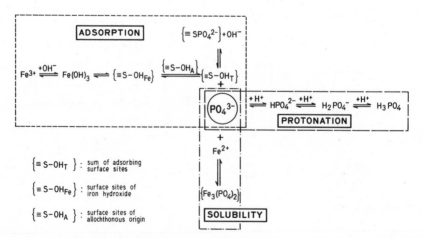

Figure 9.12. Schematic presentation of chemical equilibria determining the phosphate concentration in the boundary layer (SRP is represented by PO_4^{3-}).

**Table 9.2.   State Variables of the Two-Box Model for the
Sediment Boundary Layer**

C Total	c Dissolved	$\dot{c}$ Solid
TOC	DOC	POC
O_2	O_2	
T Fe(III)	Fe^{3+}	$Fe(OH)_3, Fe(OH)_3 SRP$
T Fe(II)	Fe^{2+}	$Fe(II)SRP$
TP	SRP	$Fe(II)SRP, Fe(OH)_3 SRP$

4.2.   Modeling the Dynamics of Phosphorus Transport

In a first approximation the physical–chemical model utilizes a *set of eight
reactions*, which are presented schematically in Table 9.1. The details on stoichiometry and kinetics are discussed elsewhere (18). The boundary conditions
for the solubility of phosphate (SRP) are given by a set of equilibria illustrated
in Figure 9.12.

The *general equation* for the transport processes of the dissolved species
was adapted from Berner (19) and simplified as follows:

$$\frac{\delta C_i}{\delta t} = D_i^* \left(\frac{\delta^2 c_i}{\delta z^2} + \frac{1}{\phi} \frac{\delta \phi}{\delta z} \frac{\delta c_i}{\delta z} \right) - \frac{1}{\phi} \frac{\delta(\phi v c)}{\delta z} + R_i (\bar{c}) \qquad (1)$$

where

C_i = total concentration of the state variable i (Table 4.2)
c_i = dissolved phase of state variable i in pore water, which is calculated
 as a function of the vector $\bar{c}$ of the state variables and the associated
 equilibria constants and solubility products (Table 9.2)
R_i = reaction term, presented in Table 9.1
ϕ = porosity
v = velocity of advection
z = depth
t = time
D_i^* = diffusion coefficient of the dissolved phase of the state variable i. For
 the OXY box, D_i is adapted empirically (see Section 3.3). For the RED
 box, only molecular diffusion is assumed.

The application of this model to the laboratory experiments (in which the
advection term of Eq. 1 is zero) is discussed elsewhere (18).

5.   CONCLUSIONS

The phenomenology of phosphate transport processes at the sediment–water
interface can be described qualitatively and eventually quantitatively by a sys-

tem approach to the boundary layer. Within this biological system three important processes determine the net flux of phosphorus from the sediment back to the hypolimnion:

1. The thickness of the aerobic zone (OXY box) in which bioturbation can increase the flux of phosphorus from the anaerobic zone (RED box) to the hypolimnion. The thickness of this layer, that is, the OXY box, is indirectly proportional to the concentration of particulate organic carbon.
2. The aerobic microbial decomposition of organic matter prevents any phosphorus transfer to the dissolved phase. The carbon/phosphorus ratio in this compartment is decreased.
3. The iron oxide barrier of the aerobic–anaerobic interface which can act as a sink for phosphorus. Its capacity depends on the iron/phosphorus ratio of the dissolved phase.

6.   APPENDIX

6.1.   Sampling Locations, Separation Techniques, and Chemical Analysis

In 1982 gravity cores in acryl resin tubes were taken in intervals of 2 months from two different basins of the Lake of the Four Cantons (Kreuztrichter, Lake Alpnach) and from Lake Zug (Fig. 9.13). Pore waters were separated in an N_2 atmosphere by filtration under pressure. A detailed description of the separation technique and the chemical methodology used will be published elsewhere (8).

6.2.   Flux Measurements Across the Sediment Boundary Layer

After recovery sediment cores with a diameter of 11 cm and an overlying water column of at least 10 cm were equipped with a polyacrylic head (Fig. 9.14). Then the sediment was pushed upward until the sediment–water interface reached the rim of the tube. The overlying water had a volume of approximately 700 mL. At the top of the head a magnetic stirrer was installed in a cylinder with three vertical slits to prevent resuspension of the surface layer and to enable a permanently homogenized water column. Details of the method are published elsewhere (8).

The cores were kept in the dark (except for a short sampling period of about 1 h per day) at 10°C. For technical reasons the temperature was not kept at the initial value of 4 to 5°C. Kamp-Nielsen (20) has shown that between 4 to 10°C microbial activity is not significantly changed if the temperature increases. The flux experiments were run from 7 to 14 days. At the end of the experiments, the cores were sliced and the pore waters separated. The same procedure was applied to a fifth analogous core at the beginning of the experiment.

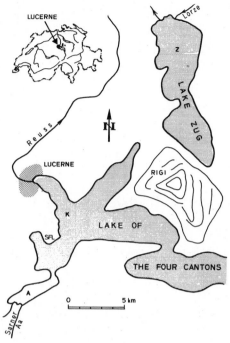

Figure 9.13. Sampling locations: (A) Lake Alpnach, depth 32 m; (K) Kreuztrichter, depth 110 m; (Z) Lake of Zug, depth 60 m; and (SFL) Lake Research Laboratory of EAWAG/ETH at CH-6047 Kastanienbaum.

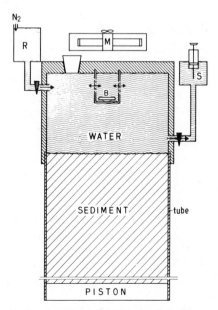

Figure 9.14. Experimental setup for flux measurement at the sediment–water interface of intact cores: (M) magnetic stirrer, (B) magnetic bar, (S) sampling device, and (R) refill device.

Thereby information on pore-water chemistry at the beginning and at the end of the experiment were available. To examine the influence of incoming organic carbon and phosphorus, dead plankton was added. To distinguish between plankton borne P and sediment P a culture of *Fragilaria crotonensis* was marked with ^{32}P and ^{14}C. The details of this method will be described elsewhere (16).

REFERENCES

1. W. Einsele, "Ueber die Beziehungen des Eisenkreislaufes zum Phosphatkreislauf im Eutrophen See," *Arch. Hydrobiol.* **29**, 664–686 (1936).
2. C. H. Mortimer, "The Exchange of Dissolved Substances between Mud and Water in Lakes," *J. Ecol.* **29**, 280–329; **30**, 147–201 (1940/41).
3. D. W. Schindler, R. Hesslein, and G. Kipphut, "Interactions between Sediments and Overlying Waters in an Experimentally Eutrophied Precambrian Shild Lake." In H. L. Goltermann (Ed.), *Interactions between Sediments and Freshwater*, Junk, The Hague, 1977, pp. 235–243.
4. G. F. Lee, "Significance of Oxic vs. Anoxic Conditions for Lake Mendota Sediment Phosphorus Release." In H. L. Goltermann (Ed.), *Interactions between Sediment and Freshwater*, Junk, The Hague, 1977, pp. 294–306.
5. O. Hoyer, H. Bernhardt, J. Clasen, and A. Wilhelms, "In Situ Studies on the Exchange between Sediment and Water Using Caissons in the Wahnbach Reservoir," *Arch. Hydrobiol. Beih.* **18**, 79–100 (1982).
6. B. Boström, M. Jansson, and C. Forsberg, "Phosphorus Release from Lake Sediments," *Arch. Hydrobiol. Beih.* **18**, 5–59 (1982).
7. R. A. Berner, *Early Diagenesis: A Theoretical Approach*, Princeton University Press, Princeton, N.J., 1980.
8. P. Baccini, E. Laczkó, J. Ruchti, and M. Sturm, "Phosphorus Transport Studies at the Sediment/Water Interface" (in preparation).
9. P. Baccini, "Untersuchungen über den Schwermetallhaushalt in Seen," *Schweiz. Z. Hydrol.* **38**, 121–158 (1976).
10. Y. H. Li and S. Gregory, "Diffusion of Ions in Seawater and in Deep Sea Sediments," *Geochim. Cosmochim. Acta* **38**, 703–714 (1974).
11. B. B. Jørgensen and N. P. Resbech, "Colorless Sulfur Bacteria, *Beggiatoa* spp. and *Thiovolum* spp., in O_2 and H_2S microgradients," *Appl. Environ. Microbiol.* **45**, 1261–1270 (1983).
12. J. J. Morgan, "Chemical Equilibria and Kinetic Properties of Manganese in Natural Water." In S. D. Faust and J. V. Hunter (Eds)., *Principles and Applications of Water Chemistry* Wiley, New York, 1967.
13. J. M. Jaquet, G. Nembrini, P. Garcia, and J. Vernet, "The Manganese Cycle in Lac Léman, Switzerland: The Role of Metallogenium," *Hydrobiol.* **91**, 323–340 (1982).
14. W. Stumm and J. J. Morgan, *Aquatic Chemistry*, Wiley, New York, 1980.

15. R. Carignan and R. J. Flett, "Postdepositional Mobility of Phosphorus in Lake Sediments," *Limnol. Oceanogr.* **26,** 361–366 (1981).

16. E. Laczkó, "Kinetics of Phosphorus Transfer by Microbial Activity in Lacustrine Sediments," Thesis, ETH Zürich (in preparation).

17. S. Emerson and G. Widmer, "Early Diagenesis in Anaerobic Lake Sediments. II. Thermodynamic and Kinetic Factors Controlling the Formation of Iron Phosphate," *Geochim. Cosmochim. Acta*, **42,** 1307–1316 (1978).

18. J. Ruchti and P. Baccini, "A Two-Box Model for the Phosphorus Transport at the Sediment/Water Interface" (in preparation).

19. R. A. Berner, "Diagenetic Models of Dissolved Species in the Interstitial Waters of Compacting Sediments," *Am. J. Sci.* **275,** 88–96 (1975.

20. L. Kamp-Nielsen, "A Kinetic Approach to the Aerobic Sediment–Water Exchange of Phosphorus in Lake Esrom," *Ecol. Modelling* **1,** 153–160 (1975).

10

THE INFLUENCE OF COAGULATION AND SEDIMENTATION ON THE FATE OF PARTICLES, ASSOCIATED POLLUTANTS, AND NUTRIENTS IN LAKES

*Charles R. O'Melia**

Department of Geography and Environmental Engineering, The Johns Hopkins University, Baltimore, Maryland 21218, USA

Abstract

Some solid particles are in all lakes. Pollutants and nutrients are associated with these particles. Collisions occur between particles and aggregates are formed. Particles and their aggregates are both subject to sedimentation. We have examined the effects of some physical factors on the origins and effects of coagulation and sedimentation in lakes. In this chapter, we focus on the influence of selected chemical factors on the transport and fate by these processes of solid particles and associated substances in lakes.

1. INTRODUCTION

Some solid particles are present in all lakes. Pollutants and nutrients are associated with these particles. Collisions occur between particles and ag-

* This work was realized in collaboration with Mark Wiesner, Ulrich Weilenmann, and Waris Ali at the Department of Geography and Environmental Engineering, The Johns Hopkins University, Baltimore, Maryland 21218.

gregates are formed. Particles and their aggregates are both subject to sedimentation. O'Melia and Bowman (1) have examined the effects of some
physical factors on the origins and effects of coagulation and sedimentation
on particles in lakes. The focus of this present work is the influence of
selected chemical factors on the transport and fate by these processes of
solid particles and associated substances in lakes.

This paper contains the following parts: (1) a brief description of the model
used in the work; (2) a selective summary of results reported previously; (3)
new results evaluating chemical factors; and (4) a discussion of the significance of these results in understanding and controlling water quality in
lakes.

2.   THE MODEL

A lake is divided into vertical boxes denoted as I, J, K, L, and so on. In
this work three boxes are used, corresponding to the epilimnion, thermocline, and hypolimnion. All particles in the lake are divided into size compartments denoted as i, j, k, l, and so on. In this work, 25 size compartments
are usually used. A general equation may be written for particles of any size
in any location. For example, for particles of size k in box I,

$$\frac{dn_{k,I}}{dt} = \tfrac{1}{2} \sum_{i+j \to k} \alpha_I \lambda(i,j)_I n_{i,I} n_{j,I} - n_{k,I} \sum_{i=1}^{\infty} \alpha_I \lambda(i,k)_I n_{i,I}$$

$$+ \frac{w_{k,H}}{z_I} n_{k,H} - \frac{w_{k,I}}{z_I} n_{k,I} \pm W_{k,I} + \frac{q_0}{z_I} n_{k,0} - \frac{q_I}{z_I} n_{k,I} \quad (1)$$

where

$n_{k,I}$ = the number concentration of particles of size k in lake compartment I

α = the collision efficiency factor reflecting the stability of the particles and the surface chemistry of the system

$\lambda(i,j)$ = a collision frequency function that depends on physical modes
of interparticle contact

$w_{k,H}$ = the settling velocity of particles of size k located directly above
box I

z_I = is the depth of box I

$W_{k,I}$ = the rate of production or destruction of particles of size k in
box I

q_0 and q_I = the areal hydraulic loadings into and out of compartment I

$n_{k,0}$ = the number concentration of particles of size k in the water
flowing into compartment I.

The left-hand side of Eq. 1 describes the rate at which the number concentration of particles of size k and location I changes with time (particles per m^{3}s). The first term on the right-hand side (rhs) expresses the rate of formation of particles of size k (or volume v_k) from smaller particles having the total volume v_k. The condition $i + j \rightarrow k$ under the summation denotes the condition that $v_i + v_j = v_k$. The factor $\frac{1}{2}$ is needed since collisions are counted twice in this summation. The second term on the rhs describes the loss of particles of size k by growth to form larger aggregates; this occurs when a size k particle collides with and attaches to a particle of any size i.

The third term on the rhs of Eq. 1 describes the addition of particles of size k to box I by settling from above (box H). The fourth term on the rhs expresses the loss of size k particles from box I by settling into box J below. For the bottom box, this corresponds to removal of particles from the lake. $W_{k,I}$ is described previously. The sixth and seventh terms on the rhs describe the input and output of the size k particles by hydraulic inflow and discharge.

Contacts between particles in water can occur by three different physical processes, namely, Brownian diffusion (thermal effects), fluid shear (flow effects), and by differential sedimentation (gravity effects). These transport processes have been studied by many investigators including Smoluchowski (2) and Friedlander (3) and are discussed in more detail elsewhere (4). The collision frequency functions or rate coefficients for these processes are as follows:

Brownian Diffusion $\qquad \lambda(i,j)^{BD} = \dfrac{2\bar{k}T}{3\mu} \dfrac{(d_i + d_j)^2}{d_i\, d_j}$ $\qquad\qquad$ (2a)

Fluid Shear $\qquad\qquad \lambda(i,j)^{SH} = \tfrac{1}{6}(d_i + d_i)^3\, G$ $\qquad\qquad\qquad$ (2b)

Differential
$\quad$ Sedimentation $\qquad \lambda(i,j)^{DS} = \dfrac{\pi(\rho_p - \rho)}{72\mu} (d_i + d_j)^3\, |\, d_i - d_j\, |$ $\quad$ (2c)

where

$\lambda(i,j)^{BD}$, $\lambda(i,j)^{SH}$, and $\lambda(i,j)^{DS}$ = bimolecular rate coefficients for interparticle collisions between particles of diameters d_i and d_j by Brownian diffusion, fluid shear, and differential sedimentation, respectively

$\bar{k}$ = the Boltzmann's constant
T = the absolute temperature
μ = the water viscosity
G = the mean velocity gradient in the water
ρ_p and ρ = the densities of the solid particles and the water, respectively.

These rate coefficients are used additively to determine $\lambda(i,j)_I$ in Eq. 1.

The settling velocity of a particle in this model is assumed to follow Stokes' equation as follows:

$$w_k = \frac{(\rho_p - \rho)g}{18\mu} d_k^2 \tag{3}$$

where g = the gravity acceleration.

Natural suspensions contain particles of many different sizes, that is, they are heterodispersive. The size distribution has been described by a particle size distribution function $n(d_p)$ [e.g., Friedlander (3)]. If dN is the number of particles per unit of fluid volume in the particle size range from d_p to $d_p + d(d_p)$, $n(d_p)$ is defined as

$$n(d_p) = \frac{dN}{d(d_p)} \tag{4}$$

Measurements of particles in ocean waters, ocean sediments, waste waters and sludge digesters have indicated that the particle size distribution function frequently follows a power law of the form

$$n(d_p) = Ad_p^{-\beta} \tag{5}$$

Here A is a coefficient related to the total concentration of particulate matter in the water and the exponent β is observed to range from 2 to 5. Equation 5 is used in this work to characterize the particle size distribution of inputs to the lake.

Characteristics of the model lake used in this research are presented in Table 10.1. These are identical to those used previously (1,4) with two exceptions. First, the collision efficiency factor α is varied from 0 to 5×10^{-2}. Second, the concentration of particles added to the epilimnetic waters (n_0) is varied from 5 to 95 ppm by volume (5.25–100 mg liter^{-1}). This increase in particle concentration is derived from observations reported by Giovanoli et al. (5). These authors report deposition rates of solids in sediment traps in Lake Zürich ranging from 1.3 to 3.2 g of freeze-dried suspended solids per meter squared per day, with an average of 2.25 g m^{-2} d^{-1}. Approximately 25% of this material was SiO$_2$ and aluminum silicates; the remaining 75% was comprised of biologically produced particles with the stoichiometric composition $\{(CaCO_3)_{108}(MgCO_3)_4(CH_2O)_{114}(NH_3)_{16}(H_3PO_4)_1\}$. If one assumes a density of 2 g cm^{-3} for inorganic silicates and carbonates and also that the organic content of the freeze-dried material was 90% water before drying, this observed settling rate corresponds to a volumetric concentration of 60 ppm introduced into the epilimnion in a river flow of 10^{-6} m s^{-1}. This estimate could be low because of dissolution and mineralization of particles during the 14-day collection period in the traps, because of ineffective collection of small particles in these traps, and because some par-

Table 10.1.   Properties of Model Lake

Parameter	Epilimnion	Thermocline	Hypolimnion
Mean depth $\bar{z}$ (m)	10	5	50
Temperature T (°C)	25	15	5
Viscosity μ (g cm^{-1} s^{-1})	0.8915×10^{-2}	1.146×10^{-2}	1.515×10^{-2}
Water density ρ (g cm^{-3})	0.997	0.999	1.000
Particle density ρ_p (g cm^{-3})	1.05	1.05	1.05
Velocity gradient G (s^{-1})	10	0.1	1.0
Areal hydraulic loading q (m s^{-1})[a]	Varies $0-10^{-6}$	0	0
Coagulation efficiency α (dimensionless)	Varies $0-5 \times 10^{-2}$	Varies $0-5 \times 10^{-2}$	Varies $0-5 \times 10^{-2}$
Particle concentration in river inflow, n_o (ppm)[b]	Varies 5–95	0	0

[a] This corresponds to the Stokes' settling velocity of a 5.7-μm particle having a density of 1.05 g cm^{-3} in water at 25°C.

[b] The waters of the river input are assumed to contain particles with a density of 1.05 g cm^{-3}. The particles are distributed over a size range of 0.3–31 μm with $n(d_p)$ described by the power law (Eq. 5) and $\beta = 4$.

ticles must be removed from the epilimnion in the downstream discharge. Based on this estimate, a maximum input concentration to the epilimnion of 95 ppm (100 mg liter^{-1} for $\rho_p = 1.05$ g cm^{-3}) was assumed. This mass of 100 mg liter^{-1} is a wet weight, and corresponds to a dry weight of about 30 mg liter^{-1}.

The dimensionless ratio α is the fraction of those collisions occurring within the lake waters that result in successful attachment and the formation of an aggregate. For $\alpha = 0$, all collisions are ineffective. The particles are perfectly stable and no coagulation occurs. If $\alpha = 1$, all collisions are suc-

Table 10.2.   Determinations of α for Loch Raven Reservoir

Date	α
June 8, 1983	0.01
June 22, 1983	0.08
July 7, 1983	0.01
July 20, 1983	0.04
August 3, 1983	0.04
August 17, 1983	0.035

cessful and the coagulation rate is the maximum that can occur. Previous work was focused on $\alpha = 10^{-3}$ (1). Larger values of α are examined here because preliminary estimates of α for particles in the Loch Raven Reservoir (Maryland, U.S.A.) suggest values of α in these waters of 10^{-2} or larger (Table 10.2). These estimates are made from laboratory measurements in which samples of reservoir water are transported to the laboratory, stirred under controlled conditions to define the rate of interparticle contacts, and the change in particle number determined using an electronic particle counter. Samples were taken at a central location from the epilimnion of the reservoir at depths from 1 to 3 m. The waters of the reservoir are low in organic carbon (2 mg liter^{-1}) and in calcium (5×10^{-4} M).

Many pollutants that affect human health or environmental quality are associated with particulate materials. This association is often described by a partition coefficient K_p defined as follows:

$$K_p = \frac{\text{concentration of substance in the solid phase (mass/mass)}}{\text{concentration of substance in the aqueous phase (mass/volume)}}$$

$$(6)$$

Typical units are liter g^{-1}, with the mass of the solid phase expressed on a dry-weight basis. In this work, values of K_p ranging from 0.1 to 10^5 liter g^{-1} are considered. Estimates of K_p for selected substances of interest are presented in Table 10.3. For the organic substances, the organic carbon content of the settling particles is assumed to be 7%, based on the results of Giovanoli et al. (5).

Table 10.3.   Partition Coefficients

Substance	K_p(liter g^{-1})	Reference
1,4-Dichlorobenzene	0.06[a]	(6)
1,2,4,5-Tetrachlorobenzene	0.54[a]	(6)
DDT	6.3[a]	(6)
2,4,5,2',4',5' PCB	15[a]	(6)
P	21	(7)
Cu	36	(7)
Zn	100	(7)
Pb	500	(7)
Fe	1300	(7)

[a] Calculated from the relationship by Schwarzenbach and Westall (6), log K_p = 0.72 log K_{ow} + log f_{oc} + 0.49, in which K_p has units of cm^{-3} g^{-1}, K_{ow} is the octanol–water coefficient of the substance of interest, and f_{oc} is the organic carbon content of the solid phase, assumed here to be 0.07.

3. PREVIOUS WORK

Earlier research (1) has evaluated coagulation and sedimentation in lakes, focusing on the effects of physical factors on the concentrations, size distributions, and fluxes of particles in the epilimnion, thermocline, and hypolimnion of a model lake chosen to have physical similarities with Lake Zurich. Selected results are summarized here.

Field observations suggest and conceptual analyses affirm that coagulation, sedimentation, and hydraulic loading exert significant and perhaps controlling effects on the transport and fate of particles and associated pollutants and nutrients in lakes. Coagulation in the epilimnion substantially reduces the number and surface area of particles in this lake compartment. Coagulation can, under some circumstances, decrease the mass concentration of particles in the epilimnion and increase the deposition of particles in the lake sediments. This increased deposition is accomplished by increasing the downward flux of particles settling by gravity into the thermocline, thereby reducing the downstream withdrawal of particles in the lake discharge and the resulting mass concentration of particles in the epilimnetic waters. Coagulation in the thermocline and hypolimnion is less extensive than in the epilimnetic waters. For the conditions and reactions considered in the model lake, coagulation in the two lower lake compartments does not affect mass deposition to the sediments. It does, however, substantially reduce the mass, surface area, and number concentrations of particles in these lake waters. The relatively slow coagulation rates considered in this previous work promote sufficient aggregation to offset the reduction in the settling velocities of the particles that occurs in these colder waters. Coagulation effects yield particle concentrations (mass, surface area, and number) that decrease downward through the lake—a trend consistent with some field observations.

4. RESULTS

The focus of the results presented here is on three areas: (1) the effects of colloid stability on the mass deposition or removal of particles from lakes; (2) the kinetics of coagulation and sedimentation in the epilimnion, thermocline, and hypolimnion of lakes; and (3) the deposition or removal from lakes of pollutants and nutrients associated with solid particles.

The responses or effects of coagulation and sedimentation in lakes are expressed in five ways:

1. Total mass, surface area, and number concentrations in lake waters expressed in units of mg liter^{-1}, m^2 liter^{-1}, and particles per liter, respectively.

2. Mass removals of particles and associated substances from a lake, expressed as a percent of the material introduced into the lake waters.
3. Distributions of particle volume (ppm) as a function of particle size (μm) in the lake waters.
4. Mass input and output fluxes by river flow, sedimentation, and coagulation processes as functions of particle size, with fluxes expressed as $mg \ cm^{-2} \ s^{-1}$.
5. The response time required for lake waters to achieve 95% of steady-state mass concentrations of particles.

4.1. Colloidal Stability

Some effects of surface chemistry (α) on particle concentrations in lakes at steady state are illustrated in Table 10.4. Steady-state concentrations of mass, surface area, and particle number in the epilimnion, thermocline, and hypolimnion of a lake are presented for three input concentrations and several values of α. In the absence of coagulation ($\alpha = 0$), concentrations are lower in the epilimnion than in the inflow due to removal by sedimentation, but then increase downward through the lake. For example, for an input concentration of 100 mg liter^{-1}, the mass concentrations in the epilimnion, thermocline, and hypolimnion are 62, 83, and 112 mg liter^{-1}, respectively. This increase in mass concentration occurs because of the effects of temperature on fluid viscosity and density and hence on the Stokes' settling velocities of particles. As the water temperature decreases from the epilimnion to the hypolimnion (from 25°C to 5°C in the case considered), the Stokes' settling velocities of particles with a density of 1.05 g cm^{-3} are reduced by 44%. The resulting steady-state particle concentration in the hypolimnion is increased by a factor of 100/(1−0.44) or 80% above the concentration in the epilimnion. The vertical distributions of particle mass, surface area, and number concentrations in the absence of coagulation ($\alpha = 0$) are a direct consequence of the balance between hydraulic loading (overflow rate) and gravity forces on the particles in the lake.

In the presence of coagulation (α from 10^{-3} to 5×10^{-2} in Table 10.4), particle concentrations decrease with depth. For example, for an input of 100 mg liter^{-1} and with $\alpha = 10^{-2}$, mass concentrations decrease by 57% from the epilimnion to the hypolimnion (from 15.6 to 6.7 mg liter^{-1}). Greater reductions are predicted for surface area (94%) and number (99.96%) concentrations.

Effects of coagulation on the mass removal of particles are presented in Figure 10.1. Mass removal or deposition of particles is plotted as a function of areal hydraulic loading for four values of α ranging from 0 to 5×10^{-2}. The curve for $\alpha = 0$ is independent of particle inputs; other curves are calculated for an input concentration (wet weight) of 100 mg liter^{-1}. For a

Table 10.4. Predicted Particle Concentrations[a]

α	Location	Mass Concentration (mg liter^{-1})	Surface Area Concentration (m^2 liter^{-1})	Number Concentration (particles per liter)
	Input	5.25	2.29×10^{-2}	34.1 $\times 10^9$
0	Epilimnion	3.28	2.14×10^{-2}	33.9 $\times 10^9$
	Thermocline	4.35	2.85×10^{-2}	45.1 $\times 10^9$
	Hypolimnion	5.89	3.83×10^{-2}	60.8 $\times 10^9$
10^{-3}	Epilimnion	3.11	1.51×10^{-2}	15.2 $\times 10^9$
	Thermocline	2.87	0.76×10^{-2}	1.8 $\times 10^9$
	Hypolimnion	2.17	0.34×10^{-2}	0.26 $\times 10^9$
10^{-2}	Epilimnion	2.26	0.74×10^{-2}	5.55 $\times 10^9$
	Thermocline	2.03	0.32×10^{-2}	0.36 $\times 10^9$
	Hypolimnion	1.25	0.11×10^{-2}	0.03 $\times 10^9$
	Input	30	13.1×10^{-2}	195 $\times 10^9$
0	Epilimnion	18.7	12.3×10^{-2}	193 $\times 10^9$
	Thermocline	24.9	16.2×10^{-2}	258 $\times 10^9$
	Hypolimnion	33.7	21.9×10^{-2}	347 $\times 10^9$
10^{-3}	Epilimnion	14.6	5.24×10^{-2}	41.3 $\times 10^9$
	Thermocline	13.3	2.40×10^{-2}	3.15 $\times 10^9$
	Hypolimnion	8.49	0.85×10^{-2}	0.30 $\times 10^9$
10^{-2}	Epilimnion	7.49	1.94×10^{-2}	13.5 $\times 10^9$
	Thermocline	6.23	0.64×10^{-2}	0.50 $\times 10^9$
	Hypolimnion	3.55	0.17×10^{-2}	0.02 $\times 10^9$
5×10^{-2}	Epilimnion	3.96	0.88×10^{-2}	6.03 $\times 10^9$
	Thermocline	2.92	0.20×10^{-2}	0.11 $\times 10^9$
	Hypolimnion	1.62	0.05×10^{-2}	0.002 $\times 10^9$
	Input	100	43.8×10^{-2}	649 $\times 10^9$
0	Epilimnion	62.4	41.0×10^{-2}	645 $\times 10^9$
	Thermocline	82.9	54.3×10^{-2}	859 $\times 10^9$
	Hypolimnion	112	73.0×10^{-2}	1.160 $\times 10^9$
10^{-3}	Epilimnion	36.1	10.8×10^{-2}	77.4 $\times 10^9$
	Thermocline	31.8	4.3×10^{-2}	4.2 $\times 10^9$
	Hypolimnion	18.8	1.3×10^{-2}	0.25 $\times 10^9$
10^{-2}	Epilimnion	15.6	3.60×10^{-2}	24.7 $\times 10^9$
	Thermocline	11.9	0.89×10^{-2}	0.55 $\times 10^9$
	Hypolimnion	6.73	0.22×10^{-2}	0.01 $\times 10^9$
5×10^{-2}	Epilimnion	7.80	1.59×10^{-2}	11.0 $\times 10^9$
	Thermocline	5.12	0.25×10^{-2}	0.10 $\times 10^9$
	Hypolimnion	2.87	0.06×10^{-2}	0.001 $\times 10^9$

[a] Based on the model lake presented in Table 10.1, with $q = 10^{-6}$ m s^{-1}.

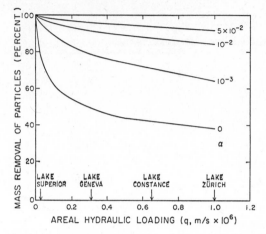

Figure 10.1. Effects of areal hydraulic loading q and particle stability α on particle retention in lakes. $n_0 = 100$ mg liter^{-1}. (See Table 10.1 for other assumed lake properties.)

lake with a hydraulic loading of 10^{-6} m s^{-1} (Lake Zürich), particle retention can vary from 38 ($\alpha = 0$) to 92% ($\alpha = 5 \times 10^{-2}$), indicating that colloid stability can significantly affect particle fluxes in lakes.

Effects of coagulation on the size distribution of particles in the lake waters are presented in Figure 10.2. The particle size distribution is represented as the volume concentration distribution [$\Delta V/\Delta \log d_p$ vs. $\log d_p$]. Here ΔV is the volume concentration (ppm) in a particular logarithmic size interval and d_p is the particle diameter (μm). The area under such a curve over any size interval is proportional to the volume concentration of particles

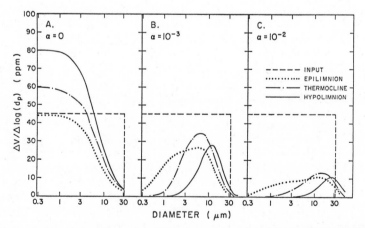

Figure 10.2. Effects of particle stability (α) on volume concentration distributions in lake waters: (A) $\alpha = 0$; (B) $\alpha = 10^{-3}$; (C) $\alpha = 10^{-2}$. $q = 10^{-6}$ m s^{-1} and $n_0 = 100$ mg liter^{-1}. (See Table 10.1 for other assumed lake properties.)

in that size interval. For spherical particles, the volume concentration equals this area (3).

The river inflow contains a total volume concentration of 95 ppm; for β = 4, this corresponds to an equal distribution of particle volume with every logarithmic size interval. Note the horizontal lines for river inputs in each case presented in Figure 10.2. Three cases are considered, corresponding to $\alpha = 0$, 10^{-3}, and 10^{-2}.

In the absence of coagulation (Figure 10.2A), particles larger than about 5.7 μm (i.e., those with settling velocities greater than 10^{-6} m s^{-1}, the hydraulic loading to the lake) are removed effectively from all of the lake compartments. Submicron particles which have settling velocities significantly smaller than the hydraulic loading are not removed effectively from any lake compartment and accumulate at steady-state to high concentrations in the thermocline and hypolimnion. These particle distributions i.e., increasing total mass, surface area, and number concentrations with depth (Table 10.4) and a selective accumulation of fine particles in the bottom waters (Fig. 10.2A), are not reported in most lake studies. Some other processes must occur. One possibility suggested here is coagulation, which increases the settling velocities of particles by forming aggregates and thereby enhances their removal by sedimentation from the lake.

Volume concentration distributions resulting when coagulation is considered are presented in Figures 10.2B and 10.2C. Significant differences from Figure 10.2A are: (1) small particles are removed from all lake compartments; (2) volume or mass concentrations (proportional to the area under the curves) are less than the input and decrease with depth; (3) particles larger than the size corresponding to the hydraulic loading can predominate in all compartments; and (4) the predominant particle size increases with depth. Data to test these predictions are not presently available; studies of the Loch Raven Reservoir (Maryland, U.S.A.) are being conducted for this purpose. It is reasonable to state, however, that the size distributions in Figures 10.2B and 10.C are consistent with present knowledge, and suggest that coagulation is an important process for particle transport in lakes.

Effects of coagulation on lakes are examined in another way in Figures 10.3 and 10.4, in which mass fluxes are plotted as functions of particle size. More specifically, $\Delta J/\Delta \log d_p$ is plotted versus $\log d_p$ for several physical transport processes in each lake compartment. J is a steady-state mass flux (mg cm^{-2} s^{-1}) by river flow, sedimentation, or coagulation. The area under a curve over any size interval in this graph is the steady-state mass flux of particles in that size interval by a particular physical process. The flux distributions in the absence of coagulation are presented in Figure 10.3; fluxes with coagulation ($\alpha = 10^{-2}$) are presented in Figure 10.4. Results are presented for a hydraulic loading of 10^{-6} m s^{-1} and an input mass concentration of 100 mg liter^{-1}.

In the absence of coagulation (Fig. 10.3), small particles are removed from the epilimnion by river discharge and large particles are transported

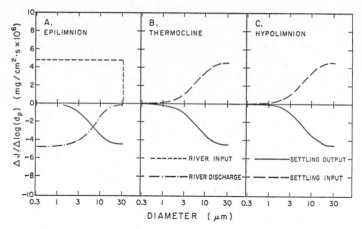

Figure 10.3. Mass flux distributions for model lake with no coagulation ($\alpha = 0$). $q = 10^{-6}$ m s^{-1} and $n_0 = 100$ mg liter^{-1}. (See Table 10.1 for other assumed lake properties.)

downward by settling. Input and output fluxes of small particles in the thermocline and hypolimnion are small, while steady-state concentrations of these particles in the water column are large (Fig. 10.2A). Inputs and outputs of large particles in the thermocline and hypolimnion are by sedimentation only. The fluxes are large (Fig. 10.3B and 10.3C), while the concentrations in the water are small (Fig. 10.2A).

When coagulation occurs (Fig. 10.4), hydraulic discharge of particles from the epilimnion can be reduced substantially. Small particles are "removed" from the epilimnetic waters by coagulation to form larger aggregates. Large particles, which include those introduced by river inflow, produced by biological or chemical processes, or formed by aggregation, are removed primarily by settling to the deeper waters.

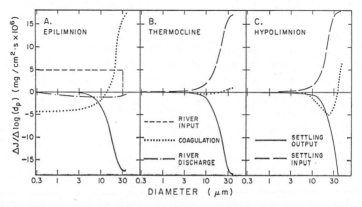

Figure 10.4. Mass flux distributions for model lake with coagulation ($\alpha = 10^{-2}$). $q = 10^{-6}$ m s^{-1} and $n_0 = 100$ mg liter^{-1}. (See Table 10.1 for other assumed lake properties.)

In the thermocline (Fig. 10.4B), input and output of particles are primarily by sedimentation, and these fluxes are almost equal. A small amount of coagulation occurs, and this small coagulation flux is sufficient to overcome the effects of temperature on Stokes' settling velocities. As stated previously, particle concentrations are reduced substantially compared to systems without coagulation (Table 10.4 and Fig. 10.2). Coagulation fluxes are larger in the hypolimnion than in the thermocline. Coagulation in the epilimnion removes small particles by forming larger aggregates and can increase particle removal from the lake. Coagulation in the thermocline and hypolimnion continues the removal of small particles, but does not increase mass deposition from the lake for the systems examined here.

4.2. Kinetics

An important subject left unresolved in previous analyses is the rate at which coagulation and sedimentation can affect water quality in lakes. This is examined in Table 10.5. The time for the particle mass concentration to reach 95% of its steady-state value (t_{95}) is presented for several conditions of colloid stability (α) for the epilimnion, thermocline, and hypolimnion of a lake with a hydraulic loading of 10^{-6} m s^{-1} and with $n_0 = 100$ mg liter^{-1}.

The following procedure was used to determine t_{95}. First, the entire lake was considered to be filled with a particle concentration equal to the input concentration, in this case 100 mg liter^{-1}. Next, the steady-state mass concentration of particles in each lake compartment was determined by integrating the model over time until the mass concentrations were constant. Except for $\alpha = 0$, those steady-state concentrations were all less than 100 mg liter^{-1}, and most were substantially less than this value (Table 10.4). Finally, the times to reach mass concentrations 5% greater than the steady-state values were determined from the change of concentration with time during the model integrations. This procedure was followed for each value of α of interest.

The response times (values of t_{95}) are rapid (Table 10.5). For $\alpha = 10^{-2}$, the preliminary estimate for the Loch Raven Reservoir, t_{95} values for the

Table 10.5. Lake Response Times (t_{95})[a]

Lake Compartment	Colloidal Stability α			
	0	10^{-3}	10^{-2}	5×10^{-2}
Epilimnion	80 days	77 days	18 days	7 days
Thermocline	9 years	2 years	6 months	3 months
Hypolimnion	20 years	4 years	1 year	100 days

[a] Calculations are for $q = 10^{-6}$ m s^{-1} and $n_o = 100$ mg liter^{-1}.

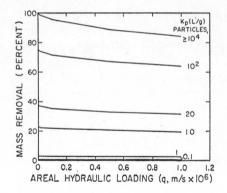

Figure 10.5. Effects of solid–solution partitioning K_p and areal hydraulic loading q on the retention of particles and particle-reactive substances in lakes. $q = 10^{-6}$ m s^{-1}, $\alpha = 10^{-2}$, and $n_0 = 100$ mg liter^{-1} (wet weight) $= 30$ mg liter^{-1} (dry weight). (See Table 10.1 for other assumed lake properties.)

epilimnion, thermocline, and hypolimnion are 18 days, 6 months, and 1 year, respectively. In the epilimnion, t_{95} is less than or equal to 80 days for all values of α, that is, from 0 to 5×10^{-2}. The hydraulic residence time of the model lake is 2 years. Since the initial conditions use a particle concentration that is substantially greater than that which occurs in a typical system (i.e., initial lake concentration assumed to equal input concentration), these calculated values of t_{95} are larger than those expected in a real lake.

4.3. Sorbed Substances

Some results obtained by adding sorption to the particle model are presented in Figure 10.5. The mass removal of a sorbable substance is plotted as a function of areal hydraulic loading for values of K_p from 0.1 to 10^4 liter g^{-1}, a range which includes the substances listed in Table 10.3. These calculations are based on an input concentration of particles of 100 mg liter^{-1} (wet weight) or 30 mg liter^{-1} (dry weight) and also an assumption of $\alpha = 10^{-2}$. For values of K_p greater than about 10^3 or 10^4 liter g^{-1} [e.g., substances such as Fe(III)], the substance of interest follows the particles and is transported effectively to the lake sediments. Removals of these substances are not distinguishable from particle removals in the lake. Substances with values of K_p less than about 10 liter g^{-1} remain predominantly in solution (e.g., tetrachlorobenzene). Factors affecting particle removal (e.g., q and α) will not influence their fate appreciably. For substances with values of K_p from 10 to 10^2 liter g^{-1} (e.g., PCBs, phosphorus, several metals), removal to the lake sediments can be substantially less than particle removals, but still significant. At present, the model does not include partitioning at the air–water interface, and this process may alter these results somewhat.

5. DISCUSSION

This discussion is focused first on the effects of coagulation and sedimentation on solid particles in lakes, and then on the effects of these processes on pollutants and other substances that may be associated with particles in lacustrine systems.

Coagulation and sedimentation can have dramatic effects on the size distribution of particles in lakes (Fig. 10.2); on the vertical distributions of particle mass, surface area, and number in lakes (Table 10.4); and on the transport of particles to lake sediments (Fig. 10.1). The kinetics of these processes are sufficiently rapid to exert these important effects (Table 10.5). The extent of these effects depends upon many factors. Three are discussed here, namely, hydraulic loading q, colloidal stability α, and particle input concentration n_0.

Particle removal can depend significantly on the areal hydraulic loading of a lake (Fig. 10.1). This is because the areal hydraulic loading is a settling velocity; it is equal to the settling velocity of a particle that can be effectively removed from a lake by gravity in the absence of coagulation. Larger particles are also removed effectively. Lake Superior, with $q = 3 \times 10^{-8}\,\mathrm{m\,s^{-1}}$, is a more effective settling basin than Lake Zürich ($q = 10^{-6}\,\mathrm{m\,s^{-1}}$). Considering water at 20°C and materials with a density of 1.05 g cm^{-3}, Lake Zurich can effectively remove particles larger than about 6 μm by gravity alone, and Lake Superior can effectively settle particles of 1.9 μm or larger in size.

Coagulation acts to increase the effects of gravity in removing particles from lakes. This is particularly significant for small particles, with settling velocities less than the areal hydraulic loading; this usually includes particles with sizes from a few microns downward through the submicron size range. Regardless of the areal hydraulic loading of a lake, coagulation will increase deposition. If particle inputs are large and the particles are somewhat unstable, coagulation will be extensive and this increase in deposition can be substantial (Fig. 10.1).

Coagulation effects depend on the colloidal stability of the particles, characterized in this work by α. Colloidal stability in lake waters may be expected to depend upon the presence of natural organic matter, the concentrations of major divalent cations (Ca^{2+} and Mg^{2+}) and iron and manganese, pH, redox conditions, and seasonal variations in physical factors, such as temperature and stratification. Gibbs (8) studied the coagulation of natural particles in the laboratory using interparticle contacts induced by fluid shear. Particles were studied as obtained from the field and after treatment with sodium hypochlorite to remove their natural organic coatings. Salinity varied from 0.06 to 20‰. Calculated values of α ranged from 0.077 to 0.82, depending upon salinity and upon the presence of adsorbed organic matter. It was concluded that natural coated samples coagulated significantly slower than samples with organic coatings removed. Tipping and Higgins (9) studied the effects of humic substances and Ca^{2+} on the coagulation of hematite particles in the laboratory at pH 6.6. Interparticle contacts were produced by Brownian motion. Colloid stability increased with increasing concentration of humic substances, and decreased with increasing dosage of $CaCl_2$. The results suggest that α varied from about 10^{-2} in the presence of 10^{-3}

M calcium and 20 mg liter^{-1} humic substances to about 1 with 0.5 M calcium and 1 mg liter^{-1} humic substances.

Surface chemistry can significantly affect the extent of particle transport in lacustrine systems. Since chemical conditions in lakes vary widely, α and hence particle deposition can be expected to vary accordingly and substantially among lakes.

Coagulation kinetics may be rapid or slow, and the rate of coagulation depends upon α. Surface chemistry thus affects both the extent of removal of particles (Fig. 10.1) and the rate at which this removal is accomplished (Fig. 10.2). Chemical factors noted previously as affecting the extent of deposition through α will also affect the rate of response of lakes to inputs of materials. Hard-water lakes, for example, may respond more rapidly in removing inputs of particles and pollutants than soft-water systems because calcium and magnesium ions may destabilize colloids, increase coagulation rates, and hasten removal.

Coagulation is a second-order reaction with respect to the number of particles in a system (Eq. 1), so that its effects in a lake can be expected to depend on the inputs of solid particles to a lake and the production of particles within the lake waters. This is examined in Figure 10.6, in which the mass removal of particles at steady state (%) is plotted as a function of particle mass input to the epilimnion for four values of α. These results are calculated for an areal hydraulic loading of 10^{-6} m s^{-1}. In the absence of coagulation ($\alpha = 0$), particle input concentrations have no effects on particle removal, which occurs by sedimentation only. For large values of α (5 $\times$ 10^{-2}), mass removal is sensitive to particle inputs at low concentrations, and rather insensitive at input concentrations greater than about 20 mg liter^{-1} (6 mg liter^{-1} dry weight).

Two proposals are drawn from the results in Figure 10.6. First, coagulation provides a feedback for limiting productivity in lakes, similar to effects

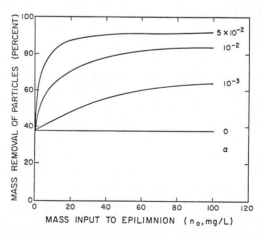

Figure 10.6. Effects of particle inputs to the epilimnion n_0 and particle stability α on particle retention in lakes. $q = 10^{-6}$ m s^{-1}. (See Table 10.1 for other assumed lake properties.)

suggested for light limitation and zooplankton feeding. As nutrient inputs to a lake increase, the production of biological particles is also increased. This, in turn, enhances coagulation and increases the removal of biomass and associated pollutants by sedimentation. Second, particle removal is more sensitive to coagulation effects in oligotrophic lakes than in eutrophic ones. In Lake Zürich, a moderately eutrophic lake with mass settling rates in excess of 2 g m^{-2} d^{-1} (5) (corresponding to input concentrations of 60 mg liter^{-1} wet weight), coagulation effects are sensitive to α but less sensitive to n_0. Deposition in a lake with a production rate of 0.5 g m^{-2} d^{-1} (corresponding to an input concentration of 15 mg liter^{-1} wet weight) would be quite sensitive to both n_0 and α.

Many substances are transported with particles in lakes. Since surface chemistry affects both the rate and the extent of particle deposition, it will also affect the rate and extent of removal of particle-reactive substances. The fate of PCBs in lakes, for example, may be expected to depend upon the concentrations of divalent metal ions through the effects of coagulation. It is also useful to quantify what is meant by "particle reactive." In this work the partition coefficient is used. Results indicate that the deposition of substances with K_p from 10 to 100 liter g^{-1} or so is quite sensitive to K_p. As a result, chemical factors affecting sorption (e.g., pH and ligand concentration) may exert important effects on trace metals and phosphorus, but are unlikely to affect the removal of synthetic organic substances with K_p values less than 10.

Chemistry can therefore affect the transport and fate of particle-reactive substances in two ways: (1) through solid–solution partitioning (K_p); and (2) through particle stability (α). For example, humic substances might increase the sorption of an organic contaminant on natural solids (increase in K_p) and also increase the stability of the sorbent particles (decrease α). Deposition of the pollutant in the lake sediments would then depend on the relative effects of increased sorption and decreased particle deposition.

6. CONCLUSIONS

Coagulation can be sufficiently rapid and extensive to affect water quality in lakes significantly. The process can increase particle deposition in bottom sediments and alter the mass, surface area, and number concentration of particles in the water column. Coagulation can also affect the transport and fate of many pollutants and nutrients. The results are sensitive to particle stability α and solute–solid partitioning K_p.

ACKNOWLEDGMENTS

The authors gratefully acknowledge the assistance of J. K. Edzwald and C.-Y. Yen, and the seminal work of Hahn and Stumm (10). This research has been supported by the National Science Foundation (U.S.A.) under Grant No. CEE81-21501.

REFERENCES

1. C. R. O'Melia and K. S. Bowman, "Origins and Effects of Coagulation in Lakes," *Schweizerische Z. Hydrol.* **46** (1984), in press.
2. M. Smoluchowski, "Versuch einer Mathematischen Theorie der Koagulations—Kinetic Kolloider Losungen," *Z. Physik. Chem.* **92,** 129 (1917).
3. S. K. Friedlander, *Smoke, Dust, and Haze,* Wiley–Interscience, New York, 1977.
4. C. R. O'Melia, "Aquasols: The Behavior of Small Particles in Aquatic Systems," *Environ. Sci. Technol.* **14,** 1052 (1980).
5. R. Giovanoli, R. Brütsch, D. Diem, G. Osman-Sigg, and L. Sigg, "The Composition of Settling Particles in Lake Zürich," *Schweizerische Z. Hydrol.* **42,** 89 (1980).
6. R. P. Schwarzenbach and J. Westall, "Transport of Nonpolar Organic Compounds from Surface Water to Groundwater, Laboratory Sorption Studies," *Environ. Sci. Technol.* **15,** 1360–1367 (1981).
7. L. Sigg, M. Sturm, L. Mart, H. W. Nürnberg, and W. Stumm, "Schwermetalle im Bodensee; Mechanismen der Konzentrationsregulierung," *Naturwissenschaften* **69,** 546 (1982).
8. R. G. Gibbs, "Effect of Natural Organic Coatings on the Coagulation of Particles," *Environ. Sci. Technol.* **17,** 237–240 (1983).
9. E. Tipping and D. C. Higgins, "The Effect of Adsorbed Humic Substances on the Colloid Stability of Haematite Particles," *Colloids and Surfaces* **5,** 85–92 (1982).
10. H. H. Hahn and W. Stumm, "The Role of Coagulation in Natural Waters," *Am. J. Sci.* **268,** 354–368 (1970).

11

THE COUPLING OF ELEMENTAL CYCLES BY ORGANISMS: EVIDENCE FROM WHOLE-LAKE CHEMICAL PERTURBATIONS

David W. Schindler

Department of Fisheries and Oceans, Freshwater Institute, 501 University Crescent, Winnipeg, Manitoba R3T 2N6, Canada

Abstract

Altering the inputs of phosphorus, nitrogen, or sulfur to lakes has had profound effects on the chemical cycles of other elements, largely by changes induced in organisms which couple the various cycles. Changes caused by phosphorus addition included enhancing photosynthesis which stimulates CO_2 invasion from the atmosphere; increased fixation of atmospheric nitrogen caused by changes to blue-green algae when nitrogen/phosphorus ratios were reduced; and replacement of oxic biogeochemical processes by anoxic pathways, including methanogenesis, sulfate reduction, production of ammonium, and release from sediments of ferrous iron.

Nitrogen additions caused changes in the acid-base balance of lakes, depending on whether a cationic or anionic form of the nutrient was used. Algal uptake, decomposition, nitrification, and denitrification all appeared to be key processes in the production or consumption of alkalinity, by altering the charge balance between cations and anions.

Addition of sulfuric acid caused an increased reduction of sulfate to sulfide in anoxic hypolimnions, and an increased sedimentation of iron sulfides. Because annual input of dissolved iron to lakes is small, it is possible that anthropogenic activities could introduce enough sulfate to lakes to totally

*deplete their reserves of iron, severely altering the end products of sulfate
reduction and disrupting cycles of phosphorus and trace elements.*

*The interfaces between air and water, epilimnion and hypolimnion, and
water and sediment are key sites of activity in the biological coupling of
geochemical cycles. These biological coupling mechanisms appear to be
unimpaired over wide ranges of acidities and nutrient concentrations, despite the numerous changes in biota which occur.*

1. INTRODUCTION

In natural lakes at steady state, the organisms which regulate various elemental cycles are presumed to operate with efficiency and regularity. Because our knowledge of biogeochemical processes under background conditions is still relatively poor, it is difficult to predict how geochemical cycles
and the organisms which couple them will respond to the stress of anthropogenic perturbations. In order to discover such responses, and to examine
how intricate and sensitive such relationships are, my colleagues and I have
deliberately and quantitatively perturbed the cycles of several elements in
natural lakes, either singly or two or three at a time. We have made detailed
studies of the "coupling" processes before, during, and after the perturbations in order to ascertain the rate and extent to which aquatic systems
can respond to chemical stresses, and the rate and degree of recovery which
can be expected when perturbation ceases.

Many of the whole-lake perturbations at the Experimental Lakes Area
(ELA) have involved changing the inputs of phosphorus, nitrogen, carbon,
and sulfur, either as neutral salts or as acids (Table 11.1). These are elements
which have been enhanced greatly in natural waters due to man's activity
in the past several decades. In this paper, I describe the biological mechanisms which "couple" or synchronize these different cycles.

2. CONSEQUENCES OF ALTERING THE INPUT OF NUTRIENTS

In lakes, most uptake of nutrients entering from outside the lake occurs in
near-surface waters, and cycling in this region is likely to be very rapid [e.g.,
see the phosphorus studies of Lean (1), and Schindler and Lean (2).]. After
being incorporated in organic tissues, some proportion of these nutrients is
lost by sinking (sedimentation). Typically, 1–5% of the mass of nutrients in
the water column of the lake is sedimented each day (3). If the lake is ther-
mally stratified, little of the nutrient sinking through the thermocline will be
available to organisms in the epilimnion until after overturn. Schindler et al.
(4) have shown that nutrients released below the thermocline are less than
10% as efficient at producing algal blooms as nutrients released into the
epilimnion. Thus, one of the most important effects of organisms on nutrients

Table 11.1   Whole-Lake Perturbations by Addition of Nutrients or Acids to Surface Waters of the Experimental Lakes Area, 1969–1983. All Additions Are Made Only During May through October.

Lake	Duration of Treatment	Treatment and Dose
		Nutrient Experiments
227	1969–1983	0.56 g phosphorus as H_3PO_4 and 6.27 g nitrogen as $NaNO_3$ m^{-2} yr^{-1} from 1970–1974; 0.54 g P as H_3PO_4, 2.24 g N as $NaNO_3$ m^{-2} yr^{-1} from 1975–1983.
226 NE	1973–1980	0.34 g phosphorus as H_3PO_4, 1.79 g nitrogen as $NaNO_3$, and 3.43 g carbon as sucrose m^{-2} yr^{-1}.
226 SW	1973–1980	1.92 g nitrogen as $NaNO_3$ and 3.67 g carbon as sucrose m^{-2} yr^{-1}.
304 (a)	1971–1972	0.38 g phosphorus as H_3PO_4, 4.92 g nitrogen as NH_4Cl, and 5.27 g carbon as sucrose m^{-2} yr^{-1}.
(b)	1973–1974	5.16 g nitrogen as NH_4Cl and 5.52 g carbon as sucrose m^{-2} yr^{-1}.
(c)	1975–1976	1.02 g phosphorus as H_3PO_4 and 14.45 g nitrogen as $NaNO_3$ m^{-2} yr^{-1}.
303	1974–1975	0.20 g phosphorus as H_3PO_4 and 3.01 g nitrogen as $NaNO_3$ m^{-2} yr^{-1}.
261	1973–1976	0.25 g phosphorus as H_3PO_4 m^{-2} yr^{-1}.
		Acid Experiments
223	1976–1983	Enough H_2SO_4 to lower the pH of the epilimnion by about 0.25 units per year. This has been 18.6–36.6 g m^{-2} yr^{-1}.
114	1979–1983	51.03 mg H_2SO_4 m^{-2}, once each month.
302 S	1982–1983	Enough H_2SO_4 to destroy most of the epilimnion alkalinity in 1982 followed by a decrease of about 0.25 pH units per year.
302 N	1982–1983	The same amount of acid (as equivalents) as applied to 302 S, except that HNO_3 is used.

is simply to assimilate them and sink, enhancing the transfer of nutrients from the epilimnion to the hypolimnion, from the euphotic to the profundal, and from the water column to the sediments. In the case of "rare" nutrients like phosphorus and nitrogen, this biological transfer causes nutrient residence times in the water column to be much shorter than for more conservative substances (Table 11.2). Similar processes affect the residence times of trace metals (5–8), except that adsorption to organic particles rather than active uptake is involved. As I shall discuss below, the organism-induced

Table 11.2.   Examples of Residence Times (Years) for Water and Several Elements in ELA Lakes

Lake	Status	Water	P	N	C	Si	Fe	SO_4	Na	Years
239	Oligotrophic reference	5.4	1.5	2.4	2.6	3.5	1.0	5.7	5.2	1970–1973
227	Eutrophic	6.1	0.6	2.5[a]	6.3[a]	1.6	13.9	4.4	4.3	1980–1981
223	Oligotrophic acidified to pH 5.1 with H_2SO_4	8.4	2.4	2.0	1.8	2.4	2.8	2.6	3.8	1979–1980

[a] Not including atmospheric invasion, which would make this time shorter.

shortages of phosphorus and nitrogen in surface waters may cause secondary reactions both in biological communities and in other chemical cycles. Likewise, enhanced inputs of nutrients and organic matter to deep waters cause dramatic changes in chemical cycles, the most important of which is a change from oxic to anoxic pathways in the lake's metabolism after oxygen has been depleted (Table 11.3).

3.   CHANGES INDUCED BY ADDITION OF PHOSPHORUS

Fertilization of lakes at ELA with phosphorus has been done both alone and with a number of combinations of nitrogen and carbon (Table 11.1). A num-

Table 11.3.   Reactions Taking Place in the Hypolimnion of Lakes Which Develop Anoxia During Stagnation. The Form of Reactions (a), (b), (d), and (e) is taken from Froelich et al. (16) with CH_2O Representing a Simplified Redfield Molecule[a] (17). Reactions (a) through (f) Are Microbially Mediated and (g) Is Dependent On (b) and (d) Illustrating the Importance of Microbial Activity in Linking the Cycles of Several Critical Elements. Phosphorus Is Ignored Here.

(a) $CH_2O + O_2 \rightarrow CO_2 + H_2O$ (early stagnation only)
(b) $CH_2O + 4Fe(OH)_3 + 8H^+ \rightarrow 4Fe^{2+} + CO_2 + 11H_2O$
(c) $CH_2O + MnO_2 + 2H^+ + CO_2 + H_2O$
(d) $2CH_2O + SO_4^{2-} \rightarrow S^{2-} + 2CO_2 + 2H_2O$
(e) $CH_2O \rightarrow CH_4 + CO_2$
(f) $CO_2 + NH_3{}^a + H_2O \rightarrow NH_4^+ + HCO_3^-$
(g) $Fe^{2+} + S^{2-} \rightarrow FeS$
The sum of (b), (d), and (g) is
(h) $3 CH_2O^a + SO_4^{2-} + 4Fe(OH)_3 + 8H^+ \rightarrow FeS + 3Fe^{2+} + 3CO_2 + 13H_2O$

[a] The full Redfield molecule is $\{(CH_2O)_{106}(NH_3)_{16}H_3PO_4\}$. The NH_3 in reaction (f) is derived from the NH_3 in the Redfield molecule.

ber of general conclusions may be drawn:

1. Primary production and phytoplankton standing crop were stimulated in all cases (9). The resulting consumption of dissolved inorganic carbon (DIC) by phytoplankton caused a manyfold decrease in the pCO_2 of surface waters, inducing CO_2 influx from the atmosphere (10,11). These atmosphere–water exchanges are affected by physical turbulence and chemical enhancement processes, as well as by the gradient in CO_2 across the air–water interface which is induced by photosynthetic uptake of CO_2 (12–14). The invading carbon is removed by photosynthesizing algae, sedimented, and released as CO_2 and HCO_3^- by catabolism in the hypolimnion. The result has been an increase in the mass of DIC in all fertilized lakes, regardless of whether or not carbon was added with fertilizer (15). The annual range in the ratio of $\Sigma C : \Sigma P$ in the waters of such lakes remains nearly unchanged, even after many years of fertilization (e.g., Fig. 11.1), regardless of the ratio supplied with fertilizer or dissolved inputs. Also, in lakes receiving phosphorus (as phosphoric acid) plus nitrate as the salt of a conservative cation, alkalinity tends to increase as a function of the loss of nitrate relative to

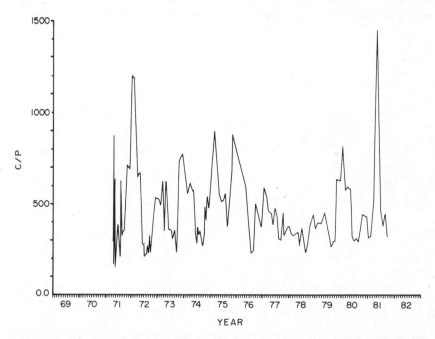

Figure 11.1. The ratio of total carbon to total phosphorus by weight in Lake 227, fertilized with phosphorus and nitrogen since 1969. Fertilizer, runoff, and precipitation together have an average $\Sigma C/\Sigma P$ ratio of about 40:1 by weight. Enhanced invasion of CO_2 from the atmosphere and recycling from sediments have maintained a much higher ratio, in the natural range observed for lakes in the area. Invasion of CO_2 is driven by photosynthesis, which is stimulated by phosphorus addition, and recycling from sediments by bacterial decomposition of the resulting organic matter.

accumulation of the cation (Fig. 11.2), as one would expect from charge-balance considerations.

2. When phosphorus inputs were high relative to nitrogen, fixation of atmospheric nitrogen also became an important source of that element to the lake (15,18,19). Only heterocystous blue-green algae are efficient fixers of N_2 in surface waters, so that a change in algal species to blue greens occurred once ionic forms of nitrogen were scarce [Fig. 11.3; see also Smith (20)]. Because of the high concentrations of N_2 in the atmosphere and relatively lower demands of algae for nitrogen, surface waters are never depleted enough in N_2 to hinder fixation, and turbulence and concentration gradients are not important controls on invasion rates as they are for CO_2. As for DIC, the photosynthetic fixation of N_2 is a light-dependent process (21,22). As found for carbon, natural compensation mechanisms were sufficient to maintain ratios of $N:P$ in lake water and algae within the range observed for background lakes and natural algal populations (Fig. 11.4) (15,20).

3. Higher oxygen consumption in the hypolimnion due to increased decomposition in fertilized lakes caused increased anoxic metabolism at the expense of oxic activity, involving a whole new suite of organisms, chemical

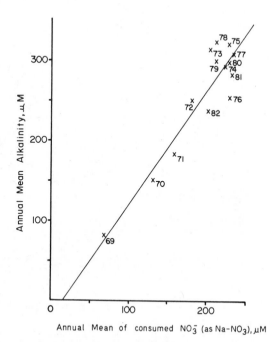

Figure 11.2. The relationship between consumed nitrate (as represented by Na–NO₃) and alkalinity in Lake 227. Na⁺ is affected only by water renewal whereas NO₃ by both biological uptake and water renewal. Points are mean annual averages for the years indicated by the numbers.

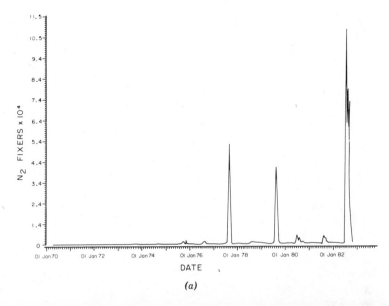

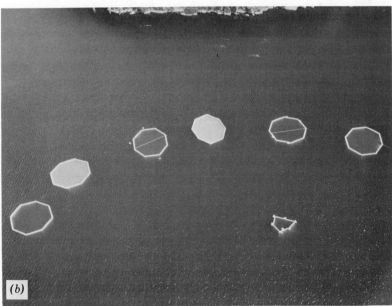

Figure 11.3. Some illustrations of how low N:P ratios in nutrient input stimulate nitrogen-fixing cyanophytes. (*a*) Biomass of nitrogen-fixing blue-green algae in the epilimnion of Lake 227 (mg m^{-3} wet weight), illustrating how this group was stimulated after the N/P ratio in input was reduced from 14:1 to 5:1 by weight in 1975, allowing atmospheric N$_2$ to become important in the aquatic nitrogen cycle. Dominant genera were *Anabaena* and *Aphanizomenon*. (*b*) A series of 10-m-diameter tubes, with N:P ratios of 15:1 and 2:1 by weight. The two green (pale) tubes had 2:1 ratios of N:P, and are dominated by *Anabaena*. The other tubes received high N:P ratios and are dominated by chlorophytes and chrysophytes (39).

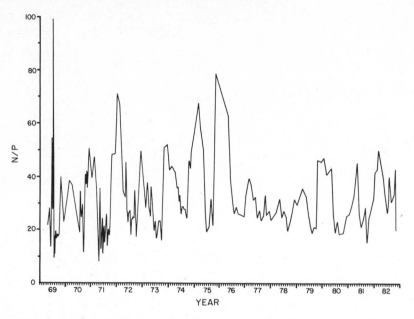

Figure 11.4.   The ratio of total nitrogen to total phosphorus by weight in Lake 227. Fertilizer, runoff, and precipitation together had an average ratio of 12.1 from 1969 to 1974 and 5.9 from 1975 on. As for carbon, fixation of atmospheric N_2 and enhanced recycling from sediments appear to maintain the natural ratio even after nitrate fertilizer additions are reduced.

products, and pool sizes. Higher concentrations of methane, ammonium, hydrogen sulfide, manganese, and ferrous iron occurred in the hypolimnion. These products are generated almost entirely at the mud–water interface in lakes with anoxic hypolimnions (23–25). As a result of fertilization, the quantity of methane in Lake 227 increased. During thermal stratification, this quantity was oxidized very slowly in the thermocline, with the rate of oxidation limited by upward diffusion of methane and downward diffusion of oxygen. In years when fall overturn is incomplete by freezeup, which is not uncommon in the extreme continental climate of the ELA, enough methane remains in the lake at freezeup to cause total anoxia by methane oxidation (26). Once the ion activity product for iron sulfides is exceeded, iron monosulfides precipitate, causing a transfer of both elements from deep water to sediments, where transformation to more stable sulfide minerals, including pyrite, occurs (24). This mechanism consumes hydrogen ion (see Eq. h, Table 11.3), which can be of considerable importance in poorly buffered aquatic ecosystems (27,28), as well as in salt marshes (29,30) and marine sediments (31). In freshwater lakes, the rate of alkalinity production can be enhanced by eutrophication, provided that nitrogen inputs are in anionic form (27,32).

4. CHANGES INDUCED BY ADDITION OF NITROGEN

In eutrophication experiments, we added neutral salts of ammonium or nitrate to several lakes, both with and without phosphorus (Table 11.1). In a recently begun acidification experiment, nitric acid has been added to Lake 302 N. In all cases, numerous effects on geochemical cycles of other elements were observed. Some have already been mentioned in the discussion of phosphorus above.

In Lake 304, there were interesting differences observed with the addition of ammonium chloride plus phosphoric acid versus ammonium chloride alone. In 1971–1972, application of phosphate plus ammonium chloride during the ice-free season caused the epilimnion of the lake to acidify rapidly, to values below pH 5. Phytoplankton increased severalfold during that period, so that conversion of ammonium to organic matter appeared to be responsible (Table 11.4) (27). The acidity produced was, however, largely reconsumed over winter as decomposition regenerated ammonium from organic matter.

In contrast, addition of ammonium chloride without H_3PO_4 in 1973–1974 had very little effect on the pH of the lake in summer. Phytoplankton standing crop remained low and ammonium accumulated in the epilimnion, presumably because the shortage of phosphorus prevented algae from consuming available nitrogen. Ammonium continued to increase through the winter months, due to decomposition in anoxic bottom waters. However, late in the winter of 1974–1975, when melt water began to enter the lake, almost all of the ammonium in the lake was oxidized to nitrate, apparently due to the sudden onset of nitrification. An increase in acidity to pH 4.6 resulted, just prior to ice-out in May.

From May of 1975 through October of 1976, the fertilization of Lake 304 was again changed, this time to H_3PO_4 plus $NaNO_3$ (Table 11.1). As expected

Table 11.4. Effects of Nitrogen Utilization by Organisms on the Acid-Base Balance of Lakes

---

Acidifying Reactions

1. Ammonium assimilation by phytoplankton:
   $$106CO_2 + 106H_2O + 16NH_4^+ \rightarrow (CH_2O)_{106}(NH_3)_{16} + 16H^+ + 106O_2$$
2. Nitrification:
   $$NH_4^+ + 2O_2 \rightarrow NO_3^- + H_2O + 2H^+$$

Alkalinizing Reactions

1. Nitrate assimilation by phytoplankton:
   $$106CO_2 + 138H_2O + 16NO_3^- \rightarrow (CH_2O)_{106}(NH_3)_{16} + 16OH^- + 138O_2$$
2. Denitrification:
   $$5C_6H_{12}O_6 + 24NO_3^- + 24H^+ \rightarrow 30CO_2 + 12N_2 \uparrow + 42H_2O$$

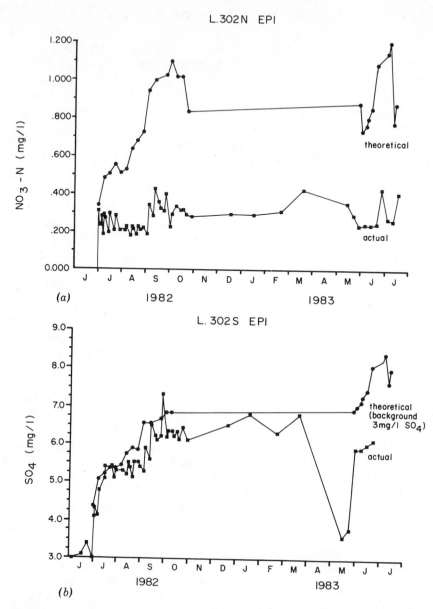

Figure 11.5. Changes in Lakes 302 N and 302 S after addition of nitric acid and sulfuric acid, respectively. (*a*) Theoretical nitrate concentration, as based on added nitrate addition, and the actual concentration of nitrate observed in the epilimnion of Lake 302 N, illustrating the rapid disappearance of the ion. Efficiency of acidification was inversely proportional to nitrate disappearance, that is, it was extremely low. The decrease in nitrate in May 1983 is due to the spring snow melt. (*b*) Theoretical sulfate concentration, based on added sulfate, and the actual concentration of sulfate observed in the epilimnion of Lake 302 S. The relatively small difference indicates high efficiency of acidification. As for nitrate, the spring decrease is due to the dilution by the spring snowmelt.

234

from the Lake 227 study, alkalinity increased rapidly. Three possible biologically mediated reactions could have been responsible: algal uptake, denitrification, or reduction to ammonium (Table 11.4). Large increases in phytoplankton suggested that algal uptake of nitrate was high. The efficiency
of alkalinity generation by nitrate addition was much higher than consumption of alkalinity by ammonium addition had been, presumably because ammonium was the primary end product of decomposition in both cases.

Addition of nitric acid to Lake 302 N has stimulated algal uptake slightly
(33) and denitrification greatly (34). These processes have proved far more
efficient at generating alkalinity than sulfate reduction, and have been sufficient to almost entirely consume the added hydrogen ion. This suggests
that nitric acid may be a less important acidifying agent than sulfuric acid
(Fig. 11.5) during periods of the year when aquatic organisms are active.
The importance of nitric acid during spring snowmelt and the direct effects
of the gaseous precursors of nitric acid on terrestrial biota are, of course,
still reason for considerable concern.

Andersen (35) reported a surprising effect of nitrate on the phosphorus
cycle of eutrophic Danish lakes. Concentrations of 0.5 to 1 mg liter^{-1} of
NO_3–N in anoxic hypolimnions prevented the release of phosphate from
sediments, whereas high phosphate releases were observed at lower nitrate
concentrations. The effect appeared to be due to the maintenance by nitrate
of high redox potentials in anoxic hypolimnions. Andersen suggested that
nitrate additions, as well as reductions in phosphorus input, might speed the
recovery of eutrophic lakes which exhibit a high return of phosphorus from
sediments. Sorensen's (36) work further elucidates this mechanism. He
showed that at high NO_3^- concentrations, Fe^{3+} reduction did not occur,
while at low concentrations of NO_3^-, Fe^{3+} was reduced by nitrate-reducing
organisms.

5.   CONSEQUENCES OF ALTERING THE INPUT OF SULFATE

Increased inputs of sulfate as H_2SO_4 to Lake 223 have caused increases in
microbial sulfate reduction with time (Fig. 11.6). The increases are proportional to sulfate concentration (37), and the rate of sulfate reduction in these
forest lakes appear to be limited by sulfate concentration instead of by organic substrate, which is abundant in sediments due to inputs of forest litter.
Sulfate reduction occurs not only in anoxic hypolimnions, but also in epilimnion sediments overlain by oxic waters which usually are depleted of
oxygen to within a few mm of the surface (Fig. 11.7) (38). As a result of
sulfate reduction, sediment pH is well buffered, even in lakes which have
been acidified by anthropogenic deposition for many years (34,39,42). As
suggested by Figure 11.7, most of the sulfate reduction in such sediments
occurs in the upper few centimeters of sediment, and the absence of sulfate
limits this activity at greater depth (41,42). In the same anoxic environments,

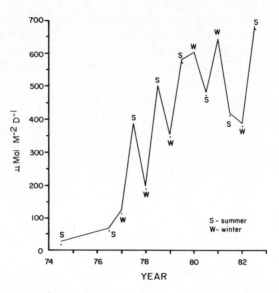

Figure 11.6. Sulfate reduction in the deep hypolimnion of Lake 223, as deduced from the rate of disappearance of sulfate from the water, plotted as a function of time in Lake 223. Acidification of the lake began in 1976. Summer data (S) are for the hypolimnion only, winter data (W) are for the whole lake.

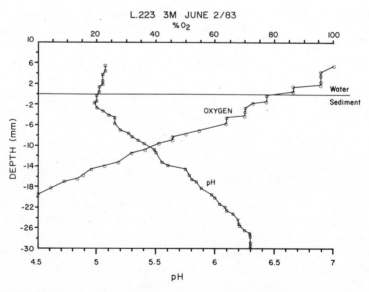

Figure 11.7. Profiles of oxygen and pH in epilimnetic sediments of Lake 223. pH and oxygen profiles were obtained at 1-mm intervals using microprobes. Sulfate concentration profiles were also obtained at 1-cm intervals using a pore water "peeper" showed considerable depletion (Kelly et al., in preparation), illustrating that high pH was maintained by sulfate reduction.

236

the rapid mobilization of ferrous iron allows the solubility product for iron and sulfide to be exceeded in both hypolimnion and pore waters, removing significant amounts of both iron and sulfur from the water column (Fig. 11.8) (24). As mentioned earlier, the formation of iron sulfides is a H^+ consuming reaction which adds substantially to the acid-neutralizing capacity of the lakes. The widespread occurrence of higher pH values in the sediments than in water columns of acidified lakes (40,42) suggests that such lakes contain internal mechanisms which are still capable of generating substantial alkalinity, which should assist in the recovery of acidified lakes once acid deposition is reduced. Once again, organisms are a key element in these reactions.

Many of the rudiments of interrelationships between the iron and sulfur cycle and the increase in alkalinity which normally accompanies increased Fe^{2+} in anoxic hypolimnions have been known for over half a century. Yoshimura (43) noted that increased iron and manganese in anoxic deep waters were always accompanied by bicarbonate. This is, of course, an important source of alkalinity, but one that is largely reversed when iron is reoxidized and reprecipitated at overturn (23,24). The precipitation of ferrous sulfides due to Fe^{2+} release from sediments and H_2S formed by sulfate reduction was recognized by several hydrobiologists, even 30–40 years ago. These early studies were reviewed by Kjensmo (44), who discussed the relationship between iron, sulfur, and alkalinity in four iron-meromictic lakes. He believed that much of the sulfur for the reduction was supplied by sedimenting organic matter rather than by sulfate, even though he correctly interprets (p. 284) an observation by Moe [cited in Kjensmo (44)] that, in the monimolimnion of Store Aaklungen, neither sulfate nor sulfide was present. Our more recent work (37) has shown that sulfate, rather than organic matter, supplies most of the sulfur to sulfate reducers.

The biological regulators linking the iron and sulfur cycle have been more precisely studied by later workers. Drabkova (46) reviews the presence and location of iron-reducing bacteria in relation to iron profiles and iron deposits in sediments in various types of lakes. Davison et al. (45) elucidated the reduction of sedimenting ferric hydroxide particles in Esthwaite Water, recording both FeS and $Fe(OH)_3$ particles present simultaneously in samples just below the oxic/anoxic boundary. They mention the presence of a layer of iron-reducing bacteria just below the maximum concentration of particulate iron in the water column.

While the oxides of iron and manganese may be reduced chemically (47), organisms capable of reducing both oxides have been identified in lakes (48). Few rate measurements have been made, so that it is impossible to generalize about the relative importance of biological versus physical mechanisms in the redox cycles. Sorensen (36) concluded that in marine systems, iron reduction was carried out by nitrate-reducing bacteria at low nitrate concentrations. However, no investigator appears to have quantitatively studied the role of bacteria in either reduction or oxidation of iron, which is rather

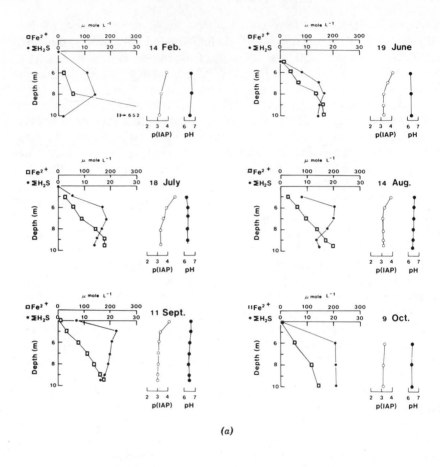

(a)

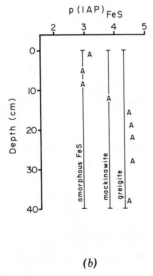

(b)

Figure 11.8. Iron and hydrogen sulfide in the Lake 227 hypolimnion (1979). (a) $\sum H_2S$ (*), Fe^{2+} ($\square$), and pH ($\bullet$) in the hypolimnion. $p(IAP)$ values indicate the degree of saturation for Fe^{2+} and HS, where the latter was calculated from $\sum H_2S$, pH and a first dissociation constant for H_2S of $10^{-7.25}$ at 5°C (9). (b) Ion activity products (A) calculated as above for sediment pore water at 10-m depth in Lake 227 (August 28, 1979). Values where formation of three forms of iron monosulfides would be expected are shown (9).

surprising in view of the fact that there are two rather convenient radioisotopes of the element which could be used. Davison and Seed (50) found that the rate of oxidation of ferrous iron in Esthwaite Water was indistinguishable from that in synthetic samples, implying that microbial activity was unimportant in the process.

Although high concentrations of iron (up to 40 mg liter^{-1}) are found in anoxic hypolimnions or monimolimnions at the ELA, annual inputs of dissolved iron to Precambrian Shield lakes are quite low. This situation is possible because oxidation and reprecipitation of iron in the upper, oxic regions of the lake prevent it from leaving via the outflow, so that its residence time in the lake is actually longer than that for water (51). Likewise, Davison (52) found that the annual accumulation of iron in the anoxic hypolimnion of Esthwaite Water could be almost entirely accounted for by fluxes from sediments, with inputs playing a minor role. At first, this seems strange, because the abundance of iron in the sediments of the ELA lakes is exceeded only by silica and aluminum, and iron is abundant in local bedrock and soil (53). Much of the input of total iron must occur in the form of large, rather insoluble particulates which fall directly to sediments, without replenishing reserves of the element in the waters of the lake. Calculations suggest that in areas subjected to high deposition of sulfate from the atmosphere, sulfate reduction followed by formation of iron sulfides may significantly reduce the dissolved-iron concentrations in hypolimnions and pore waters. In such circumstances, several effects might be seen:

1.   The above-mentioned buffering mechanism would become less important.

2.   Whereas H_2S seldom exceeds a few μg liter^{-1} when iron is abundant, it should become the major end product of sulfate reduction once iron is depleted. This seems to be the usual state in calcareous lakes, where natural sulfate concentrations are higher than in the Precambrian Shield. If degassed to the atmosphere without being oxidized, the loss of H_2S would replace sedimentation of iron sulfides as an alkalinity source, but if reoxidized, no net alkalinity would result. However, accumulations of hydrogen sulfide in the water column may cause significant taste, odor, and, in some circumstances, toxicity problems.

3.   As proposed by Hasler and Einsele (54), the removal of iron by increased sulfate reduction might "uncouple" the phosphate and iron cycles, having a fertilizing effect on lakes by allowing more phosphorus to remain in solution. Indeed, the high sulfates, low hypolimnetic irons, high H_2S, and high rate of release of phosphorus in calcareous lakes suggest that this mechanism is natural in such systems (55). Baccini et al. (Chapter 9 in this book) illustrate the key role of iron at the sediment–water interface in regulating releases of phosphate, silicate, and DOC. This interaction is also biologically affected, both by bioturbation and by the location of plates of *Beggiatoa* sp. The depletion of iron by formation of sulfides has not been observed in the

ELA lakes, and would not be expected under natural conditions because these systems contain very high concentrations of iron. In addition, phosphorus tends to be bound to organic molecules in the ELA lake sediments, rather than as iron or calcium phosphates or coprecipitates with ferric hydroxides (56). The possibility that increased sulfur inputs will disrupt the aquatic iron cycle cannot be overlooked in Precambrian Shield lakes which have sediments with low proportions of organic matter and/or iron.

4.   If sulfate reduction is insufficient to totally neutralize the additions of sulfuric acid, the balance of cations and trace metals between water and sediments is altered. In Lake 223, where the lake but not the terrestrial drainage has been acidified, acidification of surface waters to pH 5.1 from original values of 6.7–6.9 has caused significant increases in iron, manganese, zinc, and aluminum (57). For manganese, the concentration in Lake 223 at any pH value is similar to values in lakes or regions where acid deposition falls on all components of the watershed, suggesting that the effect of pH on sediment–water equilibria, rather than inputs of metals from polluted precipitation or acidified terrestrial watersheds, control the concentration of manganese in water. The same may be true for iron, though data from areas other than the ELA are too few for a meaningful comparison. Concentrations of aluminum, zinc, and copper in Lake 223 are less than 50% of those observed in other areas, indicating that the terrestrial watershed or polluted atmosphere must also be important sources (58).

As for iron, studies of the role of biota in the oxidation of manganese in ecosystems are rare. Chapnick et al. (59) illustrated that bacteria are important mediators of manganese oxidation in fresh water. They provide a review of previous studies for that element. In a recent review of microbially mediated chemical cycles in anoxic sediments, Reeburgh (61) also noted the paucity of quantitative biological studies of the aquatic iron and manganese cycles. The cycles of many trace metals and radionuclides are intimately dependent on the redox cycles of iron and manganese [e.g., Sholkovitz et al. (60)] and thus possibly on bacterial processes. It is crucial that quantitative studies of these key metals be undertaken in the near future.

6.   MULTIPLE LINKAGES BETWEEN CYCLES

There are obviously some important "multiple linkages," where two different perturbations cause the same effect on chemical processes. For example, both fertilization with nutrients and addition of sulfuric acid cause increased reduction of sulfate, the former by causing increased extent and duration of anoxic conditions in hypolimnions and surface sediments, and perhaps an increased supply of nonrefractory organic matter (32,38), and the latter by increasing the quantity of sulfate available for reduction. Another example is the acidity/alkalinity balance of lakes, which can be affected by several biological processes involving the cycles of nitrogen, sulfur, and iron as discussed above. Likewise, the simultaneous operation of two perturbations may have canceling effects, as one might expect from simultaneous acidification and eutrophication, or from simultaneous inputs of ammonium and nitrate.

Davison et al. (62) could not explain the presence of oxides of nitrogen and the absence of ferrous iron in the anoxic hypolimnion of eutrophic Rostherne Mere by strictly geochemical mechanisms. This peculiar set of circumstances was due to the predominance of mucilaginous colonial *Microcystis* and resting cysts of *Ceratium*, both of which resisted decomposition, in sedimenting organic material. Phosphorus in this material appeared to be released in near-surface sediments, but most of the organic carbon was decomposed at a depth of several centimeters in the sediment. The transfer of electrons to iron reducers was limited by the slow diffusion of ferrous iron. At this depth, the only electron acceptor abundant enough to assimilate the high production of electrons from decomposition was carbon, and methanogenesis constituted the major decomposition mechanism. Bubbling of methane from sediments was thought to constitute a rapid removal mechanism. In this way the species of algae which compose the plankton can cause decomposition and related geochemical processes to be transferred from the sediment–water interface to deeper sediments.

In summary, whole-lake experiments have shown that the cycles of phosphorus, nitrogen, sulfur, iron, and alkalinity are closely integrated, with algae and other microorganisms carrying out key functions in the interrelationships. Alteration of any one of the cycles has detectable consequences for the others, in some cases causing major metabolic and geochemical pathways to be changed dramatically. Such alterations must be predictable and quantifiable if we are to construct meaningful plans for managing the chemical inputs to lakes.

7. KEY SITES FOR THE ORGANISMS WHICH COUPLE GEOCHEMICAL CYCLES

There is considerable similarity in the sites of coupling for all of the above-mentioned cycles. With the exception of the general effect of all photosynthetic organisms on enhancing invasion of atmospheric carbon when stimulated by inputs of phosphorus, or on altering pH and alkalinity by uptake of inorganic nitrogen, critical microbiological reactions occur primarily at the interfaces between air and water, water and sediments, or between epilimnion and hypolimnion. Unfortunately, these critical zones are narrow and often contain steep chemical gradients, rendering them so difficult to study quantitatively that limnologists have traditionally focused on major, well-mixed water masses. Some examples follow.

Nitrogen fixation occurs primarily at the air–water interface, and most of the key species of algae have buoyant pseudovacuoles to help them maintain their position at or near the lake surface.

Hesslein (25) and Cook (23,24) have shown that almost all of the ammonium, methane, DIC, and iron in the hypolimnions of the ELA lakes is derived from decomposition at the mud–water interface. Comparison of rates of release from the mud–water interface with diffusive fluxes calculated by traditional flux-gradient methods shows that the surface layer of a few millimeters, containing newly sedimented material, is usually a more important site of decomposition than the entire sediment column below. Levine (19) illustrated this clearly by isolating sediments in large enclosures from the water column by covering the sediment surface with polyethylene. Releases of carbon and nitrogen from newly sedimented material falling on the upper surface of the polyethylene were over 50% of those observed in enclosures where free exchange between sediments and overlying water was permitted.

Thermoclines and chemoclines can also control both chemical and biologically mediated reactions. In Lake 227, there is a steep gradient in the concentrations of oxygen and several other substances (e.g., methane, ammonium, iron, and DIC) in the thermocline. As outlined earlier in this paper, microbiological processes in this zone are often limited by the slow vertical diffusion of chemicals in the high-density gradient at the thermocline. Quay et al. (63) have shown that in the thermocline of Lake 227, vertical diffusion is 5×10^{-5} cm^2 s^{-1} or less, approaching rates observed in sediments of the same lakes (64). Rudd and Hamilton (26,65) have shown how complicated the biogeochemical interactions in the region of the thermocline can be. They found that methane oxidation in the thermocline of Lake 227 was confined to a zone less than 1 m thick. Above that zone, the process was prevented by the availability of methane and/or ionic nitrogen, while below the zone oxygen was insufficient (Fig. 11.9).

In the epilimnion of Lake 227, methane released from sediments escaped to the atmosphere due to the inactivity of methane oxidizers. Ionic nitrogen was limiting, despite fertilizer applications of 150 μg liter^{-1} week^{-1} of NO_3^-–N. Methane oxidizers could not compete successfully with phytoplankton for ionic nitrogen. While methane oxidizers are also capable of fixing N_2, this was inhibited by high oxygen concentrations (>1 mg liter^{-1}) (66). Under winter ice, when ionic nitrogen was abundant, when phytoplankton were limited by light, and when methane was present throughout the water column, much higher overall rates of methane oxidation were observed in the lake, particularly in the upper strata (Fig. 11.10). Despite the many constraints on their activity, methane oxidizers are well adapted to life in the thermocline of Lake 227, and without their activity 50% of the carbon metabolized in the hypolimnion of Lake 227 would be lost through the thermocline, and probably to the atmosphere.

As mentioned above, Campbell and Torgersen (51) have shown that the

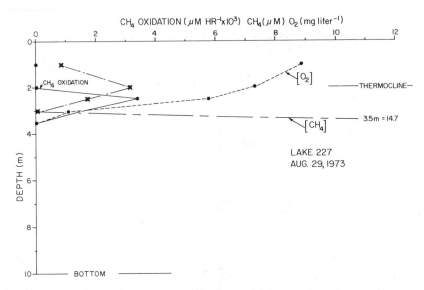

Figure 11.9. A typical profile of summer methane oxidation showing very low concentrations of methane within the zone of methane-oxidizing activity at the thermocline and higher epilimnetic methane concentrations where methane-oxidizing activity is absent (50).

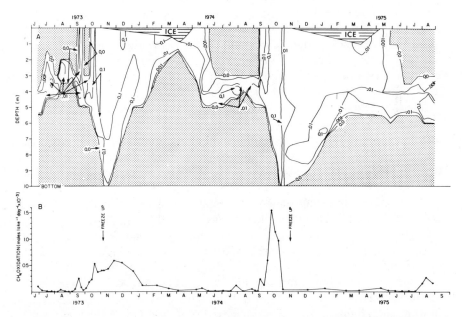

Figure 11.10. (A) Methane oxidation rates (μmol liter^{-1} hr^{-1}) in Lake 227 during a 26-month period. Stippled areas are zones of zero methane oxidation. (B) Whole-lake rates of methane oxidation (moles lake^{-1} d^{-1} $\times$ 10^{-3}) during same time period (50).

243

oxidation in the thermocline of upward-diffusing ferrous iron causes it to be retained in the anoxic zone of stratified lakes, enhancing the residence time of iron. Davison et al. (46) found that similar processes occurred in Esthwaite Water as illustrated by a middepth maximum in particulate ferric iron. Unquestionably, the trapping of iron in the thermocline enhances the availability of iron for the iron sulfide deposition mentioned earlier, and provides a replenished proton source for iron-reducing bacteria deeper in the lake. Reoxidation of methane and ammonium diffusing from anoxic sediments and denitrification of downward-diffusing nitrogen are other important biological reactions occurring in the thermocline (26,67).

8. HOW ROBUST ARE THE BIOLOGICAL COUPLING MECHANISMS?

The biological processes which couple aquatic geochemical cycles of nutrients do not appear to be easily disrupted. When phosphorus inputs to the ELA lakes were increased 10 times, mechanisms to increase nitrogen and carbon by similar amounts came into play (15). Organic carbon added to Lake 304 in excess of needs to balance the phosphorus inputs was degraded to CO_2 and lost to the atmosphere (49). Much of the nitrate added to Lake 302 N as nitric acid is denitrified. It is obvious that mechanisms exist for maintaining these nutrient balances over wide ranges, probably explaining the remarkable coordination of nutrient cycles in the sea (68,69).

Addition of nitrate or sulfate to either freshwater or marine system results in a shift from methanogenic decomposition in anoxic regions to the more energy-efficient processes of sulfate or nitrate reduction (32,70,71). This shift might also be expected to accompany the acidification of lakes, although to date, no study has confirmed such a shift.

There is, however, one exception to the chemical adaptability of ecosystems—no mechanisms appear to exist in freshwater which allow adjustments in the phosphorus cycle to maintain appropriate nutrient ratios. All of the compensation to maintain balanced $C:N:P$ ratios, whether in an upward or downward direction, must be done by the cycles of nitrogen and carbon.

Even more surprising than the adaptability of microbially mediated reactions to increased inputs of nutrients is the fact that aquatic microbiological linkages do not appear to be easily toxified. Wood and Wang (Chapter 4 in this book) provide some remarkable examples of the adaptability of microorganisms to extreme stresses. Production, respiration, and normal rates of accumulation of ammonium and methane occur in Lake 223 despite a 50-fold increase in hydrogen ion and substantial increases in the concentrations of iron, zinc, aluminum, and manganese (38,57,72). Other work at the ELA has shown that nitrification, denitrification, methanogenesis, and sulfate reduction occur at pH values as low as 4.0, both in wetlands and in laboratory experiments with lake sediments (73).

While biogeochemical cycles appear to continue with minor disruption in polluted ecosystems, biological communities normally undergo successions through many species of organisms to keep the cycles operating. The chrysophycean-dominated communities which are prominent photosynthesizers and nutrient "absorbers" in natural lakes have given way to chlorophycean and cyanophycean assemblages in Lake 227, and to chlorophycean–peridinean communities in Lake 223. Many species of zooplankton have also been displaced by others which occupy the same ecological niche (74,75). While in Lake 223 there has been dramatic elimination of the species from the food chain which have not been replaced (57,76) these are usually at the third or fourth level in the trophic pyramid, and their effects on geochemical cycling are correspondingly small.

Faced with such observations, it is paradoxical that so much effort in "ecotoxicological" studies has been devoted to metabolic parameters, that is, production, decomposition, and nutrient cycling, for it appears that ecological stresses must be very extreme before these activities are significantly disrupted. Far more sensitive and far more important in the case of organisms which are valued by humans or by higher organisms in the food chain, are the changes in species assemblages which occur. It appears that early indications of stress at the ecosystem level would be detected far more readily by changes in the species composition of the biotic community than by changes in community metabolism or chemical cycling, for the latter are among the most robust of ecosystem-level parameters.

9. CONCLUSIONS

Algae and bacteria play key roles in coupling the cycles of phosphorus, nitrogen, sulfur, carbon, and related elements in lakes. When phosphorus is added to lakes, causing them to become more eutrophic, several biological mechanisms allow corresponding increases in nitrogen and carbon, both by enhancing inputs of gaseous CO_2 and N_2 from the atmosphere to the lake and by enhancing recycling relative to phosphorus. The overall effect is a tendency for the system to maintain a "Redfield-type" balance between these elements. When ionic nitrogen is added, alkalinity balances as well as trophic state is affected. Uptake of ionic nitrogen, nitrification, denitrification, and decomposition are all microbial processes which are effective proton donors or consumers (Table 11.4).

Sulfate addition stimulates sulfate reduction in anoxic hypolimnions and sediments, causing precipitation of iron sulfides. This also affects the alkalinity of lakes. In addition, the increased sedimentation of iron as relatively insoluble sulfides has implications for the cycles of phosphorus and trace metals. Some biological coupling mechanisms depend on a delicate balance between several cycles. For example, methane oxidation requires oxygen, methane, and inorganic nitrogen for efficient operation. During summer stratification such conditions are only found in the thermocline, restricting the rates of oxidation. Under winter ice, abundance of all three substances allows high rates of methane oxidation.

Dramatic changes in species composition of algae and possibly other microorganisms allow the coupling mechanisms to adapt to wide ranges of chemical perturbations, so that nutrient cycling, production, and decomposition are not readily disrupted.

ACKNOWLEDGMENTS

G. J. Brunskill, Paul Campbell, R. B. Cook, R. H. Hesslein, C. A. Kelly, and J. W. M. Rudd provided many helpful comments on the manuscript. David Findlay, Gary Linsey, and Michael Turner provided figures from recent ELA studies. R. B. Cook, W. Davison, C. A. Kelly, and J. W. M. Rudd provided useful manuscripts and illustrations from their unpublished work.

REFERENCES

1. D. R. S. Lean, "Phosphorus Compartments in Lake Waters: A Community Excretion Mechanism," *Science* **179,** 678–680 (1973).

2. D. W. Schindler, and D. R. S. Lean, "Biological and Chemical Mechanisms in Eutrophication of Freshwater Lakes," *Ann. New York Acad. Sci.* **350,** 129–135 (1974).

3. D. W. Schindler, H. Kling, R. V. Schmidt, J. Prokopowich, V. E. Frost, R. A. Reid, and M. Capel, "Eutrophication of Lake 227, Experimental Lakes Area, Northwestern Ontario by Addition of Phosphate and Nitrate, Part 2: the Second, Third and Fourth Years of Enrichment, 1970, 1971 and 1972," *J. Fish. Res. Board Can.* **30**(10), 1415–1440 (1973).

4. D. W. Schindler, T. Ruszczynski, and E. J. Fee, "Hypolimnion Injection of Nutrient Effluents as a Method for Reducing Eutrophication," *Can. J. Fish. Aquat. Sci.* **37**(3), 320–327 (1980).

5. S. Emerson, and R. H. Hesslein, "Distribution and Uptake of Artificially-Introduced Radium-226 in a Small Lake," *J. Fish. Res. Board Can.* **30,** 1485–1490 (1973).

6. B. J. Finlay, N. B. Hetherington, and W. Davison, "Active Biological Participation in Lacustrine Barium Chemistry," *Geochim. Cosmochim. Acta* **47,** 1325–1329 (1983).

7. B. Havlik, "Radium in Aquatic Food Chains: Radium Uptake by Freshwater Algae" *Radiat. Res.* **46,** 490–505 (1971).

8. R. H. Hesslein, W. S. Broecker, and D. W. Schindler, "Fates of Metal Ra-

diotracers Added to a Whole Lake. Sediment–Water Interactions," *Can. J. Fish. Aquat. Sci.* **37**(3), 378–386 (1980).

9. D. W. Schindler, E. J. Fee, and T. Ruszczynski, "Phosphorus Input and Its Consequences for Phytoplankton Standing Crop and Production in the Experimental Lakes Area and in Similar Lakes," *Can. J. Fish. Res. Board Can.* **35**(2), 190–196 (1978).

10. D. W. Schindler, G. J. Brunskill, S. Emerson, W. S. Broecker, and T.-H. Peng, "Atmospheric Carbon Dioxide: Its Role in Maintaining Phytoplankton Standing Crops," *Science* **177**, 1192–1194 (1972).

11. D. W. Schindler, and E. J. Fee, "Diurnal Variation of Dissolved Inorganic Carbon and Its Use in Estimating Primary Production and CO_2 Invasion of Lake 227," *J. Fish. Res. Board Can.* **30**(10), 1501–1510 (1973).

12. S. Emerson, W. S. Broecker, and D. W. Schindler, "Gas-Exchange Rates in a Small Lake as Determined by the Radon Method," *J. Fish. Res. Board Can.* **30**(10), 1475–1484 (1973).

13. S. Emerson, "Gas Exchange Rates in Small Canadian Shield Lakes," *Limnol. Oceanogr.* **20**, 754–761 (1975).

14. S. Emerson, "Chemically Enhanced CO_2 Gas Exchange in a Eutrophic Lake: A General Model," *Limnol. Oceanogr.* **30**, 743–753 (1975).

15. D. W. Schindler, "Evolution of Phosphorus Limitation in Lakes," *Science* **195**, 260–262 (1977).

16. P. N. Froelich, G. P. Klinkhamner, M. L. Bender, N. A. Luedtke, G. R. Heath, D. Cullen, P. Dauphin, D. Hammond, B. Hartman, and V. Maynard, "Early Oxidation of Organic Matter in Pelagic Sediments of the Eastern Equatorial Atlantic: Suboxic Diagenesis," *Geochim. Cosmochim. Acta* **43**, 1075–1090 (1979).

17. A. C. Redfield, "The Biological Control of Chemical Factors in the Environment," *Am. Sci.* **46**, 206–226 (1958).

18. R. J. Flett, D. W. Schindler, R. D. Hamilton, and N. E. R. Campbell, "Nitrogen Fixation in Canadian Precambrian Shield Lakes," *Can. J. Fish. Aquat. Sci.* **37**(3), 494–505 (1980).

19. S. N. Levine, "Natural Mechanisms that Ameliorate Nitrogen Shortages in Lakes," Ph.D. thesis, University of Manitoba, Winnipeg, 1983, 535 pp.

20. V. H. Smith, "Low Nitrogen to Phosphorus Ratios Favor Dominance by Blue-Green Algae in Lake Phytoplankton," *Science* **221**, 669–671 (1983).

21. R. J. Flett, "Nitrogen Fixation in Canadian Precambrian Shield Lakes," Ph.D. thesis, University of Manitoba, Winnipeg, Manitoba, 197 pp.

22. S. N. Levine, Cornell University, Ithaca, N.Y., personal communication.

23. R. B. Cook, "The Biogeochemistry of Sulfur in Two Small Lakes," Ph.D. thesis, Columbia University, New York, 1981, 248 pp.

24. R. B. Cook, "Distribution of Ferrous Iron and Sulfide in an Anoxic Hypolimnion," *Can. J. Fish. Aquat. Sci.* **41**, 286–293 (1984).

25. R. H. Hesslein, "Whole Lake Model for the Distribution of Sediment-Derived Chemical Species," *Can. J. Fish. Aquat. Sci.* **37**(3), 552–558 (1980).

26. J. W. M. Rudd, and R. D. Hamilton, "Methane Cycling in a Eutrophic Shield

Lake and Its Effects on Whole Lake Metabolism," *Limnol. Oceanogr.* **23**(2), 337–348 (1978).

27. D. W. Schindler, and M. A. Turner, "Acidification and Alkalinization of Lakes by Input of Neutral Chemical Species: An Experimental, Whole Ecosystem Study of the Role of Nitrogen Compounds," (1984) (in press).

28. D. W. Schindler, R. Wagemann, R. Cook, T. Ruszczynski, and J. Prokopowich, "Experimental Acidification of Lake 223, Experimental Lakes Area: I. Background Data and the First Three Years of Acidification," *Can. J. Fish. Aquat. Sci.* **37**(3), 342–354 (1980).

29. R. W. Howarth, and A. Giblin, "Sulfate Reduction in the Salt Marshes at Sapelo Island, Georgia," *Limnol. Oceanogr.* **28**, 70–82 (1983).

30. R. W. Howarth, and J. M. Teal, "Sulfate Reduction in a New England Salt Marsh," *Limnol. Oceanogr.* **24**, 999–1013 (1979).

31. B. B. Jorgensen, "The Sulfur Cycle of a Coastal Marine Sediment (Limfjorden, Denmark)," *Limnol. Oceanogr.* **22**, 814–832 (1977).

32. C. A. Kelly, J. W. M. Rudd, R. B. Cook, and D. W. Schindler, "The Potential Importance of Bacterial Processes in Regulating Rate of Lake Acidification," *Limnol. Oceanogr.* **27**, 868–882 (1982).

33. D. W. Schindler, unpublished data.

34. J. W. M. Rudd, personal communication.

35. J. M. Andersen, "Effect of Nitrate Concentration in Lake Water on Phosphate Release from the Sediment," *Water Res.* **16**, 1119–1126 (1982).

36. J. Sorensen, "Reduction of Ferric Iron in Anaerobic Marine Sediment and Interaction with Reduction of Nitrate and Sulfate," *Appl. Environ. Microbiol.* **43**, 319–324 (1982).

37. R. B. Cook, and D. W. Schindler, "The Biogeochemistry of Sulfur in an Experimentally Acidified Lake." In R. O. Hallberg (Ed.), *Proc. 5th Int. Symp. Environ. Geochem.* **35**, 115–127 (1983).

38. C. A. Kelly, and J. W. M. Rudd, "Freshwater Sulfate Reduction and Its Relationship to Lake Acidification," *Biogeochemistry* **1**, (1984) (in press).

39. C. A. Kelly, personal communication.

40. K. Tolonen, and T. Jaakkola, "History of Lake Acidification and Air Pollution Studied in Sediments in South Finland," *Ann. Bot. Fennici* **20**, 57–58 (1983).

41. J. W. M. Rudd, C. A. Kelly, R. J. Flett, and D. W. Schindler, "Will naturally occurring bacteria facilitate the recovery of acidified lakes? Science (under review).

42. J. M. W. Rudd, C. A. Kelly, unpublished data.

43. S. Yoshimura, "Contributions to the Knowledge of the Stratification of Iron and Manganese in the Lake Water of Japan," *Japan. J. Geol. Geogr.* **9**, 61–69 (1931).

44. J. Kjensmo, "The Development and Some Main Features of Iron Meromictic Soft Water Lakes," *Arch. Hydrobiol. Suppl.* **32**, 137–312 (1967).

45. V. G. Drabkova, "Iron Bacteria in Some Lakes of the Karelian Isthmus," *Hydrobiol. J.* (English Translation of *Gidrobiol. Zh.*) **7**(1), 21–27 (1970).

46. W. Davison, S. I. Heaney, J. F. Talling, and E. Rigg, "Seasonal Transfor-

mations and Movements of Iron in a Productive English Lake with Deep-Water Anoxia," *Schweiz. Z. Hydrol.* **42,** 196–224 (1981).

47. W. Stumm, and J. J. Morgan, *Aquatic Chemistry* (2nd ed.), Wiley–Interscience, New York, 1981, 780 pp.

48. H. L. Erlich, *Geomicrobiology.* Marcel Dekker, New York, 393 pp.

49. B. M. Thompson, "Heterotrophic Utilization of Sucrose in an Artificially Enriched Lake," M.Sc. thesis, University of Manitoba, Winnipeg, Manitoba, 1972, 92 pp.

50. W. Davison, and G. Seed, "The Kinetics of Oxidation of Ferrous Iron in Synthetic and Natural Waters," *Geochim. Cosmochim. Acta* **47,** 67–79 (1983).

51. P. Campbell, and T. Torgersen, "The Ferrous Wheel: Residence Times and the Maintenance of Iron Meromixis in Lake 120, Experimental Lakes Area," *Can. J. Fish. Aquat. Sci.* **37,** 1303–1313 (1980).

52. W. Davison, "Supply of Iron and Manganese to an Anoxic Lake Basin," *Nature* **290,** 241–243 (1981).

53. G. J. Brunskill, D. Povoledo, B. W. Graham, and M. P. Stainton, "Chemistry of Surface Sediments of Sixteen Lakes in the Experimental Lakes Area, Northwestern Ontario," *J. Fish. Res. Board Can.* **28,** 277–294 (1971).

54. A. D. Hasler, and W. G. Einsele, "Fertilization for Increasing Productivity of Natural Inland Waters," Transactions of the 13th North American Wildlife Conferences. March 8–10, 1948, pp. 527–554.

55. R. E. Stauffer, "Relationships between Phosphorus Loading and Trophic State in Calcareous Lakes of Southeast Wisconsin," *Limnol. Oceanogr.* **30,** 1985 (in press).

56. T. A. Jackson, and D. W. Schindler, "The Biogeochemistry of Phosphorus in an Experimental Lake Environment: Evidence for the Formation of Humic-Metal-Phosphate Complexes," *Verh. Internat. Verein. Limnol.* **19,** 211–221 (1975).

57. D. W. Schindler, and M. A. Turner, "Biological, Chemical and Physical Responses of Lakes to Experimental Acidification," *Water, Air Soil Pollut.* **18,** 259–271 (1982).

58. D. W. Schindler, R. Wagemann, B. DeMarch, and M. Turner, "The Dynamics of Fe, Mn, Al, Cu and Ln in an Acidified Canadian Shield Lake" (in preparation).

59. S. D. Chapnick, W. S. Moore, and K. H. Nealson, "Microbially Mediated Manganese Oxidation in a Freshwater Lake," *Limnol. Oceanogr.* **27,** 1004–1014 (1982).

60. E. R. Sholkovitz, A. E. Carey, and J. K. Cochran, "Aquatic Chemistry of Plutonium in Seasonally Anoxic Lake Waters," *Nature* **300,** 159–161 (1982).

61. W. S. Reeburgh, "Rates of Biogeochemical Processes in Anoxic Sediments," *Ann. Rev. Earth Planet. Sci.* **11,** 269–298 (1983).

62. W. Davison, C. S. Reynolds, and B. J. Finlay, "Algal Control of Lake Geochemistry," *Nature* (under review).

63. P. D. Quay, W. S. Broecker, R. H. Hesslein, and D. W. Schindler, "Vertical Diffusion Rates Determined by Tritium Tracer Experiments in the Thermocline and Hypolimnion of Two Lakes," *Limnol. Oceanogr.* **25,** 201–218 (1979).

64. R. H. Hesslein, "In Situ Measurements of Pore Water Diffusion Coefficients Using Tritiated Water," *Can. J. Fish. Aquat. Sci.* **37**(3), 551–558 (1980).

65. J. W. M. Rudd, and R. D. Hamilton, "Methane Oxidation in a Eutrophic Canadian Shield Lake," *Verh. Internat. Verein. Limnol.* **19**, 2669–2673 (1975).

66. J. W. M. Rudd, A. Furutani, R. J. Flett, and R. D. Hamilton, "Factors Controlling Methane Oxidation in Shield Lakes: The Role of Nitrogen Fixation and Oxygen Concentration," *Limnol. Oceanogr.* **21**(3), 357–364 (1976).

67. Y. K. Chan, and N. E. R. Campbell, "Denitrification in Lake 227 during Summer Stratification," *Can. J. Fish. Aquat. Sci.* **37**(3), 506–512 (1980).

68. W. Broecker and T.-H. Peng, *Tracers in the Sea*, Lamont-Doherty Geological Observatory, Columbia University, Palisades, New York, 1982, 690 pp.

69. A. C. Redfield, B. H. Ketchum, and F. A. Richards, "The Influence of Organisms on the Composition of Sea Water." In M. N. Hill (Ed.), *The Sea*, Vol. 2, Wiley–Interscience, New York, pp. 26–77.

70. D. O. Mountfort, R. A. Asher, E. L. Mays, and J. M. Tiedje, "Carbon and Electron Flow in Mud and Sandflat Intertidal Sediments at Delaware Inlet, Nelson, New Zealand," *Appl. Environ. Microbiol.* **39**, 686–694 (1980).

71. M. R. Winfrey, and J. G. Zeikus, "Effect of Sulfate on Carbon and Electron Flow during Microbial Methanogenesis in Freshwater Sediments," *Appl. Environ. Microbiol.* **33**, 275–281 (1977).

72. C. A. Kelly, J. W. M. Rudd, A. Furutani, and D. W. Schindler, "Effects of Acid Precipitation on Rates of Organic Matter Decomposition in Lake Sediments," *Limnol. Oceanogr.* **29**, 687–694 (1984).

73. S. E. Bayley, J. W. M. Rudd, C. A. Kelly, personal communication.

74. D. F. Malley, D. L. Findlay, and P. S. S. Chang, "Ecological Effects of Acid Precipitation on Zooplankton." In F. M. D'Itri (Ed.), *Acid Precipitation: Effects on Ecological Systems*, Ann Arbor Science Pub., Ann Arbor, Michigan, 1982, pp. 297–327.

75. D. F. Malley, P. S. S. Chang, and D. W. Schindler, "Decline of Zooplankton Populations Following Eutrophication of Lake 227, Experimental Lakes Area, Ontario: 1969–1974," *Can. J. Fish. Aquat. Sci.* (under review).

76. K. H. Mills, "Fish Population Responses during the Experimental Acidification of a Small Lake." In G. Hendrey (Ed.), Proceedings of the Symposium of the American Chemical Society, Las Vegas, Nevada, March 1982, pp. 117–131.

12

THE GEOBIOLOGICAL CYCLE OF TRACE ELEMENTS IN AQUATIC SYSTEMS: REDFIELD REVISITED

François M. M. Morel and Robert J. M. Hudson

R. M. Parsons Laboratory, 48-425, Massachusetts Institute of Technology, Cambridge, MA 02139, USA

Abstract

Because of their isolation from allochthonous sources, the open oceans provide a convenient model system in which to examine the role of the biota in element cycles. An attempt is made here to obtain a general picture of trace-element geobiology by weaving together field data on the soluble and particulate metal concentrations in the oceanic water column, laboratory information on metal uptake by algae, and teleological reasoning. It is argued that many major and trace elements, essential and toxic, may be simultaneously controlling biological production in the oceans, and that, in turn, the cycle of these elements is controlled by the biota. A condition at the thresholds of limitation by essential elements and of toxicity by their chemical analogs is seen as the normal result of evolution in a stable environment. Conversely, the composition of the water reflects the affinity of the organisms for the various elements, such affinity controlling the partitioning into the particulate phase and the transport to the sediments.

1. INTRODUCTION

Some 50 years ago, following on earlier observations of Harvey (1), Alfred Redfield (2) noted that the variations in carbonate, nitrate, phosphate, and oxygen concentrations in seawater were highly correlated. From linear regressions he found that these variations occurred in the approximate atomic ratios $140:20:1:(-140)$, very similar to the rather constant composition of marine plankton, $140:18:1:(\cong -140)$, considering the oxygen deficit compared to CO_2), and concluded that photosynthetic and oxidative decomposition processes were responsible for the bulk of these chemical variations in the oceans. This simple model, with slightly more precise numbers for what has come to be known as the Redfield ratio, is now a central paradigm of chemical oceanography and limnology.

In the same publication, Redfield went on to wonder about the curious simultaneous disappearance of the algal nutrients, nitrate and phosphate, from surface seawater and wondered how such extraordinary coincidence could have come about. He then proposed that a colimitation of algae by the essential nutrients nitrogen and phosphorus might result from either of two processes. It could be a consequence of selective changes in the biota, the physiology of the plankton ultimately reflecting the chemistry of its surroundings. Alternatively, the composition of seawater may have resulted from long-term biological feedback processes, the nitrogen supply having been made to match that of phosphorus through nitrogen fixation and denitrification. This molding of the chemical composition of living material and seawater by each other is also part of what is now called the Redfield model.

Over the past few years, with the help of so-called "clean" sampling and measurement techniques, geochemists have begun to obtain systematic measurements of trace elements in the sea, and discovered that many trace elements display vertical concentration distributions in the ocean highly correlated with those of the major nutrients. On this basis, it is clear that the oceanographic cycles of trace elements are also under biological control— at least in part. Is this biological control similar to that of the major nutrients? To what extent is Redfield's "model" applicable to them? Such are the questions that we wish to address here in the perspective that similar processes may be applicable to lakes, as they are for major nutrients.

The principal theses of this paper are: (1) that aquatic microorganisms, chiefly algae, control the concentrations of many trace elements in surface waters; (2) that the principal carriers from surface to deep waters and to sediments are organic particles derived from phytoplankton (presumably some from direct sinking, perhaps aggregates; mostly as fragments in fecal pellets and marine snow); and (3) that indeed the Redfield concept may be extended to several trace elements, many of which are *not* essential micronutrients. Obviously, the focus here is on bioreactive elements and on the processes that incorporate them into the particulate phase, not on remobilization mechanisms.

The open oceans provide a particularly convenient model system to test these ideas because it is a geochemically stable system receiving relatively low and constant allochthonous inputs from coastal waters and the atmosphere. As a result the biological processes can clearly dominate over all others; the system can be driven to the limit of its biological logic, offering a fruitful field of application for ecological and evolutionary reasoning, that is, for teleology. Hence, even if the oceans may not serve as direct models for lakes, they provide a model for the directions in which the biological processes are pulling the chemistry of the system, for the laws that govern the geobiological cycle that is interacting with the exogenic geochemical cycle driving all elements. More pragmatically of course, the chemistry and cycle of trace elements in aquatic systems is now most easily studied in the oceans since there is at this point a trustworthy compendium of oceanic concentration data that is still lacking for freshwater systems.

Our demonstration of the role of the biota in the geochemical cycle of trace elements starts with a discussion of the characteristics of the particles that scavenge metals in the deep ocean. Focusing on a few biologically reactive metals (Fe, Mn, Zn, Cu, Ni, Cd), it is found that neither inorganic particles nor particles coated with humic material have sufficient metal binding affinities to explain observed residence times in the oceans. By default this imputes particles of rather direct biological origin. In surface waters there is indeed direct experimental evidence for the association of trace elements with biotic particles.

Laboratory data on trace metal uptake by phytoplankton confirms the great affinity of algae for trace metals; these data also demonstrate that much of the metal uptake is by binding to high-affinity surface ligands. Such binding is effectively "passive" and phytoplankton surface ligands—on live or dead organisms or incorporated in fecal pellets—may well be the source of the high metal affinity of deep-sea scavenging particles. By demonstrating competitive interactions among metals for surface and transport sites, the laboratory data also provide an explanation for the high-affinity uptake, and hence transport, of nonessential—namely, toxic—trace elements as well as essential ones.

Finally, it is observed that field data on particulate metal concentrations correspond roughly to the limits of phytoplankton deficiency for essential metals (Fe, Mn, Zn) and of toxicity for toxic ones (Cu, Ni, Cd). The same appears to be true of sedimenting particles in Lake Constance. Based on teleological reasoning, a general hypothesis of simultaneous limitation of algal growth by several major and minor nutrients and by toxicants which act as chemical analogs of the nutrients is proposed.

2. THE NATURE OF THE SCAVENGING PHASE

Perhaps the most general way to approach the issue of geochemical cycles is to assemble data for (nearly) all the elements in the Periodic Table and to correlate the solid/liquid partitioning of elements of the earth crust both with oceanic residence times and with parameters characteristic of the chemical properties of the elements. Many such correlations have been developed

(3–5) and those presented in Figure 12.1 are quite typical. In this figure an electronegativity function which measures "the power of an atom in a molecule to attract electrons to itself" (6) is used as a single parameter representing the chemical properties of the elements. Other parameters, such as hydrolysis or acidity constants, can be used for the same purpose (4,5). The general lesson from such correlations is very clear if not surprising. The chemistry of the elements is important to their geochemistry and oceanography; the more electronegative (i.e., reactive, hence insoluble) elements tend to be enriched in the solid phase and to have a relatively short residence time in the oceans. Although more can be learned by examining in detail the distributions of subgroups of related elements in such figures, one must realize that many of the parameters that enter in these correlations are very poorly known and that the trends exhibited on the log–log graphs of Figure 12.1 only demonstrate consistent behavior within some four orders of magnitude. Even in geochemistry a factor of 10,000 cannot easily be glossed over and it is difficult to propose mechanisms on the basis of such broad-brushed pictures of geochemical processes.

To advance our discussion let us then focus on six trace elements for which chemical, oceanographic, and biological information is particularly available: Zn, Cu, Cd, Ni, Mn, and Fe (7–11). Sketches of typical concentration profiles for these elements in the North Pacific are presented in Figure 12.2 along with those of phosphorus and silicon. Nanomolar and lower oceanic concentrations characterize all these trace metals as do sharp surface minima [with the notable exception of manganese which exhibits a surface maximum hypothesized to correspond to a surface water reduction and dissolution process perhaps linked to photochemical reactions (12)]. One of the most striking features of the data of Figure 12.2 is the similarity in the profiles of the algal nutrients phosphorus and silicon and those of the trace metals; various empirical correlations have been developed which predict accurately the variations of concentrations with depth (7). Since the nutrient concentration profiles are clearly controlled by biological processes—that is, incorporation into organic particles in the euphotic zone, settling, and partial remineralization and dissolution at depth—these correlations are *prima facie* evidence that the vertical profiles of the trace elements are also controlled by the biota. The question then is to understand how this biological control of vertical distributions of trace elements is achieved, the three most obvious alternatives being:

1. Generation of the vertical particle flux in which trace metals bound to inorganic particles are carried down with the organic matter (through coagulation, incorporation in fecal pellets, etc.).

2. Control of the partition between soluble and solid phases by covering inorganic particles with an organic (humic) coating that binds trace metals.

3. Direct uptake of trace elements in living matter and resulting incorporation into the organic particle flux.

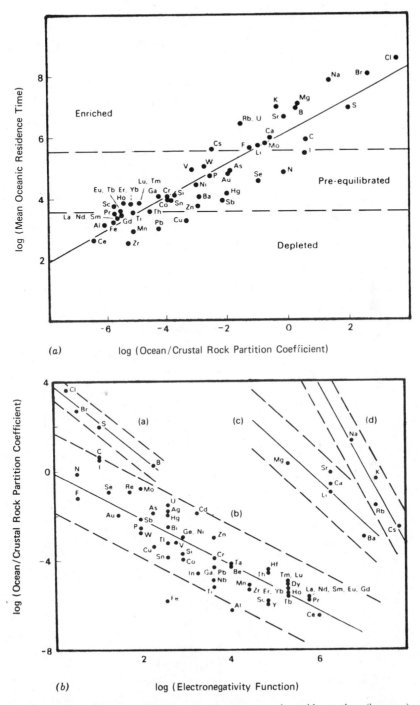

Figure 12.1. General correlations between (a) mean oceanic residence time (in years) and ocean/crustal rock partition coefficient, and (b) ocean/crustal rock partition coefficient and the electronegativity function (3).

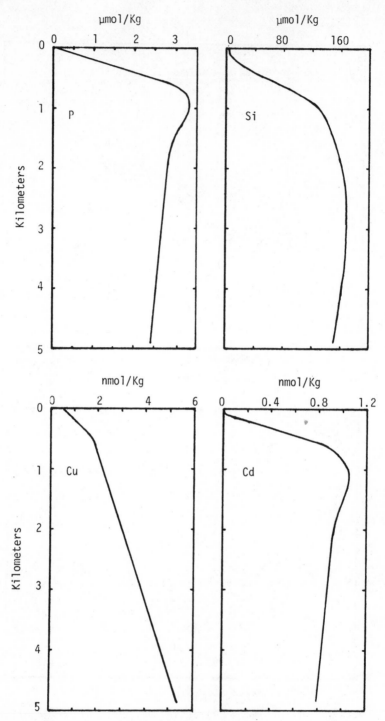

Figure 12.2. Stylized North Pacific profiles of some nutrients and trace metals (7–9). [Reprinted by permission from *Nature* **299**, 611–612 (1982). Copyright © 1982, Macmillan Journals Limited.]

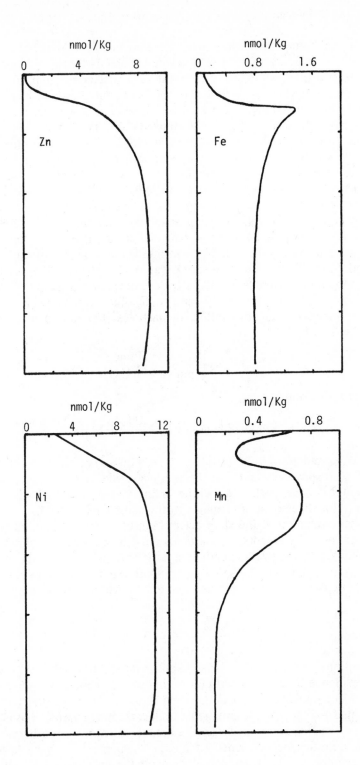

Before examining these alternatives in some detail, we must note that the shape of the vertical profiles provide evidence chiefly for the biological control of the transport out of the surface waters. The deep-ocean profiles could be explained by any type of bottom sinks and sources which may or may not be related to sedimentation and decomposition of organic matter. We must also note that this discussion is focused on the vertical distribution and transport of elements which may, in principle, be different from those processes ultimately responsible for the retention and accumulation of the elements in the sediments. Biological processes may simply distribute the concentrations of elements in the water column, while other geochemical processes control the total concentrations in the oceans.

Over the last few years, following largely on an original idea of Schindler (13), it has been popular in the geochemical literature to examine the process of vertical transport in the ocean by correlating the residence time of various elements to their tendency to adsorb on aquatic particles. [This is in effect a variation on the graph of Figure 12.1(a).] The basic idea is to consider the simplest possible one-box, steady-state model of the ocean in which removal to the sediments occur via settling of particles. For any element M, the residence time is then given by

$$T_m = \frac{M_{total} \times \text{volume}}{\text{Input rate of } M}$$

and the removal rate by

$$-\frac{dM}{dt} = \text{Removal rate of } M \doteq \frac{M_{particulate} \times \text{volume}}{T_p}$$

where M_{total} and $M_{particulate}$ are the average total and particulate concentrations of M, respectively and T_p is the average residence time of the particles in the water column. In this model the sedimentation of particles is responsible for the net removal of elements to the sediments. Obviously, this is a better approximation of the deep ocean than of the surface layers and the concentrations of elements and particles must be chosen accordingly. The applicability of the model depends to a large extent on our ability to determine average residence times that correspond to net removal processes. At steady state the input and removal rates of M must be equal, leading to the equality:

$$\frac{M_{particulate}}{M_{total}} = \frac{T_p}{T_m}$$

The partition of an element in the solid phase must thus be equal to the ratio of the residence time of the particles to that of the element. Information on the residence time of an element can then be used to estimate the necessary particulate fraction of that element and to decide among various possibilities for the scavenging solid, given some estimates of oceanic solid concentrations and affinities for the element of interest. As pointed out in the Appendix to this chapter, where a detailed account of the calculations is provided, there are three principal methods for determining residence times of elements in the oceans, relying on three different types of information: inventory of river inputs, deep-ocean profiles (i.e., the shape of the bottom of the curves in Figure 12.2), and sediment trap data. Given the uncertainties in each of these data sets and the questionable assumptions that must be made to extract residence times from some of them, the general agreement (to better than an order of magnitude) of the values presented in Table 12.1 is heartening even if somewhat surprising.

Following the work of Brewer and Hao (14), Balistrieri et al. (15), and Hunter (16), three different possible scavenging phases have been compared to the calculated residence time ratios (Table 12.5 and the Appendix): (1) an amorphous ferric hydrous oxide phase, one of the more adsorptive inorganic solids known (Table 12.5, column 2); (2) an organically coated surface with the coordinative properties of soluble marine humic substances (column 3); and (3) an organic surface with the coordinative properties of the model compound salicylic acid (column 4). According to the calculations none of these three surfaces has the required affinity for the metals of interest. This result is at variance with that of previous authors; the major causes for the discrepancies are the inclusion of both acid-base properties *and* major cation (Ca and Mg) binding to the surface ligands. One should note that the graphical presentation of such calculations can sometimes make them look better than they really are; logarithmic graphs can make a few orders of magnitude look like a small discrepancy. The choice of variables indirectly related to the partition coefficients can also obscure the quantitative differences between experimental and calculated values.

One major result on which all these calculations are in agreement is that only organic material can possibly exhibit the high affinity required for the efficient metal scavenging that is observed. In this regard it must be em-

Table 12.1.   Residence Times of Elements in Ocean Waters[a]

| | Log Residence Time (years) | | |
Metal	One-Box Model[b]	Dissolved Profiles[c]	Sediment-Trap Data[d]
Fe	1.6	—	1.1
Mn	1.7	1.7	1.4
Zn	2.9	—	—
Cu	2.8	3.2	2.8
Ni	4.6	4.2	—
Cd	4.8	5.3	—

[a] See Appendix for explanations.
[b] $T_m = 3.76 \times 10^{-4} M_{seawater}/M_{riverwater}$ (61). $M_{seawater}$ values from Quinby-Hunt and Turekian (49). $M_{riverwater}$ values from Li (48).
[c] Mn from Weiss (62); Cu from Boyle et al. (10); and Ni and Cd from Brewer and Hao (14).
[d] From Brewer et al. (52) recomputed using deep-ocean concentrations of Fe from Gordon et al. (9), of Mn from Bender et al. (11), and of Cu from Bender and Gagner (60).

phasized that humic substances may decrease as well as increase the adsorption of trace metals on inorganic solids. At the pH of seawater, a large fraction of the humates remain in solution and the soluble ligands can outcompete those on the surface for binding metals. This is the result commonly observed for copper whose adsorption to oxides at neutral pH and above is decreased in the presence of humic acids, through the adsorption of other metals, such as lead, is increased in these conditions (17,18).

In any case, it appears that the scavenging of trace metals in the deep ocean is effected by particles with much higher affinity for the metals than is observed for inorganic phases whether or not coated by an organic film. By a process of elimination we are thus led to conclude that the trace metal carrier phase in the deep ocean is an organic particle of more or less direct biotic origin—one that has not gone through the degradation–dissolution and condensation–adsorption pathway characteristic of humic substances and has very high affinity for the metals. For example, there is evidence that fecal pellets indeed have a high affinity for metals (19,20), probably reflecting in part the metal affinity of the algal material that the animals graze upon (as is discussed later). In general we should expect to find the particulate concentrations of biologically reactive trace metals in the ocean to be associated principally with organisms or their dead remains.

Direct measurements of particulate metal concentrations support that conclusion. For example, in a detailed study of surface particles in the central Pacific Ocean, Collier (21) and Collier and Edmond (22) found that Cu, Cd, Ni, and Mn were chiefly associated with biogenic organic phases, only a minor fraction ($< 10\%$) being accounted for in either aluminosilicates or biogenic inorganic material such as opal or calcium carbonate. The organic phase accounted also for about 70% of the total particulate zinc, the rest being apparently included in the opaline phase. Even for iron, which is abundant in lithogenic phases and precipitates readily as a hydrous oxide, a sizable fraction of the particulate concentration (10–70%) was estimated to be associated with the organic phase. A significant feature of the results of Collier is that a relatively large fraction of the metals associated with the particulate organic material was found to be readily leachable by seawater or distilled water rinses. Similar data have been obtained by Martin and Knauer (23), who analyzed particles from sediment traps deployed at various depths off the coast of Mexico; except for iron, which was largely present as a lithogenic aluminosilicate phase, the metals were principally associated with organic matter and large fractions of most metals were found to be labile.

3. METAL UPTAKE BY ALGAE

By a process of elimination of alternatives and from direct observations, we are led to conclude that the biota controls the vertical transport of trace elements in the ocean by direct uptake into—or onto—the organic biomass, at least for the six metals under scrutiny. The geobiological cycle of these trace elements then appears essentially similar to that of the major nutrients: they are depleted in the euphotic zone by incorporation into living particulate matter, carried out of the surface waters by settling, and partly regenerated at depth through decomposition of the organic particles. To understand these processes quantitatively, we must then examine the physiology of trace element uptake by aquatic microorganisms, particularly phytoplankton.

A few trace elements, such as iron, manganese, and zinc (in addition to Co and Mo) are essential for algal growth. These micronutrients may be limiting primary production in the open ocean, and the analogy with the major nutrients, nitrogen and phosphorus (and silica), is straightforward. Algae have no known requirements for most other trace elements however, and if anything, metals such as cadmium or nickel may act as toxicants to aquatic plants. The same is likely true of copper whose concentration in the oceans is many times over the minute nutritional requirements of algae. Why phytoplankton should then take up these trace elements and deplete surface waters is a paradox that we wish to examine here.

Though much remains to be learned about the mechanisms of trace element uptake and nutrition in phytoplankton, in the past few years laboratory and field studies have provided us with enough information to paint a reasonable if speculative picture (24–27). As illustrated in Figure 12.3, the uptake of trace metals by phytoplankton appears to be in most cases a two-stage process involving the binding of a relatively large pool of metal to the cell surface and transport through the cell membrane, presumably via metal porter molecules. The exact nature of either the surface binding groups (X) or the porter molecules (P) is not known. However, by analogy with other porter molecules (28), the latter is presumably a large protein (MW > 50,000)

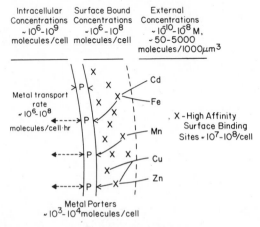

Figure 12.3. Hypothetical diagram of trace-metal transport in a typical marine phytoplankter. (Volume ≅ 500–1000 μm^3; surface area ≅ 300–800 μm^2.)

present at a concentration of 10^{-19}–10^{-21} moles ($\cong 10^4$–10^3 molecules) per cell to transport metals at rates up to 10^{-16} moles cell^{-1} hr^{-1}. (We are considering here "typical" algal cells of 500–1000 μm^3 volume.) The concentration of metals bound to the cell surface is of the order of 10^{-18}–10^{-16} moles per cell under the conditions where the metal-ion activities in the medium are maintained in the normal physiological range (corresponding to nanomolar and lower total concentrations in the absence of artificial chelating agents). Both the high concentration of these binding sites and their high affinity for metals point at substituted polysaccharides or polypeptides in the cell wall as the most likely ligand groups.

Consistent with a pseudoequilibrium model of the cell surface, both the rate of metal uptake [proportional to the metal–porter complex concentration (MP)] and the binding to the cell surface (MX) are determined by the free metal ion concentrations (or more precisely activities) in the surrounding medium. This is shown in Figure 12.4, where the rate of iron uptake (in the dark to avoid complications posed by photoreduction) and the concentration of iron released from the diatom *T. weissflogii* in the presence of 0.1 M ascorbic acid [taken to represent the total iron bound to the cell surface (FeX)] are determined by the free ferric ion concentration (Fe^{3+}), regardless of the nature of the strong complexing agents introduced in the medium. Though the exact role of the general binding to the cell surface in the metal-uptake process is unclear at this point (preliminary experiments show that

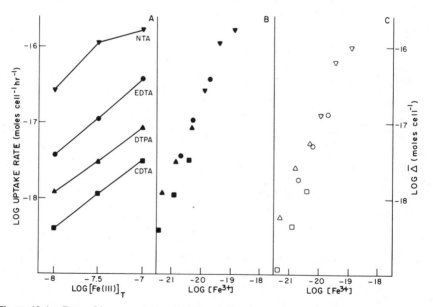

Figure 12.4. Rate of iron uptake by the diatom *T. weissflogii* in the dark in the presence of various chelators at 10^{-5} M total concentration plotted against (A) the log total iron concentration and (B) the log-free ferric ion concentration. (C) The surface bound fraction (released by suspension in 0.1 M ascorbic acid for 30 min) plotted against the free ferric ion concentration; pH 8.1; 14,000 cells mliter^{-1} (26).

this binding is not a *necessary* step for uptake), the two processes appear mechanistically linked: Seemingly a significant fraction of the surface ligands must be bound by a metal for that metal to be taken up and the extent of surface binding appears to be a good measure of the uptake rate for various necessary and toxic metals. In relation to the issue of metal transport in aquatic systems, the two important points here are that: (1) high-affinity ligands are found on algal cells; "passive" binding by these ligands [some of which are probably present more or less intact in settling particles, such as fecal pellets or marine snow (19)] may be important in the transport of metals both in surface and in deep waters; and (2) the extent of binding by a metal to surface ligands on living algae is related to the intracellular uptake of that metal.

In the prokaryotic world, a different mode of transport is often used for iron: ligands of high affinity and specificity for Fe(III)—siderophores—are released into the medium and specific receptor molecules transport the Fe siderophore complex back into the cells (29,30). This mechanism is known to be common among cyanophytes (blue-green algae) and there is evidence that it occurs in certain eukaryotic marine algae as well (31,32), particularly in dinoflagellates. However, the few existing quantitative studies on Fe transport in marine phytoplankton do not show siderophore transport to be important. Iron transport by release and uptake of a complexing agent in the open ocean seems an unlikely process in any case and we do not consider it further here. Note, however, that siderophores may be important in lakes where dilution of extracellular metabolites is not as severe a problem and that the presence of siderophores has been demonstrated in some cyanophyte blooms (33).

An interesting characteristic of marine phytoplankton in more or less natural conditions (hence excluding organisms adapted to extremes of medium composition) appears to be the relatively low specificity of the metal transport system(s). In many situations it has been observed that trace metals interfere with the transport and assimilation of other essential metals and that this is an important mode of toxicity at low external concentrations of trace metals (27,34). (This is not to say that there are not specialized porter molecules for each of the essential metals, Fe, Mn, and Zn.) For example, Figure 12.5 shows the inhibition of iron transport by cadmium in *T. weissflogii*. Direct biochemical evidence for Fe deficiency in cadmium-stressed cells has been obtained for this organism in the presence of "sufficient" iron concentrations (34). It has also been observed that a high iron concentration inhibits cadmium uptake in this diatom, as it should in the case of a competitive uptake process. Thus toxic metals seem to be transported into the cell by the same transport system as essential micronutrients, though firm experimental evidence to prove this is still lacking.

Toxic effects are observed in algae when the cellular concentration of toxic metals reaches some critical level approaching the minimum cellular concentration of essential trace metals. Table 12.2, obtained from a perusal of recent data, contains a set of reasonably consistent values for the thresh-

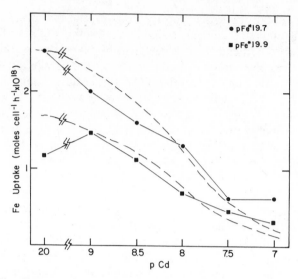

Figure 12.5. The effect of increasing free cadmium ion concentrations on iron transport in the diatom *T. weissflogii*. Iron uptake at two free ferric ion concentrations, $pFe = 19.7$ and $pFe = 19.9$, is plotted vs increasing free cadmium ion concentrations. Dashed curve represents a theoretical curve for a competitive uptake model (34).

Table 12.2. Approximate Thresholds of Metal Deficiency and Metal Toxicity in Some Phytoplankton (μmole g^{-1}) (Dry Weight)

	T. weissflogii[a]	*T. pseudonana*	*Chlamydomonas*	Reference
Deficiency				
Fe	3	—	—	(26)
Mn	0.4	2	2	Harrison (unpublished data) and (27)
Zn	0.2	—	—	Estimated from (42)
P	300	300	300	
Toxicity				
Cu	—	5–50	—	(41,43)
Ni	2	—	—	Watras (unpublished data)
Cd	0.2	—	—	(34)
Zn	—	30	15	Sunda (unpublished data) and (44)

[a] *T. weissflogii* dry weight $\cong 10^{-10}$ g cell^{-1}, hence 1 μmole g^{-1} $\cong 10^{-16}$ moles cell^{-1}. Note also that the low metal concentrations for this organism were all obtained in one laboratory using one growth medium. Other data for other organisms are from other laboratories.

olds of metal limitation and toxicity in three phytoplankton species. These data are not all obtained under similar conditions and one should be particularly aware that thresholds of deficiency or toxicity for a metal are dependent on the availability of other metals according to the aforementioned antagonistic processes. Nonetheless, the tabulated concentrations in Table 12.2 are rather constant—a remarkable fact since there is no *a priori* reason to expect that these concentrations should not vary by several orders of magnitude. Generalizing from these few data, we might say that the cellular concentrations of the essential trace metals, iron, manganese, and zinc, must be of the order of 1 μmoles g^{-1} ($\cong$ 10^{-16} moles cell^{-1} for our typical phytoplankter) for optimal growth and that the cellular concentration of toxic metals must be lower than that—say a factor of 10 lower. From the data available to date, it is observed that at such levels a major fraction (50—90%) of the toxic metals and a lesser fraction (20–30%) of the essential metals are bound to the surface of the cell. Considering a cellular phosphorus concentration of 300 μmoles g^{-1}, these critical metal concentrations correspond to an atomic ratio M/P $\cong$ 1/100. A rough stoichiometric formula for phytoplankton under optimum conditions may thus be written as $\{C_{106} H_{263} O_{110}$ $N_{16} P_1 (Fe, Mn, Zn)_{0.01} (Cu, Ni, Cd, . . .)_{0.001}\}$.

4. METAL CONCENTRATIONS IN OCEANIC PARTICLES (A TELEOLOGICAL RATIONALIZATION)

These laboratory results can be usefully compared to trace element concentrations observed in field samples of surface particulate matter. (In the open oceans surface particles are principally live and dead plankton.) The recent data obtained by Collier (Table 12.3, column 1) in the central Pacific Ocean are reasonably similar to those of Table 12.2. Older data by Knauer

Table 12.3.   Metals in Plankton and Sedimenting Material (μmoles g^{-1})

| References | Pacific Ocean | | | Lake Constance |
	(21)	(35)	(23)	Sigg[b]
Fe	3.6	11.7[a]	12–120	210
Mn	0.14	0.16	0.64–10	31
Zn	1.3	0.850	3.3–8.8	2.0
Cu	0.19	0.13	0.22–0.87	0.5
Ni	0.29	0.068	0.53–1.1	
Cd	0.27	0.021	0.075–0.29	(0.003)
Pb				0.2
P	360	—	—	50

[a] 1 high number; others approximately 4.

[b] See Chapter 13 in this book

and Martin in Monterey Bay (35) demonstrate similar results with the exception of somewhat lower nickel and cadmium concentrations, perhaps reflecting an experimental problem due to the rapid loss of metals from particles during collection and sample preparation. (In fact this lability of particulate metals, which has been realized only recently, is similar to the lability of the surface-bound metal fraction (MX) in laboratory experiments.) Martin and Knauer's more recent data from sediment traps off the coast of Mexico (23) show results generally consistent with those of Collier. With the exception of high particulate-iron concentrations which reflect terrigenous inputs, the concentrations are in the range of a few μmoles per gram for the essential metals and a few tenths of μmoles per gram for the others. (Recall also that the metals are chiefly associated with the organic particulate phase; see previous material. As a whole, the data on metal concentrations in oceanic particles (Table 12.3) thus resemble strikingly the laboratory threshold concentrations for phytoplankton limitation and toxicity (Table 12.2) and support the hypothesis that these trace elements may be controlling simultaneously algal growth in the open ocean.

There are only a few data sets available for trace metal compositions of particles in lakes. In the last column of Table 12.3, the average of Sigg's results (see Chapter 13) for sedimenting particles in Lake Constance are given for comparison. The relatively low phosphorus content is due to a high fraction of inorganic matter in the samples and the very high iron and manganese concentrations undoubtedly reflect sedimentation of allochthonous inorganic phases. However, the particulate concentration of zinc is similar to those of essential trace metals in the marine samples and those of copper and lead are in the range of particulate toxic-metal concentrations in the sea. Such data suggest that despite a much more important expected role of allochthonous inputs in lake systems than in open oceans, the same geobiological processes are at play and the biota effectively controls a major fraction of the vertical trace metal fluxes in fresh and marine waters.

The hypothesis of colimitation by several micro- and macronutrients and by toxicants, which is suggested by these data, may be arrived at by a classical teleobiological argument. That several nutrients should be colimiting is a normal consequence of either the biochemical adaptation of the biota to its environment or of the ability of the biota to control its environment. These are precisely the two explanations that Redfield offered for the apparent colimitation of marine algae by nitrogen and phosphorus. The affinity of a transport system for an essential element must at least match the availability of the element in the external milieu; but it has no reason (i.e., no evolutionary pressure) to be greater than necessary. In a system where an essential element is made scarce by biological uptake, the resulting concentration of the element precisely reflects the affinity of the biological transport system. (The proposition of colimitation by several elements is thus almost a tautology. Note, however, that it inherently does not address the issue of what ultimately limits the growth rate of autotrophic organisms.).

The same reasoning may be applied to toxic metals if one considers the issue of specificity for metal uptake. Higher specificity than is necessary for essential metals over toxic ones must be obtained at some cost and simply cannot offer "greater fitness" or "selective advantage"; however, sufficient specificity clearly does. In other words one may consider that, in the open ocean, algae *should* be growing at the limit of deficiency of several essential elements and at the limit of toxicity of several others which function as chemical analogs to the essential ones (and hence are transported in the same way but do not fulfill the requested biochemical tasks). The general descriptive word "antinutrient" (36) has been suggested for these toxic elements. Such nutrient–antinutrient relationships need not be limited to cationic metals. Oxyanions of elements such as arsenic, tin, antimony, or germanium probably act as analogs of essential nutrients such as phosphate or silicate. The wide occurrence of organic derivatives of several of these elements in seawater (37,38) may represent an important mode of control of the environment by the biota, maintaining a tenuous balance between nutrition and toxicity.

More specific biochemical mechanisms than adaptation or biological regulation of the environment may also be invoked to explain colimitation by several elements. For example, the high iron requirement of nitrogen fixers has been proposed to explain the relatively modest amount of nitrogen fixation in the present-day ocean (39), suggesting simultaneous iron and nitrogen limitation. Because trace elements are typically involved in the uptake and assimilation mechanisms of major nutrients, such interactions are generally expected and have in fact been observed in the few laboratory studies that have addressed the question (34,40,41). Collier's observation of lower metal concentrations in phytoplankton samples from the nutrient-rich upwelling water in the Antarctic (20) may represent the oceanographic consequence of these major nutrients-trace metals interactions; less metal need be taken up concurrent with the uptake and assimilation of major nutrients, in regions where major nutrients are abundant.

As is well known from the plethora of inconclusive major nutrient studies, comparisons of cellular concentrations in laboratory cultures with field data may show that several trace metals are *potential* colimiting agents for growth (as essential elements and as toxicants); but they do not provide a very convincing demonstration of actual limitation in the oceans. A more direct way to establish an actual limitation in the field is to look at the free concentrations of the metals in the water and at the apparent affinities of metal binding groups on cell surfaces. If a significant fraction of the surface ligands (X) is to be bound to essential trace metals, then the apparent affinities of these ligands for the various metals (apparent association constants) must be of the order of magnitude of the inverse of the free metal ion concentrations in the water. For toxic metals whose surface-bound concentrations must be kept a factor of 10 lower or so, the free ion concentrations must be roughly a tenth of the apparent dissociation constants.

Direct data are available for the thermodynamics of cellular metal binding of iron, manganese, and copper by marine phytoplankton. By assuming that growth responses reflect cellular binding, indirect estimations can also be made for other metals. In all cases the results are somewhat suspect because of antagonistic effects among metals and among metals and nutrients which may not be accounted for in the data. The results shown in Table 12.4 are compared to free metal ion concentrations calculated from total surface concentrations in the open ocean on the basis of an inorganic complexation model for seawater. Though the results of this comparison are not spectacular, they are certainly coherent with the general picture of metal uptake by microalgae presented so far: Apart from cadmium for which the calculated surface open-ocean free ion concentration is dramatically lower than the estimated surface affinity of the coastal diatom *T. weissflogii*, the apparent metal dissociation constants observed in the laboratory and the free metal ion concentration in the surface ocean are reasonably close to each other. Nonetheless, one would be hard pressed to consider Table 12.4 as firm evidence of metal limitation in the oceans.

One way out may be to pursue the logic of the previous reasoning: In a stable surface ocean where the nutrient concentrations reflect the affinity of the biota, evolutionary selection must favor organisms with higher and higher affinities for the essential elements in short supply. The only limit to this process is that imposed by diffusion and one would expect that, on average, over sufficient time and space scales, the supply of all limiting nutrients in the open ocean would be controlled by diffusion. (The characteristics of the uptake system must then match diffusion; this may be the reason for the two-stage uptake process described earlier.) For a spherical cell of 10 μm in diameter, an ambient concentration of 0.1 nM corresponds

Table 12.4. Free Metal Ion Concentrations in Surface Seawater and Apparent Affinities of Algal Cells

	$-\log C$				
	$M_T^{surface}$	$[M^{z+}]^a$	$+\log K_m^{apparent}$	Source of Data	References
Fe	9.9	20.8	19.0	*T. weissflogii* uptake	(26)
Mn	9.2	9.8	8.7/9.1	*T. pseudonana* accumulation	(27)
Zn	10.0	10.6	10.9	*T. weissflogii* growth	(42, 44)
Cu	9.3	10.8	8.9	Many organism growth	(43)
			11.7	*T. pseudonana* Mn competition	(27)
Ni	8.6	8.8			
Cd	11.7	14.1	8.5	*T. weissflogii* Fe competition	(34)

^{a} Ratios of free metal ion to total dissolved metal from Stumm and Morgan (63).

to a maximum diffusion rate of 2×10^{-17} moles cell^{-1} hr^{-1}. One tenth of a nmole per liter is precisely the concentration of iron and zinc in the surface open ocean and 2×10^{-17} moles cell^{-1} hr^{-1} is the minimum iron uptake rate that was observed to allow maximum growth rate in laboratory cultures of *T. weissflogii*. (On the basis of a minimum quota, approximately 3×10^{-16} moles cell^{-1}, this corresponds to a growth rate of about one doubling per day.) The concentration of dissolved manganese, which is six times higher than that of Fe and Zn at the surface (see Fig. 12.2), may not be limiting there; however, toward the bottom of the euphotic zone where the manganese concentration is lower and the others higher, Mn may also become a limiting nutrient. Note that the reasonings applied here to trace elements should also be applicable to major nutrients; on this basis one would predict ambient concentrations of approximately 10^{-8} *M* and 10^{-7} *M* for available phosphorus and nitrogen in the surface ocean (respectively 100 and 1000 times Fe and Zn concentrations) and corresponding affinities of the uptake systems (inverse of half-saturation constants) of approximately 10^8 M^{-1} and 10^7 M^{-1}.

For the sake of completeness, one should note that different approaches can be taken to study the issue of trace element limitation in the ocean. For example, Brand et al. (45) took an ecological approach and reasoned that only limiting factors could effect adaptive changes in populations. They then took advantage of the large differences in the concentrations of many trace elements in coastal and open-ocean waters and looked for differences in trace element sensitivities (essential nutrient limitation or toxicity) among neritic (i.e., coastal) and oceanic phytoplankton species. They observed, for example, that the limiting concentrations of zinc and manganese in the growth medium were much higher for neritic species than for related oceanic species. Taken at face value these results indicate that Zn and Mn somehow apply a selective pressure on marine algae and thus must in one form or another, be naturally "limiting" to the organisms.

5. CONCLUSION

In this paper, we have argued that the biota controls the vertical transport of bioreactive trace elements in the open ocean, both in surface and deep waters; that the principal mechanism of incorporation into particles is direct uptake by algae or binding to alga-derived material; that nonessential trace elements are taken along with essential ones (as chemical analogs); and that several major and minor chemical constituents, both essential and toxic, may be colimiting (synergistically) primary production. Effectively, we have proposed that the Redfield model for the biological control of major algal nutrients in the sea could be extended to many trace element using the stoichiometric formula

$$C_{106} H_{263} O_{110} N_{16} P_1, (Fe, Zn, Mn)_{0.01}, (Cu, Cd, Ni, etc.)_{0.001}$$

This formula is of course only approximate, the coefficients 0.01 and 0.001 representing the available data within a factor of 3 or so. A simple evaluation of the accuracy of the formula is provided by comparison with the one obtained from the relative depletion of surface waters—P_1, $Zn_{0.003}$, $Cu_{0.0015}$ $Cd_{0.0004}$, $Ni_{0.003}$ (see Fig. 12.2)—for those elements known to remain principally as soluble species, thus excluding Fe and Mn whose particulate fraction may contribute an unknown amount to algal nutrition. Both the differences between these two formulas and the variations in vertical profiles that are observed in different locales, reflect the specificity and the relative efficiency of the remobilization and scavenging processes as phytoplankton are transformed into settling detrital material. This important issue has been simply ignored in our discussion.

The applicability of the general approach presented here to coastal waters or even to lakes depend largely on the relative importance of the allochthonous fluxes (forcing functions) in governing the behavior of the system, and hence on the validity of the underlying physiological and ecological mechanisms. For example, various trace elements may be so abundant and the variations so rapid that the system may never evolve toward colimitation of several elements; one element may be always limiting (through deficiency or toxicity) and the relative uptake of the various elements by the algae then do not reflect the same fine balance between deficiency and toxicity hypothesized to prevail in the open oceans. In this respect it is heartening to see that the data obtained in Lake Constance (Chapter 13) do resemble those in the oceans: the dynamism of the geobiological cycle may be sufficient to dominate purely geochemical processes in many aquatic systems.

Much of the discussion in this paper consists of extrapolations and generalizations of scattered oceanographic and physiological information, and of teleological reasoning. As such, the case that is being made for biological control of the trace element composition and cycle in the oceans rests more on the coherence of the overall picture than on the careful documentation of the detailed processes. In this respect it is similar to—or rather a part of—the GAIA hypothesis which proposes that the whole chemistry of the earth surface is controlled by and for the life of the planet (46). Such a grand scheme, while it is receiving increasing attention in the geochemical literature (3,47), has the danger of bypassing necessary causal mechanisms in the evolutionary process. For example, by viewing the whole biota as a gigantic symbiotic amalgam, indeed as a single organism, evolution becomes readily unrestricted by the precept of adaptation of single species. What emerges then is a need for physiologists to explain the interdependence (and hence the adaptation) of a single organism with all the constituents of its environment. For the purpose of micronutrient uptake, maximum fitness is achieved when the affinity of the transport system matches the availability of the element in the environment (no more, no less). For "antinutrients" the selectivity must be just great enough—that is, the affinity must be just low enough—to avoid toxicity. Vice versa the elemental concentrations in surface water simply reflect the characteristics of the uptake systems of the organisms that control it.

6. APPENDIX: METAL PARTITIONING AND OCEANIC RESIDENCE TIME

In a recent paper, Balistrieri et al. (15) attempted to identify the materials that control the adsorptive properties of marine particulate matter by analyzing both field observations of trace metal scavenging and laboratory measurements of adsorption on various particulate materials. Their method derives from the work of Schindler (13), who noted that metals must occur in particulate matter in proportion to the ratio of the residence time of the particles, T_P, to the residence time of the metal with respect to removal via scavenging, T_m:

$$\frac{M_{\text{particulate}}}{M_{\text{total}}} = \frac{T_p}{T_m} \tag{1}$$

where $M_{\text{particulate}}$ and M_{total} are the particulate and total water-column concentrations of the metal. By coupling this to adsorption data, Balistrieri et al. concluded that the interaction of metals with marine particles are controlled by organic coatings analogous to humic or salicylic acid, a conclusion later corroborated by Hunter (16).

Here we reexamine this comparison of residence times and metal partitioning data to determine the validity of their conclusion. For this purpose, we consider the meaning of "scavenging residence times" in the context of removal by adsorption and take a closer look at the laboratory metal binding and adsorption data from which the partitioning estimates are made. To remain as close to measurable parameters as is practicable, we reduce both observed residence times, using Eq. 1, and adsorption data to the fraction of the total metal adsorbed rather than apparent stability constants. At equilibrium, the fraction of a metal adsorbed on particles is

$$\frac{M_{\text{adsorbed}}}{M_{\text{total}}} = \frac{K_m^c S_{\text{total}}}{1 + K_m^c \, S_{\text{total}}} \tag{2}$$

where S_{total} is the total mass or molar concentration of the adsorbing material, and $M_{\text{dissolved}}$ and M_{adsorbed} are total dissolved and adsorbed concentrations of the metal. The conditional stability constant K_m^c includes all terms necessary to adjust measured equilibrium constants for the conditions of seawater and assumes that only a small fraction of the total sites are bound to the metal of interest. As long as M_{total} remains relatively constant in the water mass under study, the results from Eq. 2 should be directly comparable to those from Eq. 1.

6.1.   Metal Residence Times

1.   *Mean Oceanic Residence Times.*   Schindler (13) originally applied the above relations to mean residence times of metals in the ocean to estimate *in situ* partitioning. By assuming that scavenging by particles was the only removal process for trace metals and that inputs balanced removal in the ocean, Schindler could use metal residence times for a well-mixed ocean calculated from input rates.

There are, however, considerable obstacles to accurately estimating input rates of trace metals to the ocean—few rivers have been sampled, actual fluxes out of estuaries are usually not known, and atmospheric contributions are largely uncertain. But, estimates of input rates based on river input only have been made and look reasonably good on log–log plots. Table 12.1 lists the estimates of mean residence times based on river inputs from Li (48) and recent measurements of trace metals in seawater (49).

2.   *Scavenging Residence Times from Dissolved Profiles.*   An alternative to averaging over the entire ocean, as in the one-box model, is to analyze a single profile of a metal's concentration over depth. Craig (50,51) nicely describes the use of a one-dimensional advection–diffusion model to infer scavenging by analyzing the deviation of an element's concentration profile from conservative behavior. This deviation is presumed to arise from scavenging by particles, which is represented as an irreversible uptake process from solution; that is,

$$-\frac{\partial(M_{\text{dissolved}})}{\partial t} = \frac{\partial(M_{\text{adsorbed}})}{\partial t} = \psi M_{\text{dissolved}} \tag{3}$$

where ψ is the first-order scavenging rate constant. If all metal removal from the system is via scavenging, the steady-state balance between diffusion, generation, advection, and scavenging is

$$K\frac{\partial^2(M_{\text{dissolved}})}{\partial z^2} + J = w\frac{\partial(M_{\text{dissolved}})}{\partial z} + \psi M_{\text{dissolved}} \tag{4}$$

where

w = the vertical advection velocity
J = the concentration-independent production rate
K = the vertical eddy diffusivity
z = the vertical dimension, positive upward

This model incorporates the assumptions that horizontal gradients do not affect the vertical profile, that the mixing parameter (K/w) is constant, and that the density and vertical advection velocity are nearly constant.

These restrictions mean the model cannot be applied to the surface layer of the ocean; rather it must be applied to a coherent mass of deep water. To do so, one first determines K and w by fitting the profiles of a conservative property, for example, salinity or potential temperature, and of a radiotracer, for example, ^{14}C. The scavenging rate constant or the *in situ* generation term for a trace metal may then be estimated by fixing the upper and lower concentrations and fitting the element's dissolved concentration profile. This process determines only the *net removal* or deviation from conservative behavior of the dissolved element; the depth-integrated removal may not represent the entire removal flux if there is an input flux to the water mass other than advection/diffusion, for example, metal in settling particles. We must then interpret the scavenging residence times (ψ^{-1}) obtained by this method as a residence time of the dissolved metal with respect to removal by adsorption, as can be readily seen from Eq. 3. This residence time can be reasonably used in Eq. 1 as long as $M_{\text{dissolved}}$ is much greater than M_{adsorbed}. Several values of scavenging residence times from dissolved profiles are listed in Table 12.1.

3. *Scavenging Residence Times from Sediment Trap Data.* A third method of estimating removal/residence times of metals involves modeling the variation over depth of metal fluxes measured with sediment traps. A one-dimensional model for particulate metal fluxes is developed assuming irreversible adsorption, Eq. 3, and neglecting diffusion and advection of the settling particles:

$$\frac{\partial(\text{flux of particulate metal})}{\partial z} = S \frac{\partial(M_{\text{adsorbed}})}{\partial z} = -\psi M_{\text{dissolved}} \qquad (5)$$

where the particle settling velocity S is assumed constant with depth. Theoretically, the value of ψ obtained from this analysis should be the same as that determined by analyzing dissolved profiles; the same limitations apply to the interpretation of each. Brewer et al. (52) used this method to determine scavenging residence times for iron, manganese, and copper (see Table 12.1). They point out that their result may be influenced by sediment trap inefficiencies and by horizontal transport of fine particles; but the residence times for Mn and Cu are reasonably close to those obtained using other methods.

The reader may have already noted the contradiction inherent in using parameters derived by assuming irreversible uptake by particles within the reversible adsorption model which serves as the basis for our analysis. In fact, Bacon and Anderson (53) found that irreversible adsorption definitely could not simultaneously explain both the dissolved and particulate profiles of ^{230}Th, a strongly scavenged element; reversible adsorption could explain both profiles. The distinctive feature of their work was the simultaneous determination of both dissolved and particulate metal profiles, a necessary condition for the application of a reversible model. Unfortunately, such data have not yet been both collected and analyzed for the trace elements of interest here. Hence we must accept the fact that the irreversible scavenging rate constant ψ has no real chemical significance other than as a ratio of a net removal flux to a storage of dissolved metal.

6.2. Particle Residence Times

In order to estimate partitioning, the metal residence times just obtained must be supplemented with particle residence times. Brewer et al. (52) obtained a particle residence time of 0.365 years from measurements of thorium isotopes caught in sediment traps. Bacon and Anderson (53) argue that this short residence time reflects rapidly sinking particles from biologically active surface waters and therefore does not reflect the slower settling velocities of the larger population of small particles. Since these smaller particles have a much greater surface area, they would be expected to determine the adsorptive behavior of deep-sea particles. By fitting observed metal partitioning data and metal residence times to Eq. 1, Bacon and Anderson (53) estimated T_p to be 5–10 years. Here we will use a value of 7.5 years as did Hunter (16).

6.3. Estimates of Partitioning

From the residence times obtained above, we can now estimate the required extent of partitioning to the solid phase assuming adsorptive scavenging is the only metal removal process. Table 12.5 shows that these values range from 0.6 for iron to 4×10^{-5} for cadmium. The required partitioning would decrease by 1.3 log units if a particle residence time of 0.365 years were used. It now remains to be seen if we can determine which solid phases control adsorption.

6.4. Partitioning onto Model Solids

1. *Amorphous $Fe(OH)_3$.* As an analog for the inorganic constituent of marine particulate matter, we first consider amorphous iron hydroxides, a solid with relatively high adsorptive affinities. Oakley et al. (54) measured conditional stability constants in seawater at pH 8.0 for the adsorption of copper and cadmium onto this material. Following Balistrieri et al. (15), we will use 2 µg liter^{-1} of particulate matter in the deep sea. If we assume two-thirds of the total particulate matter to be inorganic (i.e., not reduced carbon compounds), and all the inorganic particulate matter to bind metals as strongly as amorphous $Fe(OH)_3$, then we obtain an estimate for the fraction of these two metals adsorbed by inorganic particles in the ocean. In Table

Table 12.5.   Adsorption of Metals by Potential Carrier Phases in the Ocean[a]

Metal	T_p/T_m[b]	Amorphous Fe(OH)$_3$[c]	Humic Acid[d]	Salicylic Acid[e]
		−log (Fraction Adsorbed)		
Fe	0.2			8.5
Mn	0.5		7.3	8.5
Zn	2.0		6.5	7.8
Cu	1.9	3.6	3.0	4.7
Cd	<4.4	5.2	8.9	10.7
Ni	<3.3		5.9	7.1

[a] See Appendix for explanation.
[b] T_p = 7.5 yr (53); T_m from sediment-trap data (Fe, Mn, Cu), dissolved profiles (Ni, Cd), and one-box model (Zn).
[c] Calculated using Eq. 2. From Oakley et al. (54), values of logK_m^c are Cu (2.31) and Cd (0.64), where S_{total} is in g liter^{-1} (salinity = 32‰, T = 20°C, pH 8.0); assumes 1.3×10^{-6} g liter^{-1} amorphous Fe(OH)$_3$.
[d] Calculated using Eqs. 2 and 7; (I = 0.5 M, pH 8.0); assumes L_T = 3.4×10^{-9} moles HA liter^{-1} (5×10^{-3} moles HA g^{-1}; ratios of free metal ion to total dissolved metal from Stumm and Morgan (63). From Mantoura et al. (55), values of log K_m° are Mn (4.51), Zn (5.31), Cu (9.71), Cd (4.69), Ni (5.51), Ca (4.12), and Mg (3.98).
[e] Calculated using Eq. 2 and 8; (I = 0.5 M, pH 8.1); assumes 5×10^{-3} moles "salicylate" liter^{-1}; ratios of free metal ion to dissolved metal from Stumm and Morgan (63). From Martell and Smith (59), values of log β_{ML}° are FeL (17.6), MnL (6.8), NiL (7.8), CuL (11.5), ZnL (7.5), CdL (6.4), HL (13.74), and H$_2$L (16.71). Competition from Ca and Mg is insignificant.

12.5, we can see that the fractions of Cd and Cu adsorbed are one to two orders of magnitude too low to account for the observed partitioning. Thus, to the extent that amorphous Fe(OH)$_3$ fairly represents inorganic marine particles, we can conclude that inorganic particles do not control adsorptive partitioning in the deep ocean.

In fact, Gordon et al. (9) measured approximately 1.6×10^{-7} g liter^{-1} particulate iron in the deep ocean. If the remaining inorganic particulate matter were much less strongly adsorptive than Fe(OH)$_3$, the discrepancy between the observed partitioning and the partitioning on amorphous Fe(OH)$_3$ would be an order of magnitude larger. Oakley et al. (54) did measure conditional constants for adsorption onto MnO$_2$ that were approximately an order of magnitude larger than for amorphous Fe(OH)$_3$. The lower concentrations of particulate manganese, $\cong 8 \times 10^{-8}$ g liter^{-1} (8), in the ocean, however, make it unable to significantly scavenge these metals.

2.   *Marine Humic Acids.*   As an alternative to control of adsorption by inorganic particles, we will consider humic material as an analog for the organic coatings found on marine particles. Mantoura et al. (55) determined conditional stability constants at pH = 8 and I = 0.02 for numerous metal–humate complexes. The overall stability constants they observed are defined

by

$$K_m^{obs} = K_m^0 \, \gamma_m = \frac{(ML)}{(M^{z+}) \, (L_T - ML)} \tag{6}$$

where

(ML) = the concentration of the metal complex (moles liter^{-1})
(M^{z+}) = the concentration of the free ion (moles liter^{-1})
K_m^0 = the stability constant at zero ionic strength
γ_m = the activity coefficient of the metal ion
L_T = the total concentration of humate ligand (moles liter^{-1})

and the ratio of activity coefficients of bound and unbound ligand is assumed equal to 1. Since essentially all the humic acid is complexed with calcium and magnesium, the conditional constant in seawater is

$$K_m^c = K_m^0 [K_{Ca}^0 (Ca^{2+}) + K_{Mg}^0 (Mg^{2+})]^{-1} \alpha_m \frac{\gamma_m}{\gamma_{2+}} = \frac{(ML)}{(M)L_T} \tag{7}$$

where (M) is the dissolved inorganic metal concentration, α_m is $(M^{z+})/(M)$, and the γ_i are activity coefficients from the Davies equation ($\gamma_{Ca} = \gamma_{Mg} = \gamma_{2+}$).

Hunter (16) has noted that Balistrieri et al. (15) did not account for competition for the humate ligand by protons. His speciation calculations transformed the constants used by Balistrieri et al. back to the original K_m^{obs} reported by Mantoura et al. However, Hunter, following Balistrieri et al., did not consider competition by Ca and Mg. A comparison of computed conditional constants with values measured in seawater for copper and cadmium (54,56) suggests there is an uncertainty in the complexation constants of about the same order of magnitude as the calcium/magnesium competition term ($\cong 10^{2.4}$). The constants computed with competition lie at the low end of the observed range. The possibility that Eq. 7 overcompensates for the competitive effect of major ions in seawater must therefore be kept in mind in the following partitioning calculations.

The calculation of partitioning to humics requires that an estimate of the number of humate-like sites in marine organic matter be made. Mantoura and Riley (57) analyzed a fulvic acid with approximately 7×10^{-4} moles of sites per gram of organic matter. For a river humic material, Sunda and Hanson (58) found a total of 5×10^{-3} moles per gram, but only 4×10^{-5} moles per gram of the strongest sites. Balistrieri et al. (15) considered a range of 10^{-4}–10^{-2} moles per gram of particulate matter, which is similar to those used in the references cited here. We take a value of 5×10^{-3} moles per gram as a reasonable upper bound and assume one-third of the total particulate matter, or 0.6 μg liter^{-1}, to be organic.

As can be seen in Table 12.5, none of the metals are sufficiently complexed by humic acids to account for the observed scavenging by particles. Even considering the uncertainty in conditional constants, only copper is bound as strongly by humics as the observed residence times require. A similar analysis of other trace elements known to be scavenged—Be, Sc, ^{210}Pb, and [230]Th—yields only beryllium as a potential candidate for scavenging by humics. Most importantly, the variation in affinity for different metals does not resemble that obtained from scavenging data. These results cast serious doubt on the claim that humic-like organic coatings on marine particles control the scavenging of many or most trace metals (60–63).

3. *Salicylic Acid.* Finally, we consider salicylic acid (2-hydroxybenzoic acid) as an analog for the organic coating of marine particles. Salicylic acid is a bidentate ligand which loses two protons ($pK_{a1} = 3$, $pK_{a2} = 13.7$) upon forming a complex (59). Its structural similarity to humic acid makes it an attractive model for humic substances. Since competition from Ca and Mg is insignificant, the conditional stability constant for salicylate–metal complexes need account only for competition from protons and for metal speciation:

$$K_m^c = \frac{\beta_{ML}^0}{\alpha_m \alpha_L} \frac{\gamma_m \gamma_{2-}}{\gamma_{ML}} \tag{8}$$

where β_{ML}^0 is the complex stability constant ($I = 0$), γ_i are activity coefficients from the Davies equation, and $\alpha_L = (L^{2-})/(L_T)$. Assuming, as we did for humic materials, 5×10^{-3} moles of sites per gram of organic matter, we obtain the values of metal partitioning to "salicylate-coated" particles shown in Table 12.5. All of the partitioning values are low by three or more orders of magnitude, demonstrating that "salicylate-like" organic coatings cannot explain the observations.

The discrepancy between these results and those of Balistrieri et al. (15) and Hunter (16) arises from their assumption that salicylic acid acts as a monodentate ligand and loses only the carboxylic proton upon binding a metal. Hence, both papers greatly underestimated the effect of competition by protons on metal binding.

7. SUMMARY

Neither the inorganic nor the organic analogs of marine particulate matter considered here have a sufficient affinity for trace metals, with the possible exception of copper, to account for the adsorption onto particles required by the scavenging residence times for the metals. This difference from the conclusions of Balistrieri et al. (15) and Hunter (16) is the consequence of a more complete consideration of competition for metal binding sites from calcium, magnesium, and hydrogen ions in seawater and as comparison of observations and calculations for each metal rather than for correlations fit to many elements. Thus, there must be some other solid(s) with higher affinities for metals that is responsible for scavenging in the deep ocean.

ACKNOWLEDGMENTS

The authors are indebted to M. Bacon, S. W. Chisholm, R. Collier, W. Fish, J. Murray, L. Small, and T. Trull for helpful comments and discussions and to G. Knauer, J. Martin, and W. Sunda for access to unpublished data. This work was supported in part by NOAA Grant No. NA79AA-D-00077 and NSF Grant No. OCE-8119103.

REFERENCES

1. H. W. Harvey, "Nitrate in the Sea," *J. Mar. Biol. Assoc.* **14,** 71 (1926).
2. A. C. Redfield, "On the Proportions of Organic Derivatives in Sea Water and Their Relation to the Composition of Plankton." In *James Johnstone Memorial Volume*, Liverpool Univ. Press, Liverpool, 1934, pp. 176–192.
3. M. Whitfield, "The World Ocean: Mechanism or Machination?" *Interdisc. Sci. Rev.* **6,** 12–35 (1981).
4. M. Whitfield and D. R. Turner, "Chemical Periodicity and the Speciation and Cycling of the Elements." In C. S. Wong, E. Boyle, K. W. Bruland, J. D. Burton, and E. D. Goldberg (Eds.), *Trace Metals in Sea Water*, Plenum Press, New York, 1983, pp. 719–750.
5. Y. -H. Li, "Ultimate Removal Mechanisms of Elements from the Ocean (Reply to a Comment by M. Whitfield and D. R. Turner)," *Geochim. Cosmochim. Acta* **46,** 1993–1995 (1982).
6. L. Pauling, *The Nature of the Chemical Bond*, University Press, Cornell (1955), as quoted in reference (4).
7. K. W. Bruland, "Oceanographic Distributions of Cadmium, Zinc, Nickel, and Copper in the North Pacific," *Earth Planet. Sci. Lett.* **47,** 176–198 (1980).
8. W. M. Landing and K. W. Bruland, "Manganese in the North Pacific" *Earth Planet. Sci. Lett.* **49,** 45–56 (1980).
9. R. M. Gordon, J. H. Martin, and G. A. Knauer, "Iron in North-East Pacific Waters," *Nature* **299,** 611–612 (1982).
10. E. A. Boyle, F. R. Sclater, and J. M. Edmond, "The Distribution of Dissolved Copper in the Pacific," *Earth Planet. Sci. Lett.* **37,** 38–54 (1977).
11. M. L. Bender, G. P. Klinkhammer, and D. W. Spencer, "Manganese in Seawater and the Marine Manganese Balance," *Deep Sea Res.* **24,** 799–812 (1977).
12. W. G. Sunda, S. A. Huntsman, and G. Harvey, "Photoreduction of Mn Oxide in Seawater: Geochemical and Biological Implications," *Nature* **301,** 234–236 (1983).
13. P. W. Schindler, "Removal of Trace Metals from the Oceans: A Zero Order Model," *Thalassia Jugoslav.* **11,** 101–111 (1975).

14. P. G. Brewer and W. M. Hao, "Ocean Elemental Scavenging." In E. A. Jenne (Ed.), *Chemical Modeling in Aqueous Systems: Speciation, Sorption, Solubility and Kinetics*, ACS Symposium Series 93, American Chemical Society, 1979, pp. 261–274.

15. L. Balistrieri, P. G. Brewer, and J. W. Murray, "Scavenging Residence Times of Trace Metals and Surface Chemistry of Sinking Particles in the Deep Ocean," *Deep Sea Res.* **28A,** 101–121 (1981).

16. K. A. Hunter, "The Adsorptive Properties of Sinking Particles in the Deep Ocean," *Deep Sea Res.* **30,** 669–675 (1983).

17. J. A. Davis (submitted for publication).

18. A. C. M. Bourg, "Role of Fresh Water/Sea Water Mixing on Trace Metal Adsorption Phenomena." In C. S. Wong, E. Boyle, K. W. Bruland, J. D. Burton, and E. D. Goldberg (Eds.), *Trace Metals in Sea Water*, Plenum Press, New York, 1983, pp. 195–208.

19. S. W. Fowler, "Trace Elements in Zooplankton Particulate Products," *Nature* **269,** 51–53 (1977).

20. R. D. Cherry and J. J. W. Higgo, "Zooplankton Fecal Pellets and Element Residence Times in the Ocean," *Nature* **274,** 246–248 (1978).

21. R. W. Collier, "The Trace Element Geochemistry of Marine Biogenic Particulate Matter," Ph.D. thesis, Woods Hole Oceanographic Institution, Woods Hole, Massachusetts, 1981.

22. R. W. Collier and J. M. Edmond, "The Trace Element Geochemistry of Marine Biogenic Particulate Matter," *Prog. Oceanogr.*, in press (1984).

23. Martin and Knauer (in preparation).

24. S. A. Huntsman and W. G. Sunda, "The Role of Trace Metals in Regulating Phytoplankton Growth in Natural Waters." In I. Morris (Ed.), *The Physiological Ecology of Phytoplankton, Studies in Ecology*, Vol. 7, Blackwell Sci. Publications, Boston, 1980.

25. F. M. M. Morel and N. M. L. Morel-Laurens, "Trace Metals and Plankton in the Oceans: Facts and Speculations." In C. S. Wong, E. Boyle, K. W. Bruland, J. D. Burton, and E. D. Goldberg, (Eds.), *Trace Metals in Sea Water*, Plenum Press, New York, 1983, pp. 841–869.

26. M. A. Anderson and F. M. M. Morel, "The Influence of Aqueous Iron Chemistry on the Uptake of Iron by the Coastal Diatom *Thalassiosira weissflogii*," *Limnol. Oceanogr.* **27,** 789–813 (1982).

27. W. G. Sunda and S. A. Huntsman, "Effect of Competitive Interactions between Manganese and Copper on Cellular Manganese and Growth in Estuarine and Oceanic Species of the Diatom *Thalassiosira*," *Limnol. Oceanogr.* **28,** 924–934 (1983).

28. J. A. Raven, "Nutrient Transport in Microalgae." In A. H. Rose and J. G. Morris (Eds.), *Advances in Microbial Physiology*, Vol. 21, Academic Press, London, 1980, pp. 47–226.

29. J. B. Nielands, "Microbial Iron Transport Compounds." In G. L. Eichhorn (Ed.), *Inorganic Biochemistry*, Vol. 1, Elsevier North-Holland, 1973, pp. 167–202.

30. F. B. Simpson and J. B. Nielands, "Siderochromes in Cyanophyceae: Isolation

and Characterization of Schizokinen from *Anabaena* sp.," *J. Phycol.* **12**, 44–48 (1976).

31. C. G. Trick, R. J. Anderson, and P. J. Harrison, "Prorocentrin: An Extracellular Siderophore Produced by the Marine Dinoflagellate *Prorocentrum minimum*" *Science* **219**, 306–308 (1983).

32. C. G. Trick, R. J. Anderson, N. M. Price, A. Gilliam, and P. J. Harrison, "Examination of Hydroxamate-Siderophore Production by Neritic Eukaryotic Marine Phytoplankton," *Mar. Biol.* (in press).

33. T. P. Murphy, D. R. S. Lean, and C. Nalewajko, "Blue-Green Algae: Their Excretion of Iron-Selective Chelators Enables Them to Dominate Other Algae," *Science* **192**, 900–902 (1976).

34. G. I. Harrison and F. M. M. Morel, "Antagonism between Cadmium and Iron in the Marine Diatom *Thalassiosira weissflogii*," *J. Phycol.* **19**, 495–507 (1983).

35. G. A. Knauer and J. H. Martin, "Phosphorus-Cadmium Cycling in Northeast Pacific Waters," *J. Mar. Res.* **39**, 65–76 (1981).

36. L. F. Small, personal communication, 1983.

37. M. O. Andreae, "Arsenic Speciation in Seawater and Interstitial Waters: The Influence of Biological-Chemical Interactions on the Chemistry of a Trace Element," *Limnol. Oceanogr.* **24**, 440–452 (1979).

38. M. O. Andreae, "The Determination of the Chemical Species of Some of the "Hydride Elements" (Arsenic, Antimony, Tin and Germanium) in Seawater: Methodology and Results." In C. S. Wong, E. Boyle, K. W. Bruland, J. D. Burton, and E. D. Goldberg (Ed.), *Trace Metals in Sea Water*, Plenum Press, New York, 1983, pp. 1–19.

39. J. G. Rueter (in preparation).

40. N. M. L. Morel, J. G. Rueter, and F. M. M. Morel, "Copper Toxicity to *Skeletonema costatum* (Bacillariphyceae)," *J. Phycol.* **14**, 43–48 (1978).

41. J. G. Rueter, S. W. Chisholm, and F. M. M. Morel, "Effects of Copper Toxicity on Silicic Acid Uptake and Growth in *Thalassiosira pseudonana*," *J. Phycol.* **17**, 270–278 (1981).

42. M. A. Anderson, F. M. M. Morel, and R. R. L. Guillard, "Growth Limitation of a Coastal Diatom by Low Zinc Ion Activity," *Nature* **276**, 70–71 (1978).

43. W. G. Sunda and R. R. L. Guillard, "The Relationship between Cupric Ion Activity and the Toxicity of Copper to Phytoplankton," *J. Mar. Res.* **34**, 511–529 (1976).

44. S. S. Bates, A. Tessier, P. G. C. Campbell, and J. Buffle, "Zinc Adsorption and Transport by *Chlamydomonas variabilis* and *Scenedesmus subspicatus* (Chlorophyceae) Grown in Semicontinuous Culture," *J. Phycol.* **18**, 521–529 (1982).

45. L. E. Brand, W. G. Sunda, and R. R. L. Guillard, "Limitation of Marine Phytoplankton Reproductive Rates by Zinc, Manganese and Iron," *Limnol. Oceanogr.* **28**, (6), 1182–1198 (1983).

46. J. E. Lovelock, *GAIA: A New Look at Life on Earth*, Oxford Univ. Press, Oxford, 1979.

47. W. S. Broecker, "The Ocean," *Sci. Am.* **249**, 146–161 (1983).

48. Y. -H. Li, "Geochemical Cycles of Elements and Human Perturbation," *Geochim. Cosmochim. Acta* **45,** 2073–2084 (1981).

49. M. A. Quinby-Hunt and K. K. Turekian, "Distribution of Elements in Sea Water," *EOS*, April 15, 1983.

50. H. Craig, "Abyssal Carbon and Radiocarbon in the Pacific," *J. Geophys. Res.* **74,** 5491–5506 (1969).

51. H. Craig, "A Scavenging Model for Trace Elements in the Deep Sea," *Earth Planet. Sci. Lett.* **23,** 149–159 (1974).

52. P. G. Brewer, Y. Nozaki, D. W. Spencer, and A. P. Fleer, "Sediment Trap Experiments in the Deep North Atlantic: Isotopic and Elemental Fluxes," *J. Mar. Res.* **38,** 703–728 (1980).

53. M. P. Bacon and R. F. Anderson, "Distribution of Thorium Isotopes between Dissolved and Particulate Forms in the Deep Sea," *J. Geophys. Res.* **87,** 2045–2056 (1982).

54. S. M. Oakley, P. O. Nelson, and K. J. Williamson, "Model of Trace-Metal Partitioning in Marine Sediments," *Environ. Sci. Technol.* **15,** 474–480 (1981).

55. R. F. C. Mantoura, A. Dickson, and J. P. Riley, "The Complexation of Metals with Humic Materials in Natural Waters," *Est. Coast. Mar. Sci.* **6,** 387–408 (1978).

56. T. D. Waite and F. M. M. Morel, "Characterization of Complexing Agents in Natural Waters by Copper (II)/Copper(I) Amperometry," *Anal. Chem.* **55,** 1268–1274 (1983).

57. R. F. C. Mantoura and J. P. Riley, "The Use of Gel Filtration in the Study of Metal Binding by Humic Acids and Related Compounds," *Anal. Chimica Acta* **78,** 193–200 (1975).

58. W. G. Sunda and P. J. Hanson, "Chemical Speciation of Copper in River Water. Effect of Total Copper, pH, Carbonate, and Dissolved Organic Matter." In E. A. Jenne (Ed.), *Chemical Modeling in Aqueous Systems*, ACS Symposium Series 93, American Chemical Society, Washington, D.C., 1979, pp. 147–180.

59. A. E. Martell and R. M. Smith, *Critical Stability Constants: Other Organic Ligands*, Vol. 3, Plenum Press, New York, 1977.

60. M. L. Bender and C. Gagner, "Dissolved Copper, Nickel and Cadmium in the Sargasso Sea," *J. Mar. Res.* **34,** 327–339 (1976).

61. H. J. M. Bowen, *Environmental Chemistry of the Elements*, Academic Press, London, 1979.

62. R. F. Weiss, "Hydrothermal Manganese in the Deep Sea: Scavenging Residence Time and Mn/^{3}He Relationships," *Earth Plant. Sci. Lett.* **37,** 257–262 (1977).

63. W. Stumm and J. J. Morgan, *Aquatic Chemistry*, 2nd ed., Wiley, New York, 1983.

13

METAL TRANSFER MECHANISMS IN LAKES; THE ROLE OF SETTLING PARTICLES

Laura Sigg

Institute for Water Resources and Water Pollution Control (EAWAG), Swiss Federal Institute of Technology (ETH), Zurich, Switzerland

Abstract

Settling particles, especially biogenic organic particles, play a dominating role in binding heavy metals and transferring them into the sediments, thereby regulating the concentrations of dissolved metals in lakes. As shown by investigations in Lake Constance and in Lake Zurich, lakes, despite being much more polluted with heavy-metal ions, are nearly as much depleted in these trace-metal concentrations as are the oceans. Larger productivities and higher particle sedimentation rates are primarily responsible for the more efficient scavenging. The partition of the heavy metals between the particles and the water, is influenced by the affinity of the metal for the particle surface and the chemical speciation of the metal in solution. In the lakes investigated a large part of the settling particles consists of phytoplankton and biological debris. The mean elemental composition of these particles corresponds—within broad margins—to a reasonably constant stoichiometry of $C_{113}N_{15}P_1Si_{14}Cu_{0.008}Zn_{0.06}Pb_{0.004}Cd_{0.00005}$

Another scavenging and metals regeneration cycle operates in the deeper water layers. Mn(II) and Fe(II) are released from the sediments to the overlying water where they become oxidized to oxides of Mn(III, IV) and Fe(III) which are potential carrier phases of heavy metals.

1. INTRODUCTION

Increased amounts of heavy metals have been released into the environment over the last decades; their growing use for different industrial purposes

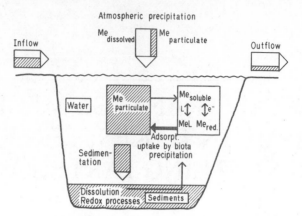

Figure 13.1. Schematic representation of the transfers and fate of heavy metals in the lake. Input is by river inflow and atmospheric precipitation and output by sedimentation and river outflow. Various processes, such as adsorption, uptake by biota, and precipitation, in the lake transfer metals from a soluble into a particulate form.

leads to an increased load of these metals to aquatic systems. In lakes, this increasing load is reflected in the sedimentary record; higher concentrations of heavy metals are found in more recent sediments (1–5). The resulting concentrations in the water column affect the biota. Heavy metals and biota are interrelated in an intricate way: Metal ions, if present in too high concentrations may adversely affect phytoplankton; settling organisms on the other hand—by taking up or adsorbing metal ions—tend to reduce their residual concentrations (6–8). To what extent have lakes been able to offset their increased heavy-metal pollution through natural regulatory mechanisms? In this contribution I intend to examine the most important processes which metals undergo in a lake and to assess the factors which influence the removal of metal ions by particle sedimentation and regulate the metal ion concentrations. Special emphasis will be given to the processes and properties that affect the partition of metal ions between the soluble and the solid phase.

Figure 13.1 gives a general idea of the processes involved. Heavy metals are carried into the lake by the river inflow and by atmospheric precipitation; the fractions of metals in particulate form is higher in the river than in rain water. In the lake different processes may lead to binding of the metals to particles, that is, adsorption on solid particles, uptake by biota, and chemical precipitation, and thus contribute to their removal by sedimentation. At the sediment–water interface, redox processes are occurring, which may change the partition between solids and solution.

This chapter is based on an investigation in Lake Constance and partly on some preliminary results obtained in Lake Zurich. These are prealpine lakes, in calcite-rich regions. The hypolimnion of Lake Constance is aerobic during the whole year, while the deepest layer of Lake Zurich becomes anaerobic during part of the year. Conditions encountered in highly eutrophic lakes with anoxic hypolimnion (H_2S production) are not considered.

Data.   The data that are presented here have been obtained in a joint project on heavy metals in Lake Constance and Lake Zurich carried out in collaboration with the Institute of Applied Physical Chemistry, Nuclear Research Center, Jülich, FRG (9,10). Data on concentrations of heavy metals in the water column and on fresh sedimenting material caught in sediment traps [the sedimentological work was carried out by Michael Sturm, EAWAG (11)] were collected during spring and summer 1981 and 1982, at the deepest point of Lake Constance (250 m), and in spring and summer 1983 at the deepest point of Lake Zurich (135 m).

2. IMPORTANT FACTORS REGULATING THE HEAVY-METAL CONCENTRATIONS IN LAKES

2.1. Speciation and Partition of Metals between Solution and Solid Phase

No analytical methods are available to determine directly the speciation of heavy metals in lake water. Ion-selective electrodes that can detect selectively free aquo metal ions are not sensitive enough for the concentrations encountered. Knowing the concentrations of ligands present in the water (CO_3^{2-}, OH^-, Cl^-), equilibrium calculations allow us to estimate the concentrations of the main inorganic species (Table 13.1). Carbonato and hydroxo species predominate under the conditions of Lake Constance. These calculations do not take into account any organic ligands for which concentrations and equilibrium constants are essentially unknown. Binding to the surface of a model solid phase can be included in such calculations. It

Table 13.1.   Main Inorganic Species for Conditions of Lake Constance

Main Inorganic Ligands pH 7.9–8.7; $I = 5.10^{-3}$		Main Inorganic Species[a]		Total Concentration (10^{-9} moles liter^{-1})
	(Moles liter^{-1})	Fe (III)	$FeOH_2^+$, $FeOOH_{(s)}$	200–400
		Mn (IV)	$MnO_{2(s)}$	20–1000
OH^-	8×10^{-7}– 5×10^{-6}	Mn (II)	$Mn^{2+}_{(aq)}$	—
HCO_3^-	2.0–2.5×10^{-3}	Cu (II)	$CuCO_{3(aq)}^0$	5–15
CO_3^{2-}	8×10^{-6}– 5×10^{-5}	Zn (II)	Zn^{2+}, $ZnCO_{3(aq)}^0$	10–30
Cl^-	1.3–1.6×10^{-4}	Cd (II)	Cd^{2+}, $CdCO_{3(aq)}^0$	0.05–0.1
SO_4^{2-}	3.6×10^{-4}	Pb (II)	$PbCO_{3(aq)}^0$	0.2–0.5

[a] Constants used for the calculations are from Smith and Martell (12) and Dyrssen and Wedborg (13).

has been shown that the binding of metals to surfaces can be described as an equilibrium reaction, in which the surface OH groups act as ligands (14,15):

$$\equiv SOH + M^{2+} \rightleftharpoons \equiv SO\,M^+ + H^+ \qquad K_1^s$$
$$2 \equiv SOH + M^{2+} \rightleftharpoons (\equiv SO)_2 M + 2H^+ \qquad \beta_2^s$$

(1)

($\equiv SOH$ represents OH groups attached to a surface S, e.g., as found on the surface of a hydrous oxide or in functional groups of organic matter present as particles or attached to a surface).

The metal binding reactions, Eq. 1, are strongly pH dependent; the pH encountered in the lake is in a favorable range where metal-ion uptake is possible. Figure 13.2(a) gives examples of speciation calculations for Pb(II) and Cu(II); SiO_2 and Al_2O_3 are used as model solid phases, representative for particles existing in natural waters. Depending on the particle concentrations assumed (and on the assumed total concentrations of surface OH groups), different fractions of surface-bound metals are obtained.

Calculations are exemplified for the case of Pb in the presence of SiO_2 as a solid phase (only the most abundant species are taken into account):

$$\beta_2^s = \frac{\{(\equiv SiO)_2\,Pb\}\,[H^+]^2}{\{\equiv SiOH\}^2\,[Pb^{2+}]}$$

(2)

where

$$\{\equiv SiOH\} = \text{concentration of OH groups at the}$$
$$\text{surface of } SiO_2 \text{ (moles kg}^{-1})$$

$$\{(\equiv SiO)_2 Pb\} = \text{concentration of Pb bound to } SiO_2$$
$$\text{(moles kg}^{-1})$$

Stability constants for different cations and surfaces are given, for example, in Schindler (15). The total concentration $[Pb]_T$ can be represented as the sum of the soluble and surface species; the concentration of each species can be calculated with the help of the stability constants and of the ligand concentrations:

$$[Pb]_T = [Pb^{2+}] + [Pb\,OH^+] + [Pb\,CO_3^0] + [Pb(OSi\equiv)_2]\ (\text{moles m}^{-3})$$

$$= [Pb^{2+}]\left(1 + K_{OH}[OH^-] + K_{CO_3}[CO_3^{2-}] + S_w\,\beta_2^s\,\frac{\{\equiv SiOH\}^2}{[H^+]^2}\right)$$

$$S_w = \text{concentration of particles (kg m}^{-3})$$

(3)

From this equation, $[Pb^{2+}]$ and the concentration of different species can be calculated. β_2^s has to be corrected for the influence of surface charge at

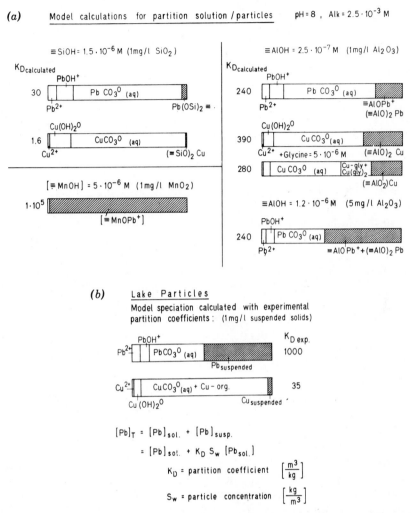

Figure 13.2. Model calculations for the partition of Pb and Cu between the solution and the particles and their speciation in solution. The different species are represented as percentages of the total concentrations in the bar diagrams; the fraction bound to the particles is indicated by hatched surfaces. (*a*) Calculations based on model oxide surfaces, using stability constants for surface complexes (15); K_D are derived from the calculations. (*b*) Calculations based on the field data; experimental K_D are used in order to calculate the fraction bound to particles.

the corresponding pH; such calculations can be done readily with a computer program such as MINEQL (16).

The partition coefficient of an element between particles and solution can be defined as:

$$K_D = \frac{C_s \ (\text{moles kg}^{-1})}{C_w \ (\text{moles m}^{-3})} = (\text{m}^3 \text{ kg}^{-1}) \tag{4}$$

where

C_s = concentration in particles (moles kg^{-1})
C_w = soluble concentration (moles m^{-3})

For the example given above, the partition coefficient can be calculated, on the basis of the equilibrium constants, as

$$K_D = \frac{\beta_2^s (\{\equiv SiOH\}^2/[H^+]^2)}{(1 + K_{OH}[OH^-] + K_{CO_3}[CO_3^{2-}])} \; (m^3 \; kg^{-1}) \qquad (5)$$

In this case, it is evident that K_D is strongly pH dependent.

The concentration of inorganic ligands is quite constant in the water column, so that little changes in the inorganic speciation can be expected. The concentrations of particles are varying, so that the fractions of metals bound to particles can vary. The difference in chemical conditions between the river inflow and the lake is small, so that little change in speciation can be expected for metals brought by the river. But there are very important differences between the conditions in rain and lake water. Rain water has a pH around 4.5; the ionic strength is very low, as is also the particulate-matter concentration. The metals carried by rain might for a large part be in dissolved form; large changes in speciation and in turn on the distribution between solid particles and solution will occur in the upper few centimeters of the lake.

2.2.   Metal Uptake by Phytoplankton

The cycles of the nutrient elements in a lake are in a general way influenced and partially regulated by aquatic organisms. Algal biomass is formed during photosynthesis by the uptake of nutrients (N, P, Si) and possibly other elements together with carbon. These elements are released again upon subsequent partial destruction (oxidative mineralization or respiration) of the settling organism-produced particles. As a consequence of photosynthesis and respiration, depletion of nutrient elements in the surface layers and enrichment in the deeper-water layers, accompanied by a depletion of oxygen, is typically observed during stagnation periods [Fig. 13.3(a)]. If the aquatic biomass has a relatively constant composition uptake, and release occurs in constant proportions, a covariance of the concentration of HCO_3^-, NO_3^-, phosphate and silicate is observed in the water column (Figure 13.3(b)). The relatively constant proportions found in the concentration differences $\Delta C : \Delta N : \Delta P : \Delta Si : \Delta O$ reflect the relative composition of the phytoplankton particles (Tables 13.2 and 13.3) and the stoichiometry of the photosynthesis and respiration process. The agreement between the stoichiometric coefficients observed in the water column and the composition

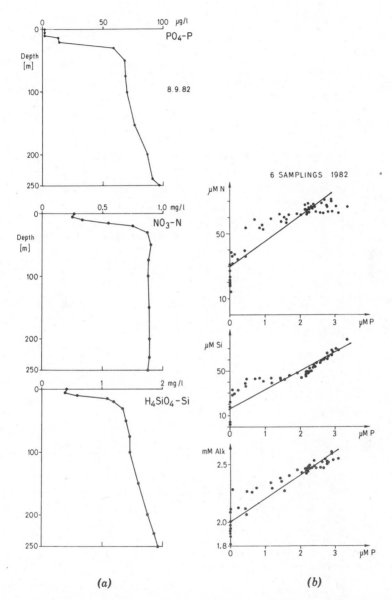

Figure 13.3. (*a*) Nutrient profiles in the water column of Lake Constance (middle of the lake, depth 250 m, 8/9/82). P, N, and Si are strongly depleted in the upper 10–20 m. (*b*) Covariance of HCO_3^-, Si, N, and P in the water column. The data are from six different sampling campaigns (1982).

Table 13.2. **Elementary Composition of Sedimentary Material from Sediment Traps (Lake Constance 1981/1982).** Values Given Are Means of the Samples Collected during a Collection Period (Maximum Eight Samples per Period) in moles g^{-1} Relative to moles phosphorus g^{-1}.

Collection Period	Depth (m)	C	N	P	Ca	Mg	Fe	Mn	Cu (× 10^{-3})	Zn (× 10^{-2})	Pb (× 10^{-3})
5/23–6/12/81	51	145	20	1	57	5.6	2.6	0.10	6	2.6	4
6/12–7/3/81	243	—	—	1	230	—	2.6	1.0	—	15	—
8/21–9/11/81	51	148	15	1	111	7.2	4.3	0.14	10	11	5
9/11–10/23/81	243	93	11	1	37	5.0	4.1	3.6	8	4.6	4
10/2–10/23/81	44	78	12	1	48	11.2	8.9	0.24	11	4.0	5
10/23–11/6/81	52	113	15	1	41	10.5	2.6	0.21	12	5.6	6
5/28–6/18/82	90	121	16	1	62	5.9	4.2	0.10	8	5.4	2
5/28–6/18/82	250	—	—	1	114	5.9	3.1	0.13	6	2.3	5
6/18–7/8/82	205	105	17	1	56	12.2	7.5	1.16	10	7.0	3
7/8–7/30/82	49	117	16	1	216	6.5	3.2	0.09	6	5.2	2
7/30–8/20/82	49	85	12	1	30	2.1	1.0	0.04	3	2.3	1
8/20–9/9/82	230	116	15	1	75	3.8	2.5	0.84	6	6.6	2
9/9–10/9/82	91	127	17	1	128	8.6	6.1	0.18	9	5.3	5
Mean composition		113	15	1	93	7	4	0.6	8	6	4

Table 13.3. Elementary Composition of Sedimentary Material from Sediment Traps: Sediment Traps at 243-m Depth: 09/11/81–10/23/81, Eight Successive Samples Taken during 2.5 Days Each.

moles g^{-1}/moles phosphorus g^{-1}	C	N	P	Ca	Mg	Fe	Mn	Cu ($\times 10^{-3}$)	Zn ($\times 10^{-2}$)	Pb ($\times 10^{-3}$)
	98	11	1	50	5.7	4.4	3.7	12	5.6	4
	85	10	1	40	4.8	3.8	2.9	8	4.9	4
	104	13	1	47	5.4	4.4	5.9	10	5.6	4
	109	13	1	49	6.0	5.3	4.1	10	5.3	5
	87	12	1	36	5.3	4.2	3.0	8	4.1	4
	122	15	1	43	6.3	5.0	3.6	10	5.5	5
	95	11	1	27	4.0	3.0	2.7	7	3.5	3
	93	11	1	28	4.1	3.3	3.4	7	4.5	3
Mean composition	99 ± 11	12 ± 1	1	40 ± 8	5.2 ± 0.8	4.2 ± 0.7	3.7 ± 0.9	9 ± 2	4.9 ± 0.7	4 ± 1

of settling particles (Table 13.2) is for the main nutrient elements quite good, especially if one considers that the settling particles contain, in addition to the biota, various proportions of $CaCO_3$ and allochthonous material, such as aluminum silicates (the biogenic particles—biota and $CaCO_3$—make up more than 50% during a large part of the summer stagnation period) and that some fluctuations in the composition of the biota with time and depth and nutrient states (17) are plausible. The mean elemental composition of the settling particles with regard to the main nutrients, $C_{113} N_{15} P_1$, is in good agreement with the marine algal composition originally given by Redfield:

$$113CO_2 + 15NO_3^- + 1HPO_4^{2-}$$
$$+ 17H^+ \quad \xrightarrow[\text{respiration}]{\text{photosynthesis}} \quad \{(CH_2O)_{113} (NH_3)_{15} (H_3PO_4)_1 \ldots\} + 130O_2$$

$$(6)$$

Can the idea of a constant elemental algal composition be extended to the heavy metals? The experimental data obtained so far are not fully conclusive. Although there is considerable scatter in the data, the results given in Tables 13.2 and 13.3 show a tendency towards a constant composition of the particles in Cu, Zn, Pb, Fe, and Mn relative to P.

In interpreting the data one needs to consider (1) that the particles contain—in addition to biota—various proportions of $CaCO_3$ and allochthonous materials; (2) that the data in Table 13.2 are from samples which were collected at different depths and at different time periods and are thus of different age and mineralization history which represents residues of different algal populations (17); and (3) that analytical difficulties, especially those caused by allowing different time periods between the retrieval of the samples and their processing (freeze drying) could not fully be avoided. That the data (Table 13.3) from individual collection periods (eight subsequent samples, each sample collected for a 60-hr interval and retrieved together after 480 hr) show a better compositional constancy is very suggestive; these are samples of the same depth from the same seasonal period and of similar proportions in organic and inorganic composition that have been analytically

Table 13.4. Particle–Water Partition Coefficients as Determined from Representative Concentrations in Water, and in Settling Particles

	Fe	Zn	Cu	Cd	Pb
Water (Total concentration C_t) (mg m^{-3})	10–30	1–2	0.3–1.0	5×10^{-3}–0.02	0.05–0.1
Fresh sediment particles C_s (mg kg^{-1})	12,000	130	30	0.3	50
Partition coefficients (m^3 kg^{-1}) $K_D = \dfrac{C_s}{C_w}$	10,000	100	35	30	1000

Table 13.5.   Soluble Concentration as Function of K_D

$$C_t = C_s \cdot S_w + C_w$$
$$= K_D \cdot C_w S_w + C_w$$
$$= C_w(K_D \cdot S_w + 1)$$

where

C_t = total concentration (mg m^{-3})
C_w = soluble concentration (mg m^{-3})
C_s = concentration in particles (mg kg^{-1})
K_D = partition coefficient = C_s/C_w(m^3 kg^{-1})
S_w = concentration of suspended particles in water (kg m^{-3})

$$\frac{C_w}{C_t} = \frac{1}{K_D S_w + 1}$$

For $S_w = 1 \times 10^{-3}$ kg m^{-3}:

$K_D = \ \ 1; C_w/C_t = 0.999$
$K_D = \ 10; C_w/C_t = 0.99$
$K_D = 10^2; C_w/C_t = 0.91$
$K_D = 10^3; C_w/C_t = 0.5$
$K_D = 10^4; C_w/C_t = 0.09$

---

processed simultaneously. That these subsequent samples, retrieved after different time periods between 2 and 20 days, show no systematic trend in composition may be taken as an indication that either little degradation took place (at 5°C) or that degradation occurred congruently.

What are the better scavengers for heavy metals, the biological surfaces, or the mineral surfaces? Various arguments suggest that the biological surface is primarily responsible for the tying up of heavy-metal ions and for the relatively constant proportion of heavy metals in relation to N or P: The inorganic part of the suspended particles consist primarily of $CaCO_3$, quartz, chlorite, illite, and feldspars (11); these minerals, also capable of adsorbing both organic substances and metal ions, have relatively small specific surface areas and thus relatively little capacity for metal ion binding. Furthermore, as shown by Balistieri et al. (18), inorganic surfaces (even if partially covered by adsorbed humates) have less affinity for metal ions than biological surfaces. Thus, for the biota of Lake Constance a hypothetical stoichiometric composition*) of

$$\{C_{113} \ N_{15} \ P_1 \ Cu_{0.008} \ Zn_{0.06} \ Pb_{0.004} \ Cd_{0.00005}\} \tag{7}$$

may as a first approximation be postulated. It is this approximate constancy that justifies the use of comprehensive distribution coefficients—conditionally valid for Lake Constance (Tables 13.4–13.6). The hypothesis of constant

---

* Ca, Mg, Fe, and Mn are excluded from the formula because these elements form precipitates. Fe is dependent on the proportion of allochthonous material; Mn is sensitive to the redox conditions. $CaCO_3$ is mostly biogenic; it may enter the formula with an approximate coefficient of 100. For Cd, where only a few data have been measured, the coefficient 5×10^{-5} is tentative.

Table 13.6. Residence Time and Steady-State Concentrations of Metals Treating the Lake as a One-Box Model (See Chapter 12 by D. Imboden in This Book)

where

$Q = Q_{in} = Q_{out}$ = water flow $\quad (m^3 yr^{-1})$

V = Volume of lake $\quad (m^3)$

A = Surface area $\quad (m^2)$

$h = \dfrac{V}{A}$ = mean depth $\quad (m)$

C_t = Total concentration in the lake $\quad (mg\ m^{-3})$

C_s = Concentration in particles $\quad (mg\ kg^{-1})$

C_w = Soluble concentration $\quad (mg\ m^{-3})$

S_w = Particle concentration in the lake $\quad (kg\ m^{-3})$

S = Sedimentation rate $\quad (kg\ m^{-2}\ yr^{-1}) = V_s S_w$

F_1 = Input from surface waters $\quad (mg\ m^{-2}\ yr^{-1})$

F_2 = Input by precipitation $\quad (mg\ m^{-2}\ yr^{-1})$

F_3 = Output by sedimentation $\quad (mg\ m^{-2}\ yr^{-1}) = C_s S = K_D C_w S$

F_4 = Output by outflowing water $= QC_t/A \quad (mg\ m^{-2}\ yr^{-1})$

K_D = Partition (or distribution) coefficient $= C_s/C_w \quad (m^3\ kg^{-1})$

f_p = Fraction in particulate form $= K_D S_w/(1 + K_D S_w)$

V_s = Settling velocity of particles $= S/S_w \quad (m\ yr^{-1})$

k_s = Rate constant characterizing sedimentation $= V_s/h \quad (yr^{-1})$

k_{mix} = Rate constant characterizing water renewal $\quad (yr^{-1})$

$\hat{C}_{t,in}$ = Average total input concentration $= (F_1 + F_2)A/Q \quad (mg\ m^{-3})$

$\hat{C}_t$ = Steady-state concentration $\quad (mg\ m^{-3})$

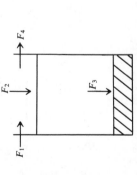

Mean residence time[a] of the water:

$$\tau_w = \frac{V}{Q} = \frac{1}{k_{mix}} \quad (yr)$$

Mean residence time of particles:

$$\tau_p = \frac{VS_w}{Qs_w + AS} = \frac{1}{k_{mix} + k_s} \quad (yr)$$

Mean residence time of metals:

$$\tau_M = \frac{VC_t}{QC_t + ASC_s} \quad (yr)$$

$$= \frac{V(1 + K_D S_w)}{Q(1 + K_D S_w) + ASK_D}$$

$$= \frac{1}{k_{mix} + f_p k_s} \quad (yr)$$

Relative residence time of metals:

$$\frac{\tau_M}{\tau_w} = \frac{k_{\text{mix}}}{k_{\text{mix}} + f_p k_s}$$

$$= \frac{k_{\text{mix}}(1 + K_D S_w)}{k_{\text{mix}}(1 + K_D S_w) + K_D S_w k_w}$$

(see Figure 13.7)[b]

Steady-state concentration (see Chapter 1 this book):

Mass Balance:

$$V\frac{dC_t}{dt} = F_1 + F_2 - F_3 - F_4$$

At steady state:

$$\frac{dC_t}{dt} = 0 \quad \text{as} \quad t \to \infty, \quad C_t = \hat{C}_t$$

$$\hat{C}_t = \frac{k_{\text{mix}}}{k_{\text{mix}} + f_p k_s} \bar{C}_{t,\text{in}}$$

Results for Lake Constance (see Figure 13.8):

Steady-state concentrations (μg liter^{-1}): Calculated—for Cu, 1; Zn, 1.5; Cd, 0.01; Pb, 0.02–0.5; Fe, 1.
Measured—for Cu, 0.5–1; Zn, 1–2; Cd, 0.01; Pb, 0.04–0.1; Fe, 10–20.

_____

[a] The residence time of a system (in hypothetical steady state) is given by the expression:

$$\frac{\text{amount present in the system}}{\text{amount delivered or removed per unit of time}}$$

[b] If f_p (or K_D) is small, $k_{\text{mix}} > f_p k_s$ and $\tau_M \approx \tau_w$; if, on the other hand, most or all of the element is removed by sedimentation (f_p (or K_D) = large), then $\tau_M \approx \tau_p$.

elemental composition allows steady-state approximations, as will be shown in Section 3.

It must be pointed out, however, that the constant proportions found in the settling particles do not appear clearly in the profiles of concentration of heavy metals versus depth of the water column as shown in Figure 13.4. Some of the profiles obtained, not depicted here, may be indicative of such a release in the deeper-water layers. But clearcut correlations between the concentrations of P and Si and those of Cd, Zn, Ni, and to some extent of Cu that have been observed in the oceans, have not been observed in lakes so far. There can be several reasons for this:

1. The lake system is less sensitive to such change than the sea. According to the composition given, a congruent release of 10 µg of phosphorus per liter corresponds to 0.2 µg liter^{-1} Cu, 2 ng liter^{-1} Cd, 1 µg liter^{-1} Zn, respectively.

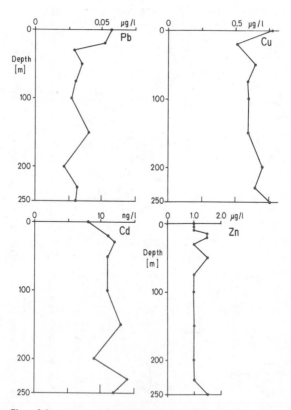

Figure 13.4. Profiles of the concentrations of heavy metals in the water column of Lake Constance (middle of the lake, depth 250 m, summer stagnation time, 9/8/82). Pb, Cu, and Cd are measurements by P. Klahre and H. W. Nürnberg, KFA Jülich, by voltametric methods (9). Concentration ranges (total concentrations) shown in these profiles were found throughout the different sampling campaigns at various times.

2. These changes in concentration may not readily be seen if blurred by other factors. In an ocean the most significant change in phosphorus (and thus in ΔMe) occurs in the top 1000 m or so, and because of much higher concentrations of algae and bacteria and other reasons, the same kind of concentration change is compressed into the top 10–20 m in lakes. Because metal pollution of the surface layers, especially due to atmospheric inputs, is much larger in lakes than in the ocean, little vertical segregation exists in the top layers between the scavenging of metals and their partial release during respiration.

3. We have few experimental data in the top 20 m; furthermore, contamination problems (proximity of the boat) are largest in these surface layers.

4. The metals that are released upon mineralization can—because of much larger particle concentrations in the lake in comparison to the sea— become readsorbed on other particles.

These complications, however, should not detract from the obvious fact that the transport of metals by biological particles is an important factor for the removal of metals from the water column. These inferences are in agreement with the conclusions drawn from the Melimex experiment [a study on the effect and fate of heavy metals in limnocorrals (6,7)] and interpretations given by Parker et al. (19) to the transport of Zn and Cd in Lake Michigan.

It is axiomatic in limnology—and well documented in other chapters of this book (Gächter and Imboden, Chapter 16; Schindler, Chapter 11; and Stabel, Chapter 7)—that the phosphorus input into the lake is the major factor in controlling the quantity of biomass synthesized and in determining the sedimentation rate of biogenic particles. The approximate constancy of the proportions of the various heavy metals relative to phosphorus in the particles implies that the removal of "biophile" heavy metals into the sediments is related to the phosphorus input into the lake. Progress in the eutrophication of lakes may therefore have alleviated to some extent the effects of progressively increased metal pollution; thus the coupling of biogeochemical cycling by organisms as described by Schindler (Chapter 11 in this book) can perhaps be extended to certain heavy-metal ions.

The interactions of algae with metal ions and other reactive elements can be thought to be of several types (20):

1. The metal ions can become surface coordinated (adsorbed) to the algal surface in accordance with a surface coordination or adsorption equilibrium. (Such an equilibrium may precede a subsequent slow, rate-determining transfer into the inside of the cell.) The quantity of metal, say Cd^{2+}, tied up by particles will be proportional to the surface-coordination constant and the maximum binding capacity of the algal surfaces; the latter in turn depends on the biological phosphorus. Thus a constant ratio of biologically bound phosphorus to surface-bound cadmium (or other metals) should result.

2. Alternatively, the metal ions like phosphorus and other micronutrients (cadmium may be mistaken as a micronutrient by the cell) become physiologically assimilated. If the metal ions are assumed to colimit the growth of algal cells, the concentration of both metal ion and phosphorus in the cell will equally depend on the generation time and on the concentrations of dissolved metal ion and phosphate, respectively. In this case the stoichiometric composition of the biotic particles would be affected by the ratios of their concentrations in solution or in turn by their inputs into the lake. Investigations in different lakes with different metal ion inputs may corroborate the uptake mechanism.

2.3. Redox Processes

Only a few elements are directly involved in redox processes; of interest here are especially manganese and iron. (In very eutrophic lakes, the formation of H_2S in deep anoxic layers would be another important point.) Indirectly, the concentrations of many elements can be influenced by the manganese and iron cycles.

Mn and Fe oxides are very good scavengers for heavy metals, as it has been demonstrated in laboratory adsorption experiments. They can thus also play this role in the lake (21,23). Coprecipitation with Mn oxides could also be a possible process (24). In Lake Constance, the oxygen concentration in the deeper-water layers does not drop below 8 mg liter^{-1}. The redox conditions just below the sediment–water interface are such that Mn is reduced and released from the sediments into the overlying water. No reduction of Fe(II) and release is observed in Lake Constance [Fig. 13.5(a)]. In contact with the oxygenated water layers, Mn(II) is reoxidized and precipitated as Mn (III, IV) oxides; the Mn oxidation is catalyzed by "Mn bacteria," that can be observed in the water column and in the fresh sediment material (11,25). Mechanisms of Mn release from sediments have been described by Davison et al. (26).

In Lake Zurich, oxygen drops to zero at the deepest point during stagnation time. The Mn release from sediments is very large. Up to 0.5 mg liter^{-1} Mn$_T$ were measured in water samples near the sediments. In this case (July 1983, depth 135 m), Mn oxide represents about 40% of the particles [Fig. 13.5(b)]. The influence of the Mn cycle on other elements is the superposition of two effects:

1. Metals bound to the Mn oxides can be released from the sediments, following the reduction to Mn(II).
2. The freshly formed Mn (III, IV) oxides are very efficient scavengers of other metals; the active cycling of Mn in the deepest layers tends to transfer efficiently other metals to the sediments. This implies that the fluxes are high, but the concentrations in the water column are low.

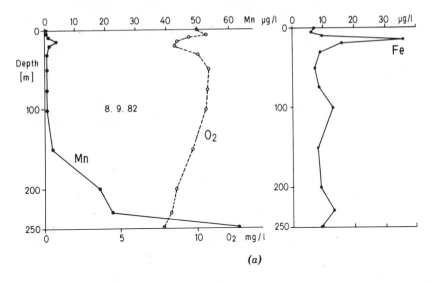

(a)

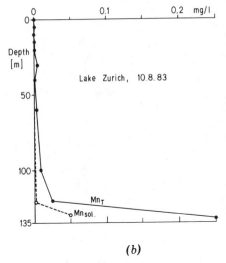

(b)

Figure 13.5. Profiles of the concentrations of Mn, O_2, and Fe in Lake Constance. (a) Mn(II) is released from sediments, although O_2 concentrations do not drop below 8 mg liter^{-1} (total Mn and Fe concentrations). Mn flux from the sides of the deep region of the lake may contribute to the high concentrations found at depths about 100 m above the sediments. Peak of Fe concentration at 15-m depth is probably from a waste-water input. (b) Profile of Mn concentration in Lake Zurich during summer stagnation time (depth 135 m, 8/10/83). Dissolved Mn(II) is only found at the lowest depth, about 2 m over the sediment–water interface.

Preliminary results from Lake Zurich support this idea: Cu and Zn show only a slight increase in concentration near the sediments, while the concentration of Pb(II) is very low near the sediments.

Studies of the sedimentary material by x-ray fluorescence (27) allow us to evaluate the elementary composition of single particles. Different types of particles can be compared. Cu and Zn were found in unusually high concentrations associated with Mn and Fe in freshly settling particles trapped at a depth of 250 m in Lake Constance (Fig. 13.6), while Cu and Zn do not appear, for example, in particles containing mainly $CaCO_3$.

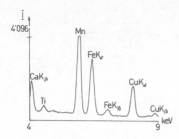

Figure 13.6. X-ray emission spectrum from sediment material (Lake Constance, 243-m depth). Single particles were selected under the electron microscope and their X-ray emission spectrum recorded. (Peak heights are related in a semiquantitative way only to the present concentrations.) In this case, an accumulation of Mn and Fe oxides was found, together with Cu [measurements by R. Giovanoli, University of Berne (27)].

2.4. Precipitation of Solid Phases

Precipitation of solid phases cannot represent an important process for the heavy metals considered here (except for Fe and Mn oxides) due to the low concentrations in the water column, as one can check by calculating the solubility products of the corresponding solid phases (28). An important process of precipitation occurring in the lake is that of calcium carbonate (17,29). Conditions of oversaturation and undersaturation can be encountered, depending on depth and seasonal changes in photosynthesis and respiration. Some evidence about the role of $CaCO_3$ in coprecipitating or adsorbing other metal ions, can be won from the composition of the fresh sediment particles. The calcium carbonate content varies widely in these sediments, depending on the time and on the depth, due to the precipitation and dissolution processes. The comparison of different series of sediment material shows quite constant heavy-metal contents, as related to phosphorus, in spite of varying $CaCO_3$ contents; no tendency to higher heavy-metal concentrations associated with high $CaCO_3$ contents appears (Tables 13.2 and 13.3). In x-ray emission spectra, no accumulation of heavy metals appears together with high Ca concentrations. Collier (30) found that the calcium carbonate fraction in marine particles contained very small concentrations of metal.

2.5. Quantitative Description of the Distribution of Heavy Metals between Dissolved and Particulate Phase

The different processes described all lead to a transfer of heavy metals from the dissolved into the particulate phase. In order to evaluate quantitatively the transfer of metals to the sediments, an overall evaluation of the distribution of metals between dissolved and particulate phase can be used, by calculating partition coefficients:

$$K_D = \frac{C_s}{C_w} \quad (\mathrm{m}^3 \ \mathrm{kg}^{-1}) \tag{4}$$

They indicate the result of all the transfer processes from the dissolved to the particulate phase and they reflect thus the general tendency of elements to be bound to particulate material, as resulting from solubility, tendency for adsorption on organic and inorganic surfaces, uptake by biota, complex formation, and so on. Partition coefficients derived from field data are conditional for given ligand concentrations, pH, type of particles, and so on; they should not be used uncritically for other systems.

Well-defined partition coefficients can be obtained in laboratory experiments with defined particles; in the case of the adsorption of metals on surfaces, they can be related to the surface complexation constants [Fig. 13.2(a)]. There are several ways of deriving partition coefficients from the field data:

1.   The concentration of metals in the dissolved phase and the particulate material can be measured in samples from the water column, respectively; this would give values for the actual partition in the water column for different times and depths. This implies the use of filtration (or another technique) in order to separate the dissolved and the particulate fraction; such an operation brings possibilities for errors at the very-low-concentration levels encountered in lakes.

2.   The concentrations in the water column and in fresh settling material can be determined separately, and partition coefficients can be derived from mean values of the measured concentrations. This method has the advantage that the concentrations in the settling material can be determined on representative samples with sufficient material, and that mean values of partition coefficients can be obtained. The results may differ from those obtained with method 1, because less small particles are obtained in sediment traps than in the water column. For the settling flux, the material in the sediment traps is representative, as far as no selective dissolution occurs before the material reaches the sediments. This method implies that the particulate matter and the conditions in the water column are sufficiently homogeneous, so that constant K_D can be assumed.

The partition coefficients given in Table 13.4 were determined by method 2. For Zn, Cu, Cd, and Pb, the measured total concentrations in the water column were used; in this calculation for Fe, the concentration measured in 0.45-μm filtrated samples was used. The assumption that the total concentrations can be used in this calculation can be verified by calculating the soluble concentration for different K_D (Table 13.5) with a particle concentration of 1 mg liter^{-1}. The soluble concentration is greater than 90% of the total concentration for K_D up to 100. Some direct measurements of Cu and Cd in filtrated solutions confirmed that the dissolved concentrations were nearly equal to total concentrations. With $K_D = 10^3$, soluble concentration is 50% of the total concentration; the uncertainty in the Pb partition coefficient is thus especially large, as the true soluble concentration is not known.

For Fe, the soluble concentration has to be used, and the partition coefficient may also be affected by a large error. In Figure 13.2(b), a model calculation, using the experimentally derived partition coefficients, is illustrated. In comparison with the model calculation derived from surface complexation constants, the distribution coefficient is found to be low for copper; this does not reflect that the binding of "free" Cu^{2+} to the surface is weak, but supports the idea that organic ligands are important—much more important than for other heavy-metal ions—in influencing the Cu(II) speciation (14,23,31,33).

The efficiency of the transport to the sediments and thus of the removal from the water column is closely related to the partition coefficients. The quantitative relationships, assuming a one-box model (see Chapter 1 in this book), are summarized in Table 13.6. Calculations with the values of Lake Constance and experimental data for several elements are given in Figure 13.7. Relative residence times have been evaluated on the basis of the input data, so that they are independent of the calculated K_D. Elements with small partition coefficients, that are thus mainly in soluble form, have residence times in the lake approaching that of the water. Elements with high partition coefficients, mainly in particulate form, have residence times approaching that of the particles. Assuming different residence times of the particles leads to different residence times of the metals associated with them. In the calculations of Figure 13.7, the total sedimentation rate was assumed to be constant, so that different residence times of particles correspond to different particle concentrations in the water column and different sinking velocities.

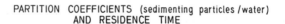

PARTITION COEFFICIENTS (sedimenting particles/water)
AND RESIDENCE TIME

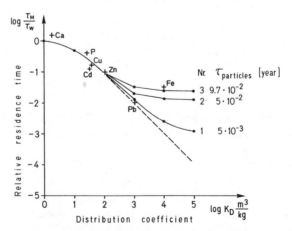

Figure 13.7. Relative residence time of elements in the lake as a function of the partition coefficients K_D. Data for the different elements are derived from the field data; relative residence times are based on the input data and concentrations in the lake. Drawn lines are calculated for the conditions of Lake Constance, with different particle residence times.

In reality, the situation is of course more complicated, due to the different dissolution and mineralization processes acting on the sinking particles, different sinking velocities due to different particle sizes, and so on.

Similar concepts on the relationship of partition coefficients and residence times have been applied to the oceans (18,34); in the case of the oceans, only removal by sedimentation has to be taken in account.

3. MASS BALANCES, RESIDENCE TIMES, AND STEADY STATE

In order to calculate a mass balance of the heavy metals in the lake, the inputs by river inflow and atmospheric precipitation and the outputs by sedimentation and river outflow have to be considered. Under the assumption that the inputs equal the outputs, steady-state concentrations can be calculated, assuming a one-box model (see Chapter 1 in this book). This model considers only mean inputs and outputs but no variations on a small time scale or local variations. It implies full mixing of the lake and thus cannot predict any temporal or local variations. In the one-box model, the sedimentation rate is assumed to be constant.

The calculation of the steady-state concentration is outlined in Table 13.6 [see Chapter 1 in this book and Imboden et al. (3)]. It is evident that sedimentation rate and steady-state concentration are closely related.

The mass balances for several elements have been roughly evaluated for Lake Constance, on the basis of the available experimental data. The atmospheric input and the sedimentation rate have been measured only at one pelagic location. The fluxes obtained are given in Figure 13.8; due to the uncertainties in the determination the fluxes do not exactly match.

The calculated steady-state concentrations of the metals are compared with the experimentally determined concentrations in Table 13.6. They agree within a factor of 2, which is reasonable considering the assumptions and the limitations of the precipitation and sedimentation data. According to these calculations, the retention of the metals in the lake is 90% for Pb and Fe; 75% for Cu; 80% for Cd; and 85% for Zn.

The evaluation of the different inputs allows an estimation of the importance of atmospheric inputs for the different metals. Although there are some uncertainties in the data used, it is evident that the atmospheric input is very important for Pb (>50% of total input), and for Zn and Cd. Atmospheric inputs have been shown to be relevant for the Great Lakes (1,36–38), especially for Pb and Zn.

The effect of the sedimentation on the steady-state concentrations is very important; due to the high sedimentation rates, the concentrations in the water column are kept low (Table 13.7) (39,40–42). Using a similar approach, the effect of different sedimentation rates on the Cd concentration in Lake Michigan has been estimated by Muhlbaier and Tisue (38).

A short comparison with small acidic mountain lakes illustrates this point.

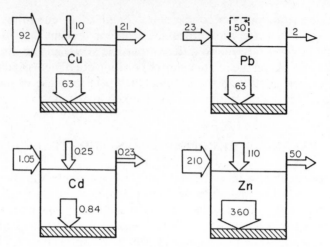

Figure 13.8. Mass balance for some heavy metals in Lake Constance. All fluxes are given in mg m^{-2} yr^{-1}. River input data are based on measurements in the Rhine near Schmitter (1981/ 1982). Total precipitations (dry + wet) were collected by EAWAG in the years 1975–1979 [Zobrist (35)] at one point on the lake shore. Sedimentation data are based on measurements 1981/1982 (10,11). Due to the imprecisions in the experimental determinations of the various fluxes, the mass balances do not exactly match.

Some small lakes in Tessin have no inflow; they are only fed by snowmelt and precipitation. They have very low-nutrient concentrations and productivity. The pH is between 4.8 and 6.6 due to acidic precipitations. Heavy-metal inputs occur by precipitation and—possibly to a small extent—by leaching of the rocks. Heavy-metal concentrations measured in one of these

Table 13.7. Heavy Metal Concentrations in Different Aquatic Systems

	Cu (10^{-9} moles liter^{-1})	Zn (10^{-9} moles liter^{-1})	Cd (10^{-9} moles liter^{-1})	Pb (10^{-9} moles liter^{-1})
Lake Constance	5–15	10–30	0.05–0.1	0.2–0.5
River Rhine[a] (Schmitter)	90	150	0.6	7
Laghetto Inferiore[b]	3	120	0.4	2
Pacific Ocean (39,40)	0.5–5	0.1–10	0.01–1	0.01–0.1
Mediterranean coastal waters (41)	—	—	0.1	0.6
Baltic Sea (42)	10–15	20–50	0.3–0.5	0.2–1

[a] Inflow to Lake Constance; mean values over 1 year.
[b] Mountain lake in Tessin (Switzerland), pH 5.5–6.5.

lakes—Laghetto Inferiore—are given in Table 13.7. The concentration of Pb, Zn, and Cd is higher than in Lake Constance and in Lake Zürich. The sedimentation rate is very low and the acidic pH keeps the metals in solution (see Eq. 1), so that removal of the metals from these lakes is inefficient.

4. COMPARISON OF HEAVY-METAL CYCLES IN OCEANS AND LAKES

Many publications have appeared in the last few years about the fate of heavy metals in the oceans (39,40,43–45), since sufficiently sensitive analytical methods and appropriate sampling techniques have been developed. There is much interest in the appearing correlations between heavy metals and nutrients that indicate a concomitant uptake by marine organisms. It may be useful to compare the lake and the oceanic system in order to understand better the involved processes.

In Table 13.7 the concentrations of Cu, Zn, Cd, and Pb in Lake Constance are compared with literature values for the Pacific Ocean and for coastal seawaters. The concentrations in the lake are in a similar order of magnitude as in marine environments, although the inputs to the lake by rivers and by the atmosphere are much higher. This must be explained by higher removal rates in the lake. Sedimentation rates in Lake Constance are about 10^3 g m^{-2} yr^{-1}; particle concentrations in the water column range from 0.1 to 5 mg $liter^{-1}$, depending strongly on depth and time. For comparison, particle concentrations are about 0.005–0.02 mg $liter^{-1}$ in the open ocean, and in turbid layers and in productive areas they can amount to 0.3 mg $liter^{-1}$; higher values are found near the coast and large rivers. Sedimentation rates vary from about 2 to 10 mm per 1000 years to 140 cm per 1000 years (in productive areas). This means that the removal efficiency of metals by sedimentation is much higher in lakes, due to the much higher sedimentation rates of particles and to larger particulate fractions of the metals. Two effects act in favor of a larger particulate fraction of metals in the lake: high particle concentrations and low chloride concentrations.

Lead profiles from lakes and from the ocean (40) are compared in Figure 13.9. Although the depth scales of the lake and of the ocean differ by orders of magnitude, the shapes of the lead depth profiles are very similar; because of much lower particle concentrations, similar processes may occur in the ocean over kilometers that would occur in a lake in the upper 20–50 m. This is observed for the uptake and release of nutrients that can be followed in the ocean over a depth of 1 km and in a lake in the upper 50 m (Figure 13.5). In the case of lead, it can be supposed that similar processes are responsible for the observed higher concentrations in the upper layers. This may indicate the influence of the atmospheric input to the upper layer in both cases, as an important source of lead. This atmospheric input is higher in the lakes that are situated near densely populated and industrialized regions. The

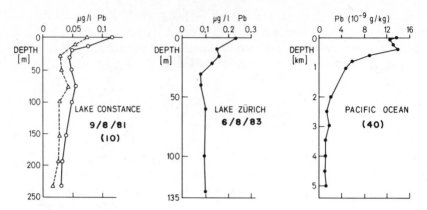

Figure 13.9. Lead profiles in Lake Constance (9/8/81) (10), Lake Zurich (6/8/83), and in the Pacific Ocean (1981). [Used with permission of Schaule and Patterson (40)]. The similar shape of these profiles, despite the difference in length scales (kilometers for the ocean and meters for the lakes), may indicate the influence of the atmospheric deposition on the upper layers.

oceans are sensitive to the widely dispersed lead inputs, far away from direct sources.

The striking correlations of metals with nutrients, for example, of cadmium with phosphate (39), found in oceans, do not seem to appear clearly in Lake Constance, although the biogenic particles appear to play an important role in the transport of the metals. Several reasons can be put forward that can explain this:

1. Only few measurements have been made in the upper layer (0–10 m) of the lake.

2. Contamination risks (boat) are especially high here.

3. Inputs of particulate heavy metals are much higher in Lake Constance than in oceanic environments.

4. Even in the middle of the lake, the influence of the Rhine can sometimes be noticed by a massive input of allochthonous material.

5. Metals that are released during mineralization of algal material may be adsorbed in the lake on other particles and transported to the sediments, so that the release of metals into the water column cannot be observed as in the oceans.

6. The scavenging by Mn and Fe oxides is an additional removal process in the lake.

5. CONCLUSIONS

As shown by the results obtained in Lake Constance and in Lake Zürich, heavy metals, such as Cu, Zn, Cd, and Pb are continuously removed from lake waters by becoming attached to settling particles; although lakes are much more polluted with heavy metals than the oceans, the residual concentrations of heavy metals in the water column of the lake are of a similar order of magnitude as those found in the marine environment. The partition of the heavy metals between the settling particles and the water determines the relative residence time of these metals in the lake and thus in turn their residual concentrations. The overall partition coefficient as determined experimentally from the field data is influenced by various factors, especially by the affinity of the free metal ion to the particle surface, by the type of solid surface present (allochthonous aluminum silicates, $CaCO_3$, MnO_2, and biota), and by the speciation of the metal ion in solution. The biological surface is most likely the better scavenger for heavy metals than the inorganic surfaces. A large part of the fresh sedimenting particles consists of particles of biogenic origin—organic material and freshly precipitated calcium carbonate. The mean elemental composition of these particles corresponds to a rather constant stoichiometry that may include the heavy metals: $\{C_{113} N_{15} P_1 Si_{14} Ca_{93} Mg_7 Cu_{0.008} Zn_{0.06} Pb_{0.004} Cd_{0.00005}\}$.

Although the removal by biological particles is certainly an important factor, the metal concentration profiles in the water column do not show large differences between the upper and deeper layers; this is probably due to high particulate inputs and incomplete release during mineralization of the algal material. An additional removal mechanism is provided by the Mn cycle in the deep layers of the lake; there the freshly produced Mn oxides are efficient scavengers of heavy metals. High sedimentation rates and favorable partition of the metals between particles and solution are the determining factors that keep the metal concentrations in the water column low, although metal pollution (inputs to the lake from atmospheric precipitation and from river inflow) is high. It appears that a productive lake with a high sedimentation rate represents a situation which is favorable for an efficient removal of metals. On the other hand, heavy metals tend to accumulate in acidic lakes of low productivity.

ACKNOWLEDGMENTS

I thank Werner Stumm for his continuous support and many discussions. Michael Sturm carried out the sedimentological part of this project; I am grateful for his permission to use some of his data and for his advice. Special appreciation is due to D. Kistler and P. Klahre for technical assistance in sampling and for analytical work.

REFERENCES

1. J. O. Nriagu, A. L. W. Kemp, H. K. T. Wong, and N. Harper, "Sedimentary Record of Heavy Metal Pollution in Lake Erie," *Geochim. Cosmochim. Acta* **43**, 247 (1979).

2. D. N. Edgington and J. A. Robbins, "Records of Lead Deposition in Lake Michigan Sediments Since 1800," *Environ. Sci. Technol.* **10,** 266 (1976).

3. D. M. Imboden, J. Tschopp, and W. Stumm, "Die Rekonstruktion früherer Stofffrachten in einem See mittels Sedimentuntersuchungen," *Schweiz. Z. Hydrol.* **42,** 1 (1980).

4. G. Müller, "Heavy Metals and PAH in a Sediment Core from Lake Constance," *Naturwissenschaften* **64,** 427 (1977).

5. B. Rippey, R. J. Murphy, and S. W. Kyle, "Anthropogenically Derived Changes in the Sedimentary Flux of Metals in Lough Neagh, Northern Ireland," *Environ. Sci. Technol.* **16,** 23 (1982).

6. P. Baccini, J. Ruchti, O. Wanner, and E. Grieder, "Regulation of Trace Metal Concentrations in Limno-Corrals," *Schweiz. Z. Hydrol.* **41,** 202 (1979).

7. R. Gächter and A. Máreš, "Effects of Increased Heavy Metal Loads on Phytoplankton," *Schweiz. Z. Hydrol.* **41,** 228 (1979).

8. W. Sunda and R. R. L. Guillard, "The Relationship between Cupric Ion Activity and the Toxicity of Copper to Phytoplankton," *J. Mar. Res.* **34,** 511 (1976).

9. L. Sigg, M. Sturm, L. Mart, H. W. Nürnberg, and W. Stumm, "Schwermetalle im Bodensee; Mechanismen der Konzentrationsregulierung," *Naturwissenschaften* **69,** 546 (1982).

10. L. Sigg, M. Sturm, J. Davis, and W. Stumm, "Metal Transfer Mechanisms in Lakes," *Thalassia Jugoslav.* **18,** 293 (1983).

11. M. Sturm, U. Zeh, J. Müller, L. Sigg, and H. H. Stabel, "Schwebstoffuntersuchungen im Bodensee mit Intervall-Sedimentationsfallen," *Eclogae Geol. Helv.* **75,** 579 (1982).

12. R. M. Smith and A. E. Martell, "Critical Stability Constants," in *Inorganic Complexes*, Vol. 4, Plenum Press, New York and London, 1976.

13. D. Dyrssen and M. Wedborg, "Major and Minor Elements, Chemical Speciation in Estuarine Waters." In E. Olausson and I. Cato (Eds.), *Estuaries*, Wiley, New York 1980.

14. W. Stumm and J. J. Morgan, *Aquatic Chemistry* (2nd ed.), Wiley–Interscience, New York, 1982.

15. P. Schindler, "Surface Complexes at Oxide/Water Interfaces." In M. A. Anderson and A. J. Rubin, (Eds.), *Adsorption of Inorganics at the Solid/Liquid Interfaces*, Ann Arbor Science Pub., Ann Arbor, Michigan, 1981.

16. J. C. Westall, J. L. Zachary, and F. M. M. Morel, Technical Note Number 18, Water Quality Laboratory Department of Civil Engineering, MIT, 1976.

17. H. H. Stabel, Chapter 7 in this book.

18. L. Balistieri, P. G. Brewer, and J. W. Murray, "Scavenging Residence Times of Trace Metals and Surface Chemistry of Sinking Particles in the Deep Ocean," *Deep Sea Res.* **28A,** 101 (1981).

19. J. I. Parker, K. A. Stanlaw, J. S. Marshall, and C. W. Kennedy, "Sorption and Sedimentation of Zn and Cd by Seston in Southern Lake Michigan," *J. Great Lakes Res.* **8,** 520 (1982).

20. J. Wood and H. K. Wang, Chapter 4 in this book.

21. J. W. Murray, "The Interaction of Metal Ions at the Manganese Dioxide-Solution Interface," *Geochim. Cosmochim. Acta* **39,** 505 (1975).

22. L. S. Balistieri and J. W. Murray, "Metal–Solid Interactions in the Marine Environment: Estimating Apparent Equilibrium Binding Constants," *Geochim. Cosmochim. Acta* **47,** 1091 (1983).

23(a) P. Baccini and U. Suter, "Melimex, An Experimental Heavy Metal Pollution Study: Chemical Speciation and Biological Availability of Copper in Lake Water," *Schweiz. Z. Hydrol.* **41,** 291 (1979).

23(b) P. Baccini and T. Joller, "Transport Processes of Copper and Zinc in a Highly Eutrophic and Meromictic Lake," *Schweiz. Z. Hydrol.* **43,** 176 (1981).

24. R. Giovanoli, P. Bürki, M. Giuffredi, and W. Stumm, "Layer Structured Manganese Oxide Hydroxides, IV: The Buserite Group; Structure Stabilization by Transition Elements," *Chimia* **29,** 517 (1975).

25. D. Diem, "Die Oxidation von Mangan (II) im See," Ph.D. thesis, ETH Zürich No. 7359 (1983).

26. W. Davison, C. Woof, and E. Rigg, "The Dynamics of Iron and Manganese in a Seasonally Anoxic Lake; Direct Measurements of Fluxes using Sediment Traps," *Limnol. Oceanogr.* **27,** 987 (1982).

27. R. Giovanoli, University of Berne, unpublished results.

28. J. Tschopp, "Die Verunreinigung der Seen mit Schwermetallen," Ph.D. thesis, ETH Zürich Number 6362 (1979).

29. K. Kelts and K. Y. Hsu, "Freshwater Carbonate Sedimentation." In A. Lerman (Ed.), *Lakes, Chemistry, Geology, Physics*, Springer-Verlag, New York, 1978.

30. R. W. Collier, "Trace Element Geochemistry of Marine Biogenic Particulate Matter," Ph.D. thesis, MIT, 1980.

31. P. M. Stokes, "Copper Accumulations in Freshwater Biota." In J. O. Nriagu (Ed.), *Copper in the Environment*, Part I, Wiley–Interscience, New York, 1979.

32. B. Imber, M. G. Robinson, and F. Pollehne, "Complexation of Trace Metals in Natural Waters," International Symposium, Texel, 1983.

33. W. Sunda, P. J. Hanson, "Chemical Speciation of Copper in River Water." In E. A. Jenne (Ed.), *Chemical Modeling in Aqueous Systems*, ACS Symposium Series 93, 1979.

34. M. Whitfield, "The Mean Oceanic Residence Time (MORT) Concept—A Rationalisation," *Mar. Chem.* **8,** 101 (1979).

35. J. Zobrist, "Die Belastung der Schweizerischen Gewässer durch Niederschläge," In *Acid Precipitation—Origin and Effects*, VDI International Conference, Lindau, 1983.

36. S. J. Eisenreich, "Atmospheric Input of Trace Metals to Lake Michigan," *Water, Air Soil Pollut.* **13,** 287 (1980).

37. E. R. Christensen and N.-K. Chien, "Fluxes of Arsenic, Lead, Zinc and Cadmium to Green Bay and Lake Michigan Sediments," *Environ. Sci. Technol.* **15,** 553 (1981).

38. J. Muhlbaier and G. T. Tisue, "Cadmium in the Southern Basin of Lake Michigan," *Water, Air Soil Pollut.* **15,** 45 (1981).

39. K. W. Bruland, "Oceanographic Distributions of Cd, Zn, Ni and Cu in the North Pacific," *Earth Planet. Sci. Lett.* **47,** 176 (1980).

40. B. K. Schaule and C. C. Patterson, "Lead Concentrations in the Northeast Pacific: Evidence for Global Anthropogenic Perturbations." *Earth Planet. Sci. Lett.* **54,** 97 (1981).

41. H. W. Nürnberg, "Voltametric Trace Analysis in Ecological Chemistry of Toxic Metals," *Pure Appl. Chem.* **54,** 853 (1982).

42. B. Magnusson and S. Westerlund, "The Determination of Cd, Cu, Fe, Ni, Pb and Zn in Baltic Sea Water," *Mar. Chem.* **8,** 231 (1980).

43. E. A. Boyle et al., "Cu, Ni, and Cd in the Surface Waters of the North Atlantic and North Pacific Oceans," *J. Geophys. Res.* **86,** 8048 (1981).

44. P. Buat-Menard and R. Chesselet, "Variable Influence of the Atmospheric Flux on the Trace Metal Chemistry of Oceanic Suspended Matter," *Earth Planet. Sci. Lett.* **42,** 399 (1979).

45. P. A. Yeats and J. A. Campbell, "Nickel, Copper, Cadmium and Zinc in the Northwest Atlantic Ocean," *Mar. Chem.* **12,** 43 (1983).

14

ACIDIFICATION OF AQUATIC AND TERRESTRIAL SYSTEMS

Jerald L. Schnoor

Department of Civil and Environmental Engineering, University of Iowa, Iowa City, Iowa 52242, USA

Werner Stumm

Institute for Water Resources and Water Pollution Control (EAWAG), Swiss Federal Institute of Technology (ETH), Zurich, Switzerland

Abstract

Our global environment is on the average, with regard to a proton and electron balance, in a stationary state. Schematically, the oxidation states and the H^+ reservoirs of the weathering sources equal those of the sedimentary products. Obviously, such balances are locally and regionally upset. In addition, they have become significantly disturbed by the combustion of fossil fuels which generates a net production of H^+ ions in atmospheric deposition.

In this chapter, we intend to show how this disturbance is transferred to the terrestrial and aquatic environment and to review the major H^+ yielding and H^+ consuming processes occurring in the watersheds, paying special attention (1) to the disturbance of the H^+ balance resulting from temporal or spatial decoupling of the production and mineralization of the biomass, and (2) to the H^+ ion consumption by the weathering of rocks. The aggradation of vegetation (productive forests) may be accompanied by acidification of surrounding waters. In some instances, the acidic atmospheric *deposition may be sufficient to disturb the existing H^+ balance between aggrading vegetation and weathering reactions.*

Chemical weathering is shown to be a key process in neutralizing the internal production and the anthropogenic input of acids to a watershed. The lakes which have been acidified by acid precipitation are those lacking carbonate minerals and characterized by sensitive hydrological settings.

1. INTRODUCTION

What controls the pH and the pε of our present-day environment*? Considering the schematic reaction,

$$\text{igneous rock} + \text{volatile substances} \rightleftarrows \text{air} + \text{seawater} + \text{sediments} \quad (1)$$

Sillén (1) gave the picture that volatiles (H_2O, CO_2, HCl, SO_2), the acids of volcanoes (i.e., the acids that have leaked from the interior of the earth), have reacted in a gigantic acid–base reaction with the bases o rocks (silicates, oxides, and carbonates). Similarly, he calculated from the model system the quantities of redox components that have participated in a redox titration. On a global average, the environment is, with regard to a proton and electron balance, in a stationary situation which corresponds to a present-day atmosphere of 20.9% O_2, 0.03% CO_2, 79.1% N_2 and a world ocean with pH $\approx$ 8 and pε $\approx$ 12.5.

This steady state is achieved by a global balance of oxidation and reduction and H^+ production and consumption. One can infer that above all, the rate of oxidation of reduced S, Fe, and C, which occurs during the weathering of continental rocks, is balanced by the rate of reduction of the higher valent S, Fe, and C into reduced forms during and after deposition in the sediments (2). Because transfer of electrons is coupled with the transfer of protons (to maintain charge balance), oxidation and reduction are accompanied by proton release and proton consumption, respectively. Furthermore, the dissolution of rocks and the precipitation of minerals are accompanied by H^+ consumption and H^+ production, respectively. Thus, the pε and pH of our present global environment reflect the levels where the oxidation state and the H^+ ion reservoirs (cf. alkalinity or acidity definitions given below) of the weathering sources equal those of the sedimentary products.

Obviously, in local environments, H^+ and e^- balances are upset and significant variations in pH and pε occur. The weathering cycle is also affected at least locally and regionally by our civilization. Redox conditions

---

* pε, a parameter for redox intensity, gives the (hypothetical) electron activity at equilibrium and measures the relative tendency of a solution to accept or transfer electrons. It is related to the equilibrium redox potential E_H (volts, hydrogen scale) by pε $= -\log \{e^-\} = E_H/2.3\,RTF$ (where R = gas constant, T = absolute temperature, and F = the Faraday Constant; RTF^{-1} = 0.059 V mole^{-1} at 25°C).

in the atmosphere are disturbed by enhanced rates of artificial weathering of fossil fuel. The combustion of these fuels leads to a disturbance in the e^- balance. The reactions of the oxidation of C, S, and N exceed reduction reactions in these elemental cycles. A net production of H^+ ions in atmospheric precipitation is a necessary consequence (3). The disturbance is transferred to the terrestrial and aquatic environment, where either oxidized species are reduced or acidity is neutralized via acid–base reactions. Oxidation reactions in the terrestrial and aquatic environment and assimilation processes by aggrading vegetation can produce local acidification in watersheds which may rival acid loading from the atmosphere.

It is the objective of this discussion to briefly review the major H^+ yielding and H^+ consuming processes occurring in watersheds; specifically, to elaborate the influence of chemical weathering, sediment formation, and biological assimilation (and humus formation) on the proton balance. An attempt will be made to give unifying definitions of acid- or base-neutralizing capacity (alkalinity and acidity, respectively) with respect to some reference level for the various subsystems of a watershed environment.

2. GENESIS OF ACID PRECIPITATION

Figure 14.1 illustrates the genesis of acid precipitation. Most of the major and minor gaseous components of the atmosphere (O_2, N_2, CO_2, CO, NO, NO_2, SO_2, CH_4) participate in elemental cycles which are governed by oxidation–reduction reactions mostly of biological origin; photosynthesis and respiration are major reactions in this cycle. The atmosphere is more susceptible to anthropogenic impacts than the terrestrial or the aqueous environment because, from a quantitative point of view, the atmosphere is much smaller than the other reservoirs. Furthermore, the time constants concerning atmospheric alterations are small in comparison to those of the seas and the lithosphere. When the rate of oxidation of carbon, nitrogen, and sulfur increases relative to the rate of reduction of CO_2, nitrogen, oxides of nitrogen, SO_2, and H_2SO_4 (which is caused by the activities of civilization during the fossil-fuel age), then the delicate balance is disturbed. Accordingly, the concentration of CO_2 has increased globally and the concentrations of SO_2, H_2SO_4, NO, NO_2, HNO_2, and HNO_3 have increased regionally. Buffering of these interactions in other reservoirs is slow. For example, the mixing time of the ocean is about 1000 years. Since a modification of the redox balance corresponds to a modification of the acid–base balance the potential acidity of the atmosphere increases vis-à-vis the bases (Fig. 14.1, upper part). Resulting from the oxidation of sulfur and nitrogen in the atmosphere, the strong acids H_2SO_4 and HNO_3 are the main participants in the formation of acid rain.

A considerable portion of the strong acids originate from the oxidation of sulfur in the combustion of fossil fuels and the "fixation" of nitrogen from

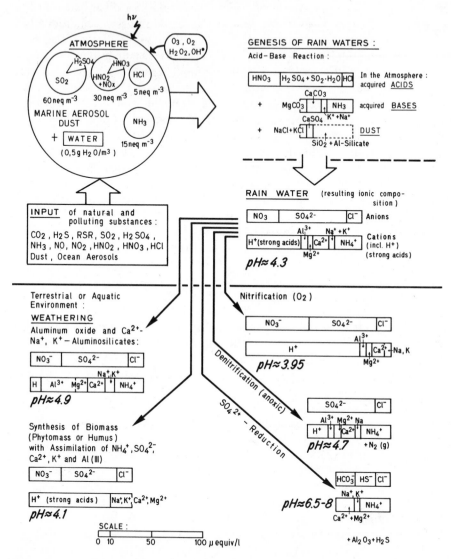

Figure 14.1. Depicts genesis of acid rain (3). From the oxidation of S and N during the combustion of fossil fuels there is a buildup in the atmosphere (in the gas phase, aerosol particles, raindrops, snowflakes, and fog) of CO_2 and the oxides of S and N, which leads to acid–base interaction. The importance of absorption of gases into the various phases of gas, aerosol, and atmospheric water depends on a number of factors. (In this figure, we assume a yield of 50% for SO_4^{2-} and NO_3^-, of 80% for NH_3, and of 100% for HCl.) The genesis of acid rain is shown on the upper right as an acid–base titration. The data given are representative of the situation encountered in Zürich, Switzerland. Various interactions with the terrestrial and aquatic environment (see Table 14.1) are shown in the lower part of the figure.

the atmosphere to NO and NO_2 (during the combustion of gasoline in motor vehicles or other combustion processes of sufficiently high temperature). HCl results from the combustion and decomposition of organochlorine compounds, for example, in polyvinyl chloride plastics during incineration. But it should also be mentioned that there exists natural sources resulting from volcanic activity and the oxidation of hydrogen sulfide from anaerobic sediments, as well as dimethyl sulfide and other volatile organic sulfur compounds originating from the ocean. Bases originate in the atmosphere as the carbonates of wind-blown dust and from ammonia, generally of natural origin. The NH_3 originates from NH_4^+ ions and from the decomposition of urea found in soil environments. Redox and acid–base reactions occur in the gas phase, in aerosols, in raindrops, cloud droplets, and fog.

The reactions of oxidation of SO_2 in the atmosphere yield a sulfur residence time of several days at the very most; this corresponds to a transport distance of hundreds to a thousand kilometers. The formation of HNO_3 oxidation is more rapid, and compared to H_2SO_4, this results in a shorter travel distance from the emission source. H_2SO_4 can also react with NH_3 to form NH_4HSO_4 or $(NH_4)_2SO_4$ aerosols. In addition, NH_4NO_3 aerosols are in equilibrium with $NH_3(g)$ and $HNO_3(g)$. The importance of gas and aerosols scavenging by atmospheric condensates and raindrops depends on many factors.

Subsequent to the deposition, there are various processes in the terrestrial and aquatic environment that can neutralize or enhance the acidity of the precipitation. Of the processes depicted in Figure 14.1, denitrification, sulfate reduction, and chemical weathering decreases acidity, while photosynthetic assimilation and nitrification increase the acidity of the waters (Table 14.1). For a better understanding of the factors that affect the H^+ balance, we first need to review some definitions.

Table 14.1.   Processes Which Modify the H^+ Balance in Waters (3)

Processes	Changes in Alkalinity $\Delta[Alk]^a = -\Delta[H-Acy]^b$ [Equivalents Per Mole Reacted (Reactant Is Underlined)]^c
1. Weathering reactions:	
$\underline{CaCO_3(s)} + 2H^+ \rightleftarrows Ca^{2+} + CO_2 + H_2O$	$+2$
$\underline{CaAl_2Si_2O_8(s)} + 2H^+ \rightleftarrows Ca^{2+} + H_2O +$ $Al_2Si_2O_5(OH)_4(s)$	$+2$
$\underline{KAlSi_3O_8(s)} + H^+ + 4\frac{1}{2}H_2O \rightleftarrows K^+ +$ $2H_4SiO_4 + \frac{1}{2}Al_2Si_2O_5(OH)_4(s)$	$+1$
$\underline{Al_2O_3 \cdot 3H_2O} + 6H^+ \rightleftarrows 2Al^{3+} + 6H_2O$	$+6$
2. Ion exchange:	
$2\underline{ROH} + SO_4^{2-} \rightleftarrows R_2SO_4 + 2OH^-$	$+2$
$NaR + \underline{H^+} \rightleftarrows HR + Na^+$	$+1$

Table 14.1. (*continued*)

Processes		Changes in Alkalinity $\Delta[Alk]^a = -\Delta[H-Acy]^b$ [Equivalents Per Mole Reacted (Reactant Is Underlined)]^c
3. Redox processes (microbial mediation):		
Nitrification	$\underline{NH_4^+} + 2O_2 \rightleftarrows NO_3^- +$ $H_2O + 2H^+$	-2
Denitrification	$1\frac{1}{4}CH_2O + \frac{1}{5}\underline{NO_3^-} + H^+ \rightarrow$ $1\frac{1}{4}CO_2 + \frac{1}{2}N_2 + 1\frac{3}{4}H_2O$	$+1$
Oxidation of H_2S	$\underline{H_2S} + 2O_2 \rightarrow SO_4^{2-} + 2H^+$	-2
SO_4^{2-} reduction	$\underline{SO_4^{2-}} + 2CH_2O + 2H^+ \rightarrow$ $2CO_2 + H_2S + H_2O$	$+2$
Pyrite oxidation	$\underline{FeS_2}$ (s) $+ 3\frac{3}{4}O_2 + 3\frac{1}{2}H_2O \rightarrow$ $Fe(OH)_3 + 2SO_4^{2-} + 4H^+$	-4
4. Synthesis ($\rightarrow$) and decomposition ($\leftarrow$) of biomass and of humus:		
Photosynthesis with NO_3 assimilation ($\rightarrow$), aerobic respiration ($\leftarrow$):		
$a\,CO_2(g) + b\,\underline{NO_3^-} + c\,HPO_4^{2-} + \cdots +$ $g\,Ca^2 + h\,Mg^{2+} + i\,K^+ + f\,Na^+ + x\,H_2O$		$+ b + 2c + 2d$ $-(2g + 2h + i + f)$
$+ (b + 2c + 2d - 2g - 2h - i - f)\,H^+$ $\rightleftarrows \{C_a\,N_b\,P_c\,S_d \cdots Ca_g\,Mg_h\,K_i\,Na_f\,H_2O_m\}_{biomass}$ $+(a + 2b)\,O_2\,(g)$		
NH_4 assimilation ($\rightarrow$); anaerobic mineralization (back reaction):		
$a\,CO_2(g) + b\,\underline{NH_4^+} + \cdots \rightleftarrows \{Ca \cdots N_b\}_{biomass}$ $+ a\,O_2\,(g)$		$-b$
N_2 fixation (half-reaction):		
$\cdots + \frac{1}{2}b\underline{N_2}(g) + \cdots \rightarrow \{\cdots N_b \cdots\}_{biomass}$		0

^{a} [Alk] = alkalinity = acid-neutralizing capacity.
^{b} [H − Acy] = mineral acidity.
^{c} In the buildup of biomass (or the exchange of ions) the uptake of each equivalent of conservative anions causes an equivalent increase in alkalinity, and each equivalent of base cations which is taken up results in an equivalent decrease of alkalinity. One comes to the same conclusion as long as the biomass or humus is formed of neutral components $\{(CH_2O)_{2a}(NH_3)_b(H_3PO_4)_c(H_2CO_4)_d \cdots (Ca(OH)_2)_g(Mg(OH)_2)_h(KOH)_i(NaOH)_f(H_2O)_m\}$ and provided that one has written the corresponding stoichiometric equations for formation and decomposition.

3. ACID- OR BASE-NEUTRALIZING CAPACITY

One needs to distinguish between the H^+ concentration (or activity) as an intensity factor and the availability of H^+, the H^+ ion reservoir, as given by the base-neutralizing capacity, BNC, or acidity[H-Acy]. The base-neutralizing capacity ($= $[H-Acy]) may be defined by a net proton balance with regard to a reference level, that is, the sum of the concentrations of all the species containing protons in excess minus the concentrations of the species containing protons in deficiency of the proton reference level. For natural waters a convenient reference level (corresponding to an equivalence point in alkalimetric titrations) is H_2O and H_2CO_3:

$$[H\text{-}Acy] = [H^+] - [HCO_3^-] - 2[CO_3^{2-}] - [OH^-] \qquad (2)$$

The acid-neutralizing capacity, ANC, or alkalinity [Alk] is related to [H-Acy] by

$$-[H\text{-}Acy] = [Alk] = [HCO_3^-] + 2[CO_3^{2-}] + [OH^-] - [H^+] \qquad (3)$$

Considering a charge balance for a typical natural water,

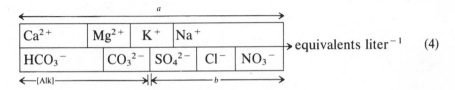

we realize that [Alk] (and [H-Acy]) can also be expressed by a charge balance, the equivalent sum of conservative cations minus the sum of conservative anions* ([Alk] $= a - b$):

$$\begin{aligned} [Alk] &= [HCO_3^-] + 2[CO_3^{2-}] + [OH^-] - [H^+] \\ &= [Na^+] + [K^+] + 2[Ca^{2+}] + 2[Mg^{2+}] - [Cl^-] \qquad (5) \\ &\quad - 2[SO_4^{2-}] - [NO_3^-] \end{aligned}$$

The [H-Acy] for this particular water, obviously negative, is defined ([H-Acy] $= b - a$) as

$$\begin{aligned} [H\text{-}Acy] &= [Cl^-] + 2[SO_4^{2-}] + [NO_3^-] - [Na^+] \\ &\quad - [K^+] - 2[Ca^{2+}] - 2[Mg^{2+}] \end{aligned} \qquad (6)$$

The definitions given for proton balance and electroneutrality agree rigorously because the reference state has been defined in terms of uncharged species. If the water under consideration contains other acid- or base-consuming species one needs to extend the proton reference level to the other components. If a natural water contains, in addition to the species given,

* The conservative cations are the so-called base cations (historically, the cations of the strong bases $Ca(OH_2)$, KOH, etc.) and the conservative anions are those anions which are the conjugate bases of strong acids (SO_4^{2-}, NO_3^-, Cl, etc.).

organic molecules such as humic or fulvic acids (which may be represented by the summarizing formula HOrg) and ammonium ions, we can extend the reference level to include these species in the form of neutral molecules HOrg, NH_3. [Alk] would now be defined as

$$
\begin{aligned}
[\text{Alk}] &= [HCO_3^-] + 2[CO_3^{2-}] + [Org^-] + \overbrace{[OH^-] - [H^+]}^{\text{negligible}} \\
&= 2[Ca^{2+}] + 2[Mg^{2+}] + [NH_4^+] + [K^+] + [Na^+] \\
&\quad - 2[SO_4^{2-}] - [NO_3^-] - [Cl^-] \\
&= -[\text{H-Acy}]
\end{aligned} \tag{7}
$$

Similarly, for an acidic water the charge balance may be

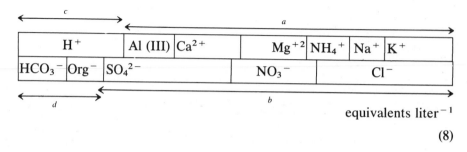

$$
\text{equivalents liter}^{-1} \tag{8}
$$

$$
\begin{aligned}
[H - Acy] &= b - a = c - d \\
&= [H^+] - [HCO_3^-] - 2[CO_3^{2-}] - [Org^-] \\
&= 2[SO_4^{2-}] + [NO_3^-] + [Cl^-] - 3[Al^{3+}] - 2[Ca^{2+}] \\
&\quad - 2[Mg^{2+}] - [NH_4^+] - [Na^+] - [K^+]
\end{aligned} \tag{9}
$$

If NH_3 is chosen as a zero-level reference state, $[NH_4^+]$ is counted (as we have done here) as a cation of a strong base to belong to the excess positive charge from strong bases. Alternatively, one could have chosen NH_4^+ as a reference state in the proton balance (because at the CO_2 equivalence point, NH_4^+ is the principal component). Then the rigorous identity between charge balance and H^+ balance no longer exists. Since we are here primarily concerned with *change* in alkalinity or acidity, it is preferable to choose as a zero-proton reference state the uncharged species, that is, H_2O, CO_2, HOrg, H_3BO_3, NH_3, and so on.

The overall change in alkalinity or [H-Acy] of a water can then readily be calculated from the stoichiometry of weathering reactions, ion exchange, redox processes, and biological production and respiration. A simple accounting can be made: every base cationic-charge unit (Ca^{2+}, K^+, NH_4^+, etc.) that is removed from the water by whatever process is equivalent to a proton added to the water, and every conservative anionic-charge unit (from anions of strong acids) (NO_3^-, SO_4^{2-}, Cl^- etc.) removed from the water corresponds to a proton removed from the water; or generally,

$$\Sigma\Delta[\text{base cations}] - \Sigma\Delta[\text{conservative anions}] = +\Delta[\text{Alk}] \qquad (10)$$
$$= -\Delta[\text{H-Acy}]$$

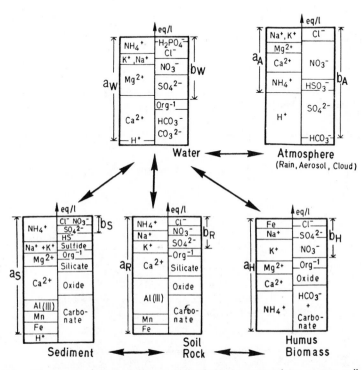

Figure 14.2. The transfer of alkalinity or acidity between atmosphere, water, soil (rocks), sediments, and biota (humus). In a (hypothetically closed) large system of the environment (atmosphere, hydrosphere, lithosphere) a proton and electron balance is maintained. Temporal and spatial inhomogeneities between (and within) the individual reservoirs may cause significant shifts in electron and proton balances so that subsystems contain differences in alkalinity and acidity. Any transfer in ions from one substance to another—however caused, by transport, chemical reaction, or redox processes—causes a corresponding transfer of [Alk] (or [H-Acy]). For each subsystem, the acid-neutralizing capacity [Alk] or the base-neutralizing capacity [H-Acy] can be defined in terms of the difference between the equivalent sum of base cations and that of conservative anions:

$$[\text{Alk}]_{\text{water}} = a_W - b_w = -[\text{H-Acy}]_{\text{water}}$$

$$[\text{Alk}]_{\text{soil, rock}} = a_R - b_R = -[\text{H-Acy}]_{\text{soil, rock}}$$

$$[\text{Alk}]_{\text{sediments}} = a_S - b_S = -[\text{H-Acy}]_{\text{sediments}}$$

$$[\text{Alk}]_{\text{humus, biomass}} = a_H - b_H = -[\text{H-Acy}]_{\text{humus, biomass}}$$

$$[\text{Alk}]_{\text{atmosphere}} = a_A - b_A = -[\text{H-Acy}]_{\text{atmosphere}}$$

As Table 14.1 illustrates, every reaction can be written by a balanced stoichiometric equation, and the change in [Alk] can be read either from Eq. 10 or the H^+ per mole of reactant needed to balance the reaction. Note that the uptake (or release) of uncharged species such as O_2, CO_2 in photosynthesis and respiration, N_2, NH_3 in N_2 fixation and denitrification, H_2S, or CH_4 has no effect on changes in the alkalinity or [H-Acy] (cf. Table 14.1).

The advantage of balancing conservative-charge units is that the method can readily be extended to other systems: to sediments, soils, rocks, and biota to account for changes in acid- or base-neutralizing capacity of these systems (Fig. 14.2).

4. PROCESSES AFFECTING ACID-NEUTRALIZING CAPACITY (ANC)

4.1. ANC of Soil

In weathering reactions alkalinity is added from the soil–rock system to the water:

$$\Delta[Alk]_{water} = -\Delta[Alk]_{soil,rock} \tag{11}$$

As Figure 14.2 illustrates, the acid-neutralizing capacity of a soil is given by the bases, carbonates, silicates, and oxides of the soil system. In a similar way as for an aquatic system, alkalinity and acidity can be defined. If the composition of the soil is not known but its elemental analysis is given in oxide components, the following kind of accounting is equivalent to that given in Figure 14.2.

$$\begin{aligned} ANC = [Alk] = {} & 6(Al_2O_3) + 2(CaO) + 2(MgO) + 2(K_2O) \\ & + 2(Na_2O) + 4(MnO_2) + 2(FeO) + (\text{``}NH_4OH\text{''}) \\ & - 2(SO_3) - (HCl) - 2(N_2O_5) \end{aligned} \tag{12}$$

Sulfates, nitrates, and chlorides incorporated or adsorbed are subtracted from ANC. On the basis of acid balance alone, one cannot distinguish whether uptake by the soil of H^+ and the release of cations is caused by weathering or by ion exchange. Similarly, the loss of SO_4^{2-} from the water accompanied by OH^- or [Alk] release may be caused by sulfate reduction (see Table 14.1) or by ion exchange.

4.2. Chemical Weathering

Figure 14.3 shows the processes which affect the acid-neutralizing capacity of soils. Ion exchange occurs at the surface of clays and organic humus in

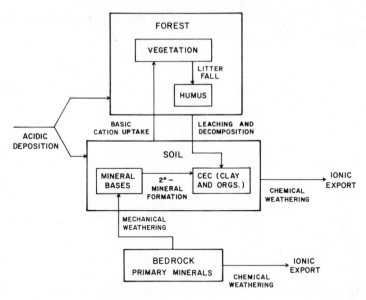

Figure 14.3. Process affecting the acid-neutralizing capacity of soils (including the exchangeable bases, CEC, and mineral bases).

various soil horizons. The net effect of ion-exchange processes is identical to chemical weathering (and alkalinity), that is, hydrogen ions are consumed and basic cations (Ca^{+2}, Mg^{+2}, Na^+, K^+) are released. However, the kinetics of ion exchange are rapid relative to those of chemical weathering (taking minutes compared to hours or even days). In addition, the pool of exchangeable bases is small compared to the total ANC of the soil (Eq. 12). Thus there exists two pools of bases in soils—a small pool of exchangeable bases with relatively rapid kinetics and a large pool of mineral bases with the slow kinetics of chemical weathering. If chemical weathering did not replace exchangeable bases in acid soils of temperate regions receiving acidic deposition, the base exchange capacity would be completely diminished over a period of 50–200 years (see Table 14.2).

In the long run, chemical weathering is the rate-limiting step in the supply of basic cations for export from watersheds. Surface waters usually are not at chemical equilibrium with regard to common minerals in soils or sediments with the possible exception of amorphous aluminum hydroxide (4). The chemistry of natural waters is predominantly kinetically controlled.

The genesis of podzolic acid soils stems from a chromatographic effect whereby bases are depleted from the upper soil profile and iron and aluminum are mobilized. Soil profiles become stratified into horizons, for example, O, A, B, and C. In podzolic soils the organic "O" horizon is typically 0–10 cm deep and contains the humus from leaf and litter decomposition. The "A" horizon is the uppermost "inorganic" soil horizon. In podzolic soils, it is depleted of basic cations and appears as a gray-to-white leached layer. The "B" horizon is classified as "spodic." It appears somewhat red-

Table 14.2.   Time Required to Titrate Exchangeable Bases in a Podzolic Soil with Acid Precipitation (100 cm yr^{-1}) at pH 4

Horizon[a]	Depth (cm)	Sum of Bases[b] (meq/100 g)	Cation Exchange Capacity (meq/100 g)	Percent Base Saturation	Soil[c] pH	Time to Titrate Bases (years)
A2	0–5	1.8	9.7	19	4.4	12.0
B21H	5–13	0.9	16.7	5	4.9	9.7
B22I	13–33	0.6	8.0	7	5.3	16.0
B23I	33–43	0.4	6.4	6	5.4	5.4
B3	43–63	0.2	2.9	7	5.6	5.4
						$\Sigma = 48.5$ yr

[a] Soil is a brown podzolic soil (Haplorthod) from north central Wisconsin. Pence, sampled by USDA S.C.S. on 5/8/80.
[b] Bases were extracted with ammonium acetate.
[c] Soil pH is 1:1H$_2$O extract.

dish and contains much of the iron and aluminum which were mobilized from above due to acidic conditions. As soil water percolates, it is somewhat neutralized by exchange reactions and chemical weathering, and the iron and aluminum precipitate or adsorb from solution.

Even in acid podzolic soils, there exist significant amounts of "fresh" minerals or weatherable bases available for dissolution. One reason is mechanical weathering, a physical process. Solifluction (the process of bedrock crack and surfacing on steep slopes) is an important mechanism in mountainous terrain. Freezing, thawing, and cracking of rocks also avails fresh minerals, a form of mechanical weathering. Especially in pit and mound topographies, tree throw provides a significant amount of fresh minerals from the B and C soil horizons. Burrowing animals can do likewise. Finally, forest canopies provide a "base pump" whereby basic cations taken from lower soil horizons are deposited in a more weatherable form on the surface of the soil during litter fall.

There are several factors which strongly affect the rate of chemical weathering in soil solution. These include:

Hydrogen ion activity of the solution.

Ligand activities in solution.

Dissolved CO$_2$ activity in solution.

Temperature of the soil solution.

Mineralogy of the soil.

Flow rate through the soil.

Grain size of the soil particles.

For a given silicate mineral, the hydrogen ion activity contributes to the formation of surface-activated complexes which determine the rate of mineral dissolution at pH < 6. Also since chemical weathering is a surface reaction-controlled phenomenon, organic and inorganic ligands (e.g., oxalate, formate, succinate, humic and fulvic acids, fluoride, and sulfate) may form other surface-activated complexes which enhance dissolution (5–7). (Some organics are known to form surface complexes which "block" dissolution.) Dissolved carbon dioxide is known to accelerate chemical weathering presumably due to its effects on soil pH and the aggression of H_2CO_3. Increases in temperature generally decrease the solubility of minerals and increase the rate of chemical weathering (8). Because mineral dissolution is kinetically controlled, increases in the rate of flow and decreases in the particle size can strongly increase the rate of chemical weathering (9).

Mineralogy is of prime importance. The Goldich (10) dissolution series, which is roughly the reverse of the Bowen crystallization series, indicates that chemical weathering rates should decrease as we go from carbonates $\rightarrow$ olivine $\rightarrow$ pyroxenes $\rightarrow$ Ca,Na plagioclase $\rightarrow$ amphiboles $\rightarrow$ K-feldspars $\rightarrow$ biotite and muscovite $\rightarrow$ aluminum oxides. If the pH of the soil solution is low enough (pH < 4.5), aluminum oxides provide some measure of neutralization to the aqueous phase along with an input of monomeric inorganic aluminum. The key parameter which affects the activated complex and determines the rate of dissolution of the silicates is the Si/O ratio. The less the ratio of Si/O is, the greater its chemical weathering rate is. Anorthite and forsterite have Si/O of $1:4$, while quartz has Si/O of $1:2$, the slowest to dissolve in acid.

4.3. Aluminum Dissolution

When soils are quite acid, as in the Lange Bramke watershed of the Harz Mountains of West Germany near Göttingen, aluminum is not only liberated from the soil in the rooting zone, but it is at high concentrations on the fine roots (and mycorrhizal fungal associations) due to a sorptive equilibrium of Al^{+3} with negatively charged root surfaces (Fig. 14.4). The importance of aluminum uptake by trees in acid soils and sorbed-Al effects on active and passive transport of nutrients to the tree are the subject of much current debate (11–14). Like chemical weathering, Al sorption is a surface-controlled phenomenon.

The regions where acid deposition has been reported to affect lakes (Adirondacks, New York, southern Norway, and Sweden) also have acid soils. These soils developed due to the leaching of basic cations over recent geologic time. Acidity of the upper soils is natural and principally caused by three phenomena: (1) a relatively low ANC of the soil; (2) an aggrading canopy whereby synthesis of biomass exerts an acidifying influence (Figure 14.1); and (3) the production or organic acids due to humics formation and decomposition. The last two are internal processes as opposed to acidic deposition which adds an additional source of acid to the system. Its mag-

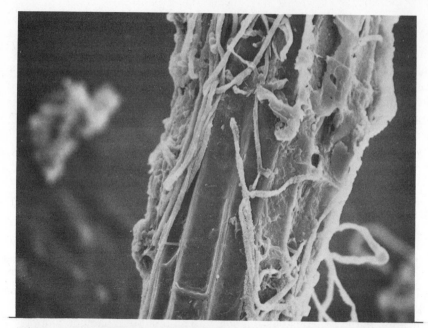

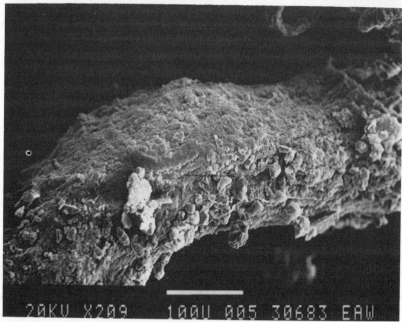

Figure 14.4. Scanning electron microphotographs of spruce roots. Fine roots from Norway spruce trees (*Picea abies*, Karst.) in Harz Mountains, Lange Bramke watershed, West Germany; soil pH = 3.8. Top: A healthy root. Energy dispersive spectroscopy (EDS) spectra showed qualitatively that aluminum is the predominate cation on the surface of the root and mycorrhizal fungi. Bottom: Root from dying tree. Fine, healthy roots were not observed. Surface of this root was scaly, and EDS spectra indicated the presence of Si, Al, Fe, and K cations, in rank order, on the surface of the root.

nitude in the hydrogen ion budget of the soils must be assessed on a case-by-case basis. Acidic deposition at the Hubbard Brook watershed, Hovattn watershed in southern Norway, and the Gardsjon watershed in Sweden, have contributed a large percentage of the total acid flux to the soil environment (15), thus decreasing the pH of the soil below "natural" levels and liberating additional Al^{+3} via chemical weathering and ion exchange.

Feldspars and other aluminum silicate minerals provide the bulk of the chemical neutralization in most watersheds affected by acidic deposition. The kinetics of weathering is of a fractional order in hydrogen ion activity (Table 14.3). Fractional-order kinetics necessitate that decreases in pH of the soil solution are partially compensated by increases in base cation weathering, but not totally. At pH < 4.5, aluminum is the principal cation, which together with hydrogen ion, makes a charge balance against increases in sulfate from acid deposition. At pH > 4.5, it is the basic cations (Ca^{+2}, Mg^{+2}, Na^+, K^+) which increase as a result of acidic deposition. To the extent that the sulfate anion can be sorbed by the soil matrix (an alkalinity anion must be exchanged into solution), this process also serves to neutralize acidic deposition, much as compensatory chemical weathering. Except in the southeast United States, most soils receiving acidic deposition are in a rough equilibrium with regard to input sulfate anions (16,17).

4.4. Biomass Synthesis

Assimilation by vegetation of an excess of cations can have acidifying influences in the watershed which may rival acid loading from the atmosphere. The synthesis of a terrestrial biomass, for example, on the forest and forest floor, could be written with the following approximate stoichiometry:

$$
\begin{array}{l}
800CO_2 \\
\;6NH_4^+ \\
\;4Ca^{2+} \\
\;1Mg^{2+} \\
\;2K^+ \\
1Al(OH)_2 \\
1Fe^{2+} \\
2NO_3^- \\
1H_2PO_4^- \\
1SO_4^{2-}
\end{array}
\xrightarrow[+\,\text{Photosynthesis}]{H_2O}
\boxed{
\begin{array}{l}
(CH_2O)_{800} \\
(NH_3)_8 \\
(H_3PO_4)_1 \\
(H_2SO_4)_1 \\
(CaO)_4 \\
(MgO)_1 \\
(Al(OH)_3)_1 \\
(FeO)_1 \\
(H_2O)_m
\end{array}
}
\; +\; 16H^+ \;+\; 804O_2 \qquad (13)
$$

Biomass

From a point of view of acid-neutralizing capacity, the biomass may also be represented in terms of corresponding cations, anions, and bases (Fig. 14.2). It is evident from this equation and from (15) that the aggrading production of terrestrial vegetation is accompanied by an acidification of the

Table 14.3. Reaction Order for the Rate Law for the Dissolution of Minerals

Mineral	Formula	Solution	Reaction Order	Reference
Dolomite	$(Ca,Mg)CO_3$	HCl	$\{H^+\}^{0.5}$	(23)
Bronzite	$(Mg,Ca)SiO_3$	HCl	$\{H^+\}^{0.5}$	(24)
Enstatite	$MgSiO_3$	HCl	$\{H^+\}^{0.8}$	(25)
Diopside	$CaMgSi_2O_6$	HCl	$\{H^+\}^{0.7}$	(25)
K-feldspar	$KAlSi_3O_8$	Buffer	$\{H^+\}^{0.33}$	(26)
Iron hydroxide	$Fe(OH)_3$-Gel	Various acids	$\{H^+\}^{0.48}$	(27)
Aluminum oxide	$\delta\text{-}Al_2O_3$	HCl	$\{H^+\}^{0.4}$	(6)

surrounding waters; correspondingly the ANC of the biomass increases. The ashes of trees are alkaline. Every temporal or spatial decoupling of the production and mineralization of biomass leads to a modification of the H^+ balance in the environment. This is the result of intensive agricultural and forestry practices and seasonal fluctuations. The release of H^+ by intensive growing vegetation (especially by forests used for wood exploitation), leads to leaching of base cations from the soil.

The interaction between acidification by an aggrading forest and the leaching (weathering) of the soil is schematically depicted in Figure 14.5. If the weathering rate equals or exceeds the rate of H^+ release by the biota, such as would be the case in a calcareous soil, the soil will maintain a buffer in base cations and residual alkalinity. On the other hand, in noncalcareous "acid" soils, the rate of H^+ release by the biomass may exceed the rate of H^+ consumption by weathering and cause a progressive acidification of the soil. In some instances, the acidic atmospheric deposition may be sufficient to disturb an existing H^+ balance between aggrading vegetation and weath-

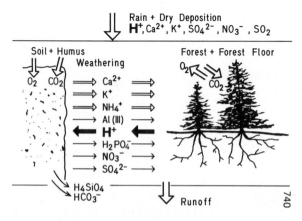

Figure 14.5. Competition between H^+ release by growing trees (aggrading forest) and H^+ consumption by soil weathering.

ering reactions. A quantitative example is provided by the carefully established acidity budget in the Hubbard Brook Ecosystem, Figure 14.6 (18), where, because of acid deposition, the H^+ ion sources nearly balance the H^+ ion sinks.

Humus and peat can likewise become very acid and deliver some humic or fulvic acids to the water. Note that the release of humic or fulvic material (HOrg, Org^-) to the water in itself is not the cause of resulting acidity, but rather the aggrading humus and net production of base-neutralizing capacity (see Gorham et al., Chapter 15 in this book).

4.5. Redox Processes

Table 14.1 lists some changes in the proton balance resulting from redox processes. Alkalinity or acidity changes can be computed as before: any addition of NO_3^- or SO_4^{2-} to the water as in nitrification or sulfur oxidation

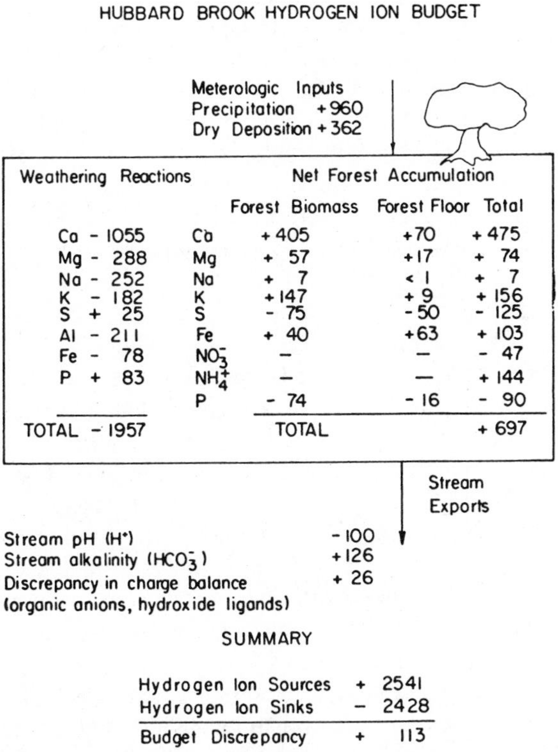

HUBBARD BROOK HYDROGEN ION BUDGET

Meterologic Inputs
Precipitation +960
Dry Deposition +362

Weathering Reactions   Net Forest Accumulation

                        Forest Biomass Forest Floor Total

Ca – 1055	Ca	+ 405	+70	+ 475
Mg – 288	Mg	+ 57	+17	+ 74
Na – 252	Na	+ 7	< 1	+ 7
K – 182	K	+147	+ 9	+ 156
S + 25	S	– 75	–50	– 125
Al – 211	Fe	+ 40	+63	+ 103
Fe – 78	NO_3^-	–	–	– 47
P + 83	NH_4^+	–	–	+ 144
	P	– 74	– 16	– 90

TOTAL – 1957   TOTAL   + 697

Stream
Exports

Stream pH (H^+)   – 100
Stream alkalinity (HCO_3^-)   +126
Discrepancy in charge balance   + 26
(organic anions, hydroxide ligands)

SUMMARY

Hydrogen Ion Sources   + 2541
Hydrogen Ion Sinks   – 2428

Budget Discrepancy   +   113

Figure 14.6. Budget of H^+ Sources and H^+ Sinks in Hubbard Brook Ecosystem (18). H^+ ion released or consumed in eq ha^{-1} yr^{-1} as a consequence of ion uptake by the aggrading forest biomass (+ forest floor) and ion release by weathering reactions, respectively. Because of atmospheric deposition of acid, H^+ sources nearly balance H^+ sinks.

increases acidity, while NO_3^- reduction (denitrification) and SO_4^{2-} reduction causes an increase in alkalinity. As shown by Schindler (Chapter 11 in this book), in experimental lakes, the nitrate removal caused by denitrification is linearly related to an increase in alkalinity. Incipient decreases of pH resulting from the addition of sulfuric acid or nitric acid to a lake is reversed by subsequent denitrification or SO_4^{2-} reduction.

Sediments in a water–sediment system are usually highly reducing environments. Electrons, delivered to the sediments by the "reducing" settling biological debris, and H^+ are consumed; thus in the sediments (including pore water) alkalinity increases, rendering the overlying water more acidic (19,20).

5.   LAKE RESPONSE

The reaction of free acidity with crystalline rocks (granite, gneiss, and mica schists) is much slower than, the dissolution of carbonates. Therefore acid lakes are more likely to be found in crystalline rock areas. This is also true in Switzerland where about 20 small acid lakes are found in the alpine southern part. There, nature makes some very interesting experiments on the weathering of these minerals. These lakes are acid because the residence time of acid precipitation in soils and the watershed is relatively short. The soils in question are very thin with exposed and weathered rocks and only little fine materials available. The water has little time to react with the minerals. Also, trees or thick vegetation is lacking at these altitudes. Some typical waters in the Maggia Valley are schematically shown in Figure 14.7. It illustrates the successive reactions and neutralization of the source water as it flows; the concentrations of cations (Al^{3+}, Ca^{2+}, K^+, and Na^+) and silicic acid increases as the dissolution process occurs.

In a first approximation, acidity is highest in those lakes with very small surface and catchment areas and the shortest residence time in the watershed. If the lakes receive runoff and snowmelt water very rapidly, then the acidity of its water is not greatly different from that of the snow.

The concentration of free aluminum can only increase the solubility of aluminum oxide; the lower the pH, the higher is the soluble Al(III) concentration. Fish cannot reproduce in acid lakes; this is probably due to the high concentrations of soluble Al(III) rather than to the concentrations of free acids. For the most part, the lakes are not sterile, but the supply of nutrients is extremely small. The water contains less than 5 µg phosphorus liter^{-1}, and thus the productivity is small too. It is interesting to note that the residual concentration of some heavy-metal ions in the acid mountain lakes, far away from streets and civilization, is much higher than in "more polluted" lakes in the lowland (Sigg, Chapter 13 in this book). The removal of these metals in the acid lakes is less efficient because of the lower sedimentation rates and because the lower pH of these waters reduces the tendency of heavy metals to become bound to the settling particles.

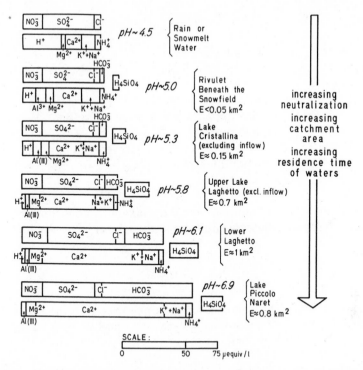

Figure 14.7. Chemical composition of several waters in a drainage pattern in the area of the upper Maggia Valley (in the vicinity of the Naret Dam, at 2100–2500 m). The extent of acid neutralization of rain or snowmelt water (chemical weathering) increases with the relative catchment area E and thus the residence time of the waters. This neutralization is accompanied by increases in pH, silicic acid concentration (H_4SiO_4), and by the liberation of cations from dissolution. Because of hydrologic changes from snowmelt, there are seasonal variations. This figure gives the condition from measurement at the end of July 1982 (21).

The rate of chemical weathering in the watersheds of these lakes can be calculated on the basis of the difference between precipitation acidity and that remaining in the lake; it amounts to 500 eq ha^{-1} yr^{-1} and is only about one-half of the rate of chemical weathering of aluminosilicate lminerals in the alpine valley of the Rhine.

5.1. Trickle-Down Model

A description of the time-variable version of the trickle-down model has been presented in Schnoor et al. (22) and the steady-state version has been given in Stumm et al. (3). The model is based on a mass balance for alkalinity in the watershed and lake.

An input/output analysis for a lake can be expressed mathematically as

$$V\frac{d(\text{Alk})}{dt} = -IL_{\text{acy}}(1 - f) - Q(\text{Alk}) + \frac{PV}{H} \qquad (14)$$

where

V = water volume of the lake (m^3)
Alk = alkalinity concentrations (μeq liter^{-1})
t = time (yr)
I = precipitation (inflow) volume per time (m^3 yr^{-1})
L_{acy} = total acidity of precipitation (dry + wet) (μeq liter^{-1})
Q = outflow or seepage rate (m^3 yr^{-1})
f = degree of acid neutralized in the watershed ($-$)
P = alkalinity production by sediment (μeq m^{-2} yr^{-1})
H = mean depth of lake (m)

Under steady-state conditions and if the alkalinity contribution from sediments is negligible, the concentration in the lake is

$$(\text{Alk}) = \frac{I}{Q} L_{acy}(f - 1) \tag{15}$$

The degree of acidic deposition that is neutralized in the watershed can be defined in terms of a watershed weathering rate W:

$$f = \frac{W}{RL_{acy}(0.1)} \tag{16}$$

where

W = weathering rate (eq ha^{-1} yr^{-1})
R = precipitation rate (cm yr^{-1})
L_{acy} = total acidity input (μeq liter^{-1})

The total acid input concentration L_{acy} must include any additional acid sources in the watershed which are significant relative to acidic deposition from the atmosphere. The rate of chemical weathering and leaching of soils is determined by the acidity of the soil water and the presence of organic and inorganic ligands. In sensitive watersheds, where water has little time to come into contact with soil minerals, the weathering of primary minerals in exposed rocks and sediments is sufficient to provide some neutralization.

Stumm et al. (6) have shown that the rate of weathering of δ-Al_2O_3 is proportional to the degree of surface protonation and to the surface complexation of oxalate in laboratory experiments:

$$R = k_1\{ \exists - OH_2^+ \}^3 + k_2\{ \exists - Ox \} \tag{17}$$

where R is the rate of dissolution of aluminum oxide. Equation 17 results in a dissolution rate that is proportional to $\{H^+\}$ to the 0.4 power. Table

14.3 shows that many common minerals have been reported to undergo fractional-order dissolution in acids.

An expression for chemical weathering (that is analogous to Eq. 17) is

$$W = k_h\{H^+\}^m + k_0 \tag{18}$$

where

W = chemical weathering rate
k_n = rate constant for acid hydrolysis
m = fractional-order constant
k_0 = rate constant in the absence of free acidity (due to CO_2 and ligand weathering)

The degree of acid neutralized in the watershed is related to the chemical weathering rate when the precipitation acidity penetrates the soil. From Eq. (16):

$$f = \frac{k_h L_{acy}^m + k_0}{R L_{acy}(0.1)} \tag{19}$$

By substituting Eq. 19 into Eq. 15, one obtains the following expression for steady-state alkalinity as a function of total precipitation acidity:

$$(Alk) = \frac{A_d(k_h L_{acy}^m + k_0)}{Q(0.001)} - \frac{I}{Q} L_{acy} \tag{20}$$

where

A_d = drainage area of the watershed (ha)

5.2. Evaluation of Coefficients

Equations 15 and 19 can be used to predict the alkalinity of lakes based on their total acidity input concentration (L_{acy}). It gives the steady-state re-action progress for a lake at various acid loadings. The coefficient k_0 can be determined from water-chemistry survey data for lakes of sensitive min-eralogy and hydrology that receive very low acid loadings. It is approxi-mately 100 eq ha^{-1} yr^{-1} for lakes in the most sensitive settings. The coef-ficients m and k_h can be determined based on two criteria: (1) field data to calibrate the model at a known acid loading and lake alkalinity; and (2) the mineralogy of the watershed compared to the reaction order for dissolution in laboratory studies (Table 14.3). Table 14.4 gives a number of sensitive watersheds which receive varying amounts of acid deposition. The range of

Table 14.4.   Watershed Measurements and Calculated Weathering Rates for Sites Reported in the Literature

Lake	Alk (μeq liter^{-1})	R (cm yr^{-1})	Q/I	L_{acy} (μeq liter^{-1})	$\frac{I}{Q}L_{acy}$ (μeq liter^{-1})	f	W (eq ha^{-1} yr^{-1})
Northcentral Wisconsin	0	76	0.45	25	56	1.0	190
Northeast Minnesota	+54	71	0.30	14	47	2.1	230
Star., Switzerland	−25	150	0.77	39	50	0.50	290
Langtjern, Norway	−20	88	0.66	52	80	0.75	340
Birkenes,[a] Norway	−33	140	0.77	57	74	0.55	440
Storgama, Norway	−30	125	0.76	60	79	0.63	470
Gardsjön, Sweden	−25	113	0.59	63	107	0.77	550
Lange Bramke,[a] FRG	−32	104.5	0.6	100	167	0.81	840
Woods, New York	−16	130	0.62	79	130	0.90	900
Hubbard Brook,[a] New Hampshire	− 6.3	130	0.62	74	120	0.95	910
Solling,[a] FRG	−10 to −30	104.5	0.6	163	272	0.93	1600

[a] Denotes a stream as opposed to a lake setting.

chemical weathering rates (200–1600 eq ha^{-1} yr^{-1}) is typical, and in general, the rates increase with increasing acid deposition. These watersheds were chosen as among the most sensitive in their region.

A reaction pathway for a lake beginning at point p (with 15 μeq liter^{-1} alkalinity) is given in Figure 14.8 using the steady-state model. Equation 20 was used for the calculation of the reaction pathway with $k_0 \simeq 30$ eq ha^{-1} yr^{-1}, $m = 0.4$, and $k_h \simeq 86$ eq ha^{-1} yr^{-1}.

5.3.   Time to Steady State

Equation 20 is actually a pseudo-steady-state relationship, since it assumes that the soil neutralizing capacity is in excess. The time to achieve such a pseudo steady state is relatively short (about three hydraulic detention times of the lake). However, the pseudo-steady-state alkalinity can decrease over a long time as minerals in the soil are titrated.

For situations where the watershed has extremely thin soils and very large acid deposition rates, the mineral phase may become exhausted. It can no longer supply basic cations to replenish the base exchange capacity of the soil, nor can it neutralize incident acid deposition. In this case, one must make a mass balance on the total acid neutralizing capacity (ANC) of both the water and the soil during their progressive titration.

The total ANC of the soil decreases over time:

$$\frac{dr}{dt} = -kr = \frac{-W}{Md}(10^7) \tag{21}$$

where

k = rate constant for soil neutralization (yr^{-1})
d = effective depth of soil (m)
M = bulk density of the soil (kg m^{-3})
W = weathering (neutralization) rate ($\text{eq ha}^{-1} \text{ yr}^{-1}$)
r = total ANC of the soil (meq kg^{-1})

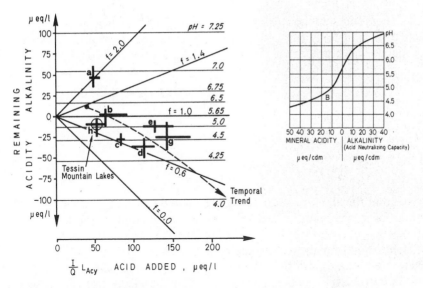

Figure 14.8. Alkalinity of natural waters as a function of acid loading (modified from Schnoor et al. (22). The acid deposition in the drainage area of lakes is neutralized by chemical weathering (the reaction with bases contained in rocks and minerals). f is the equivalents of acid neutralized by weathering ($f = 0$ signifies no neutralization; $f = 1$ corresponds to a perfect buffer; $f > 1$ signifies an alkalization of the water, e.g., the dissolution of $CaCO_3$ by CO_2). The difference between the acidity added and the resulting alkalinity corresponds to the degree of neutralization. Each lake would follow a titration curve (similar to the one on the upper right) with progressive acid loadings. The dashed line represents the temporal trend or reaction pathway as more acid is added to the system. It corresponds to the titration curve. a and b are lakes in Minnesota and Wisconsin; c represents the Birkenes stream and two other lakes in southern Norway; d signifies five lakes in southwest Sweden; e are waters in New York and New Hampshire; g represents six lakes in the La Cloche Mountains of southern Ontario, Canada; and h indicates the Swiss Lakes in the upper Maggia Valley of Tessin.

Rewriting Eq. 14 as two equations:

Lake Alkalinity $\dfrac{d(\text{Alk})}{dt} + \dfrac{Q}{V}\,(\text{Alk}) = -\dfrac{I}{V}\,L_{\text{acy}} + k\,\dfrac{V_{\text{soil}}}{V}\,Mr$ (22)

Soil ANC $\dfrac{dr}{dt} = -kr$ (23)

where

$$k = k' \left(\dfrac{[\text{H}^+]}{[\text{H}^+]_{\text{ref}}}\right)^m \left(\dfrac{\delta}{\delta_{\text{ref}}}\right)^s \left(\dfrac{Q_{\text{soil}}}{Q_{\text{ref}}}\right)^q \qquad (24)$$

k' = reference rate constant (yr^{-1})
δ = effective surface area of soils (particle size related) (m^2)
Q_{soil} = flow rate through the soil that reaches the lake (m^3 yr^{-1})

The theoretical range of exponents m, s, and q is from 0.0 to 1.0. In practice, m is usually 0.3–0.5, s is 0.4–0.7, and q = 0.5–0.8. Solving simultaneously, we obtain the solution for lake alkalinity through time:

$$(\text{Alk}) = (\text{Alk})_0 e^{-Qt/v} + \dfrac{k(V_{\text{soil}}/V)Mr_0}{Q/V - k}\,(e^{-kt} - e^{-Qt/V})$$

$$-\dfrac{I}{Q}\,L_{\text{acy}}\,(1 - e^{-Qt/V}) \qquad (25)$$

$$r = r_0 e^{-kt} \qquad (26)$$

Note that the alkalinity equation 25 consists of a first term describing the washout of initial conditions, a second term for alkalinity production in the watershed, and a third term related to the acid loading. After an increase in acid deposition, the solution to the equation 25 goes through a plateau at a pseudosteady state, and then deteriorates to the input acidity IL_{acy}/Q as $t \to \infty$. In extremely sensitive systems, the time to steady state ($t_{95\% \text{ ss}}$) can be as short as 100 years, but in most watersheds, $t_{95\% \text{ ss}}$ is hundreds of years of more. The half-life of the soil acid neutralization capacity (ANC) is given by Eq. 27:

$$t_{1/2} = \dfrac{0.69}{k} \qquad (27)$$

Figure 14.9 gives the results for a hypothetical lake receiving acid precipitation at pH 4 for an extended period of time. It is an extremely sensitive system with shallow soil and low ANC. Using Eqs. 22 and 23, the results indicate a relatively rapid response followed by a slow soil titration over hundreds of years. Hence even in very sensitive settings, there exist enough

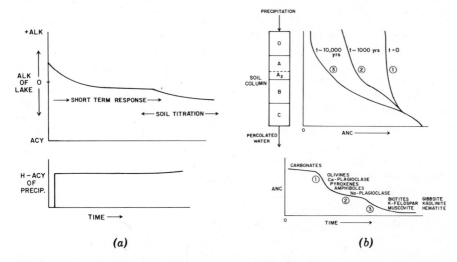

Figure 14.9.   Temporal response of a lake to acid precipitation (pH = 4). (*a*) Acidification of lake over long time. (Short-term response is over a period of a few years. Long-term soil titration takes from a few decades to centuries). (*b*) Corresponding soil titration.

mineral bases to partially neutralize acid deposition for a very long time. However, the additional input of acid deposition may be sufficient to depress soil pH.

It is possible to make the soil equations (23) and (26) more mechanistic by including two terms: a term for exchangeable bases and one for aluminosilicate weathering. The weathering of aluminosilicate minerals serves as a source term for the exchangeable base component of the soil ANC. This modifies the kinetics of soil titration somewhat by introducing a fast neutralization reaction with a relatively small buffer capacity (ion exchange) and a slower neutralization reaction with a relatively large buffer capacity (mineral weathering). If the addition of fresh mineral bases to the soil (by solifluction or other processes) is small compared to acidic deposition, then the soil would lose all of its ANC over some period of time. In such a case, mineral weathering would be the rate-determining step and Eq. 27 would give the half-life for the soil processes and the long-term resulting water chemistry. For the vast majority of cases within the time horizon of interest (decades to a few centuries), the amount of minerals in the soil and contributing bedrock can be considered to be in excess, and Eq. 20 gives the steady-state alkalinity concentration in the lake after only a few water residence times.

6.   SUMMARY

The pH and pϵ of the terrestrial and aquatic environment is determined by coupled reactions of oxidation–reduction and acid–base. If disturbances are created in elemental cycles of the environment (whereby oxidation of C,S, and N exceeds reduction reactions), a net production of H^+ ions is a necessary consequence. We have shown that aggrading biomass and humus and oxidation reactions (nitrification, sulfur oxidation) serve to add protons to aqueous systems, while chemical weathering, ion exchange, and reduction reactions (denitrification, H_2S production) serve to consume protons (add ANC to the water). Bar diagrams are a convenient means to examine changes in ANC or BNC in water and soils.

Atmospheric acid deposition creates an additional input of hydrogen and sulfate ions (H_2SO_4) to the terrestrial and aquatic ecosystem which is partly neutralized by increased weathering and cation export. It is balanced, in part, by aluminum dissolution and causes the negative effects in aquatic ecosystems on fish and possibly on forests.

The lakes which have been acidified by acid precipitation are those with extremely sensitive hydrologic settings and with watersheds lacking carbonate minerals. They tend to be small lakes. They respond relatively rapidly to changes in acid loading (on the order of a few hydraulic detention times). The soils of these watersheds have not been greatly acidified by acid precipitation nor has podzolization occurred due to anthropogenic acid deposition. However, this does not imply that more subtle changes in nutrient cycling or forest production could not have occurred. Soil solution pH could be reduced by the anthropogenic input of acids. We have proposed a quantitative framework with which to examine these questions.

ACKNOWLEDGMENTS

The authors are indebted to James J. Morgan, Laura Sigg, Jürg Zobrist, and Gary Glass who participated in the research reviewed here and provided data and ideas. Our work on atmospheric deposition is being supported by the Swiss National Foundation (National program 14 on air quality) and the U.S. Environmental Protection Agency. We thank Michael Hauhs and Bernard Ulrich for fruitful discussions and for access to the Harz Mountain watersheds. Special thanks are due to Michael Sturm for microphotographs and EDS spectrometry.

REFERENCES

1. L. G. Sillén, The Physical Chemistry of Sea Water. In *Oceanography,* M. Sears (Ed.), American Association for the Advancement of Science, Washington, D.C., 1961.

2. W. S. Broecker, "A Kinetic Model for the Chemical Composition of Sea Water," *Quarternary Res.* **1,** 188 (1971).

3. W. Stumm, J. J. Morgan, and J. L. Schnoor, "Saurer Regen, eine Folge der Störung hydrogeochemischer Kreisläufe," *Naturwissenschaften* **70,** 216 (1983).

4. N. M. Johnson, C. T. Driscoll, J. S. Eaton, G. E. Likens, and W. H. McDowell, "'Acid Rain', Dissolved Aluminum and Chemical Weathering at the Hubbard Brook Experimental Forest, New Hampshire," *Geochim. Cosmochim. Acta* **45,** 1421 (1981).

5. P. W. Schindler, "Surface Complexes at Oxide–Water Interfaces." In *Adsorption of Inorganics at Solid-Liquid Interfaces,* M. A. Anderson and A. J. Rubin (Eds.), Ann Arbor Science Publishers, Ann Arbor, Michigan, 1981.

6. W. Stumm, G. Furrer, and B. Kunz, "The Role of Surface Coordination in Precipitation and Dissolution of Mineral Phases," *Croat. Chem. Acta* **56,** 585 (1983).

7. R. Kummert and W. Stumm, "The Surface Complexation of Organic Acids on Hydrous γAl_2O_3," *J. Colloid Interface Sci.* **75,** 373 (1980).

8. E. A. Fitzpatrick, *Soils—Their Formation, Classification and Distribution,* Longmans, New York, 1980.

9. C. W. Correns, "The Experimental Chemical Weathering of Silicates," *Clay Min. Bull.* **4,** 249 (1961).

10. S. S. Goldich, "A Study in Rock-Weathering," *J. Geology* **46,** 17 (1938).

11. B. Ulrich, R. Mayer, and P. K. Khanna, "Chemical Changes due to Acid Precipitation in a Loess-Derived Soil in Central Europe," *Soil Sci.* **130,** 193 (1980).

12. B. Ulrich, "Eine ökosystemare Hypothese über die Ursachen des Tannensterbens (Abies alba Mill.)," *Forstwiss. Centralbl.* **100,** 228 (1981).

13. A. H. Johnson and T. G. Siccama, "Acid Deposition and Forest Decline," *Environ. Sci. Technol.* **17,** 294A (1983).

14. E. C. Krug and C. R. Frink, "Acid Rain and Acid Soil: A New Perspective," *Science* **221,** 520 (1983).

15. N. Van Breemen, C. T. Driscoll, and T. Mulder, *Nature* **307,** 599 (1983).

16. D. W. Johnson, G. S. Henderson, D. D. Huff, S. E. Lindberg, D. D. Richter, D. S. Shriner, D. E. Todd, and J. Turner, "Cycling of Organic and Inorganic Sulphur in a Chestnut Oak Forest," *Oecologia* **54,** 141 (1982).

17. J. N. Galloway, S. A. Norton and M. R. Church: "Freshwater Acidification from Atmospheric Deposition of Sulfuric Acid: A Conceptual Model," *Environ. Sci. Technol.* **17,** 541A (1983).

18. C. T. Driscoll and G. E. Likens, "Hydrogen Ion Budget of an Aggrading Forested Ecosystem," *Tellus* **34,** 283 (1982).

19. S. Emerson and M. Bender, "Carbon Fluxes at the Sediment–Water Interface of the Deep-Sea: Calcium Carbonate Preservation," *J. Mar. Res.* **39**(1), 139 (1981).

20. K. M. Kuivila and J. W. Murray, "Organic Matter Diagenesis in Freshwater Sediments: The Alkalinity and Total CO_2 Balance in the Sediments of Lake Washington," *Limnol. Oceanogr.* (in press).

21. J. L. Schnoor, L. Sigg, W. Stumm, and J. Zobrist, "Acid Precipitation and Its Influence on Swiss Lakes," *EAWAG News* **14/15,** 6 (1983).

22. J. L. Schnoor, W. D. Palmer, Jr., and G. E. Glass, "Modeling Impacts of Acid Precipitation for Northeastern Minnesota." In *Modeling of Total Acid Precipitation Impacts,* J. L. Schnoor (Ed.), Butterworth, Boston, Massachusetts, 1984.

23. E. Busenberg and L. Plummer, "The Kinetics of Dissolution of Dolomite in CO_2–H_2O Systems at 1.5 to 65°C and 0 to 1 atm pCO_2," *Amer. J. Sci.* **282**, 45 (1982).

24. D. Grandstaff, "Some Kinetics of Bronzite Orthopyroxene Dissolution," *Geochim. Cosmochim. Acta* **41**, 1097 (1977).

25. J. Schott, R. Berner, and E. Sjöberg, "Mechanism of Pyroxene and Amphibole Weathering—I. Experimental Studies of Iron-Free Minerals," *Geochim. Cosmochim. Acta* **45**, 2123 (1981).

26. R. Wollast, "Kinetics of Alteration of K-Feldspar in Buffered Solutions at Low Temperature," *Geochim. Cosmochim. Acta* **31**, 635 (1967).

27. R. Furuichi, N. Sato, and G. Okamoto, "Kinetics of $Fe(OH)_3$ Dissolution," *Chimia,* **23**, 455 (1969).

15

THE CHEMISTRY OF BOG WATERS

Eville Gorham, Steven J. Eisenreich, Jesse Ford, and Mary V. Santelmann

University of Minnesota, Minneapolis, Minnesota 55455, USA

Abstract

The transformation of fens receiving water from mineral soil to bogs receiving only atmospheric deposition is marked by a sharp decline in pH from above 6 to about 4 as calcium declines below 350 μeq liter^{-1}. The chemistry of bog waters is influenced chiefly by atmospheric deposition of sea spray in coastal areas (increasing sodium, chloride, and magnesium), and by dustfall from cultivated prairies in continental areas (increasing calcium, magnesium, and in lesser degree, potassium). In some locations air pollution appears to increase sulfate levels. Sulfate, nitrate, and ammonia are all much lower than in atmospheric precipitation, due presumably to plant uptake and microbial reduction. Sulfate reduction is particularly marked in oceanic sites. The low pH in bog waters is owed chiefly to yellow-brown organic acids, as indicated by highly significant interrelationships among anion deficit, hydrogen-ion concentration, absorbance, and "dissolved" organic carbon; pH also rises sharply upon photo-oxidation of water samples. Bog drainage is a significant input to the waters of many lakes and streams, and may predispose them to the further effects of acid deposition from the atmosphere.

1. INTRODUCTION

Bogs are peatlands whose vegetation and peats are commonly dominated by several species of the moss *Sphagnum*. They may have a coniferous tree cover in continental sites [Fig. 15.1(*a*)] but are generally open in maritime areas [Fig. 15.1(*b*)], with a very different surface pattern. Pools of open water are very scarce in continental bogs but are a striking feature in maritime areas. Bog surfaces receive their mineral supply wholly from the atmosphere, and are therefore said to be "ombrotrophic." Fens, on the other hand, are peatlands that receive some part of their mineral supply from waters that have percolated through mineral soil, and are said to be "minerotrophic." They are often dominated by sedges and reeds, with or without a cover of shrubs and trees. Bog waters are strongly acid (pH usually < 4.5), whereas fen waters are commonly circumneutral, although they range from moderately acid (pH rarely < 4.5) to strongly alkaline.

In wet climates and waterlogged situations the processes of peat accumulation often lead to the transformation of fens to bogs, as the surface of the peatland rises due to accumulation of peat, and becomes progressively

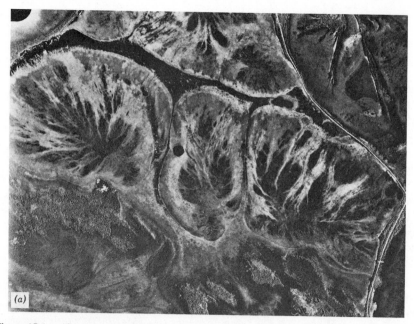

Figure 15.1. Air photos of bogs in (*a*) Caribou Cluster, Manitoba and (*b*) Gros Morne, Newfoundland. North is toward the top of each photo. The distance across Figure 15.1(*a*) is 12 km (E–W) and from top to bottom (N–S) of Figure 15.1(*b*) is 4 km. In Figure 15.1(*a*) the bogs are separated by narrow, very dark fen water tracks, and each bog displays a radiating pattern of dark forested strips alternating with light, open bog drains. In Figure 15.1(*b*) the bogs are not forested, and exhibit a pattern of alternating light ridges and dark pools oriented across the lines of water flow. Figure 15.1(*a*) is photo no. A23704-13, © (1974) Her Majesty the Queen in Right of Canada, reproduced from the collection of the National Air Photo Library with permission of Energy, Mines and Resources Canada. Figure 15.1(*b*) is photo no. 20553-87 (Nfld, © (1974) Her Majesty the Queen in Right of Canada, reproduced with permission from the collection of the Air Photo and Map Library of the Newfoundland Department of Forest Resources and Lands.

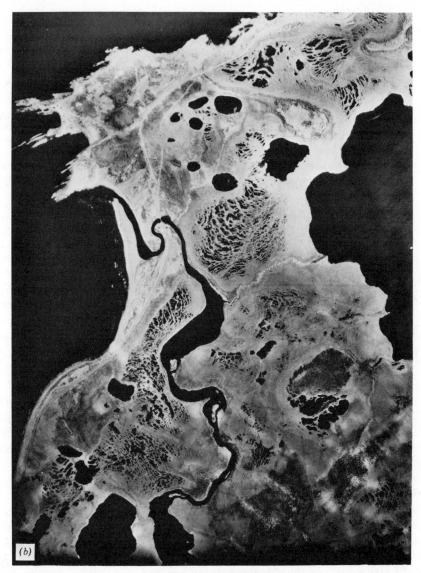

(b)

Figure 15.1. (*continued*)

more isolated from the regional ground water. In such cases the major alterations of surface-water chemistry involve a marked decline in the concentrations of calcium and bicarbonate ions and an equally marked increase in hydrogen ions. Because of lowered rates of water flow as the peat surface is transformed from a discharge to a recharge area, the concentration of "dissolved" (filter-passing) organic matter also tends to rise (1). Such a change is illustrated in Figure 15.2, which shows the relationship of pH to

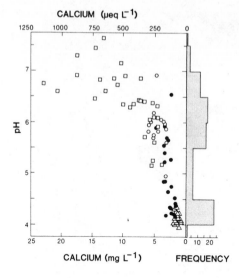

Figure 15.2. The relationship between pH and calcium in the waters of fens and bogs in the English Lake District. Open squares: fens without *Sphagnum*; open circles: fens with *Sphagnum* of the groups Squarrosa and Subsecunda but without *S. recurvum* (agg.) or *S. papillosum*; solid circles: fens with *S. recurvum* (agg.) or *S. papillosum*; open triangles: ombrotrophic bogs. [Data from Gorham and Pearsall (3) and Gorham (4)].

calcium in the surface waters of bogs and fens in the English Lake District. It is evident that the pH of fen waters declines slowly from circumneutral to faintly acid as the concentration of calcium falls below about 500 μeq liter^{-1}. The pH then falls sharply below about 350 μeq liter^{-1} of calcium as the fens give way to bogs with less than 100 μeq liter^{-1} of calcium. Bicarbonate alkalinity exhibits a similar relationship to pH (2).

The acidification of the fen sites as peat accumulates and calcium concentration declines is often associated with the invasion of species of the bog moss *Sphagnum*, first patchily by the less acidophilic members of the groups Subsecunda and Squarrosa (Ca range 160–310 μeq liter^{-1}) and later by the more acidophilic, carpet-forming species *S. recurvum* (agg.) and *S. papillosum* (Ca range 80–240 μeq liter^{-1}) that begin to form layers of *Sphagnum* peat. Finally, strongly acidophilic hummock-forming species such as *S. magellanicum* and *S. rubellum* (Ca range 35–90 μeq liter^{-1}) invade, deposit more peat, and convert the minerotrophic fen into an ombrotrophic bog, often with *S. cuspidatum* in the hollows among the hummocks. These *Sphagnum* mosses are capable of acidifying sites both by cation exchange and by the acidity of their decomposition products, as will be discussed in Section 5. Peatlands acidify rather rapidly once they reach critically low concentrations of calcium (and bicarbonate) ions, changing from fens around pH 6 to bogs below pH 4.5. The site frequency distribution at the right of Figure 15.2 illustrates this situation, with very few sites in the pH range 4.5–5.5 where bicarbonate buffering is low [cf. Gorham et al. (2)].

Processes analogous to those taking place in the surface waters of peatlands occur in the peats themselves. As peat deposits grow in height, inputs of mineral silt and clay decline and the peats become more organic. Consequently, cation-exchange capacity increases. As long as the inorganic fraction of the peat remains above about 15% dry weight metal cations predominate on the exchange complex and pH remains close to neutrality. However, as fens turn into bogs the organic content of the peat comes to exceed 85% dry weight, metal cations decline sharply, and exchangeable hydrogen ions rise equally sharply (5,6). This causes a drop in the percentage base (metal cation) saturation of the peat exchange complex from levels above 50% in fens to between 2 and 25% in bogs (5–8).

1.1. Early Work on the Chemistry of Bog Waters

The ecological significance of peat deposition above the level of standing ground water, so that the peat surface is fed only by atmospheric deposition, was appreciated as long ago as the beginning of the nineteenth century (9–11). Naismith (9) pointed out in 1807 that if a peat surface is lower than the surrounding ground, water from the mineral soil will overflow in time of flood, bringing "earthy particles" and dissolved minerals that retard the growth of bog plants and favor "esculent herbage." According to his account the water clears as it moves across the surface, so that its effect decreases toward the center of the peatland. There the upward growth of peat will be greatest and the peat will be mostly organic—with an ash content of less than 1% in one tested case.

At about this same time the acidity of bog water was thought by the Rev. Dr. Rennie (12) to be caused by a mixture of vegetable (gallic) and mineral (sulfuric) acids, which were neutralized in calcareous sites.

Actual chemical differences in the waters of different types of peatland plant communities at Lake Plager in Germany were pointed out as long ago as 1895 by Ramann [cited in Kivinen (13)]. He showed that bog waters are unusually low in silica, iron, sodium, calcium, and magnesium (Table 15.1).

Table 15.1.   Water Chemistry in a German Peatland[a]

	Total inorganic	Si	Fe	Na$^+$	K$^+$	Ca^{2+}	Mg^{2+}
	(mg liter^{-1})			(μeq liter^{-1})			
Reeds at lake edge	77	3.2	1.5	504	96	1100	303
Fen	71	3.8	9.5	239	47	953	174
Cotton–grass zone	63	5.7	1.8	397	53	689	204
Edge of bog	50	6.8	4.3	178	62	282	214
Middle of bog	20	1.5	0.9	132	47	46	75

[a] Data of Ramann as given by Kivinen (13).

1.2.   The 1940s through the 1960s

Much later, in 1947–1949, Witting (14,15) investigated the major ions in waters of four different types of plant community in Swedish peatlands, in a sequence from strongly minerotrophic, alkaline fens to ombrotrophic and strongly acid raised bogs (Table 15.2). Calcium dominated the cations of the calcareous fens, but with declining supply of calcium from mineral substrates it was exceeded (as a percentage of total cations) by sodium and magnesium—the former chiefly and the latter partly from sea spray. In the ombrotrophic bogs hydrogen ions were distinctly predominant.

Witting appreciated that the bog waters she studied owed their mineral supply wholly to atmospheric deposition. A direct comparison of rain water and bog water was made by Gorham (16), who showed a close resemblance in the ionic composition of rain and bog waters collected during wet weather in northern England (Table 15.3). During dry weather chloride concentration increased 1.84 times owing to evaporative concentration. Other ions increased more strongly, probably owing to oxidative acid production as the water table fell, with consequent exchange of some of the hydrogen ions (2.39 times) for metal cations, particularly divalent calcium (3.87 times) and magnesium (4.19 times), adsorbed on the peat. Adsorbed cations greatly outnumber cations dissolved in the interstitial waters of peats; in surface peats from Irish bogs the dissolved cations accounted for the following percentages of dissolved plus adsorbed forms: Na^+, 30%; K^+, 8%, Ca^{2+}, 1.4%, Mg^{2+}, 0.9%; and H^+, 0.03% (6).

Gorham (6) also demonstrated the great variability of sea-spray influence upon the chemistry of bog waters, with sodium and chloride ranging from 9 and 8 μeq liter^{-1}, respectively, in continental Polish bogs to 1678 and 2031 μeq liter^{-1}, respectively, in oceanic bogs of the Falkland Islands. Calcium, magnesium, and potassium were similarly affected, but to a lesser degree.

The acidity of bog waters was also a focus of interest during this period. In 1951 Villeret (18) claimed that the acidity of bog waters was caused largely by dissolved carbon dioxide, but Gorham (4) demonstrated negligible changes in English bog waters upon 3–5 min of blowing out the CO_2 with

Table 15.2.   Major Ions in Surface Waters of Swedish Fens and Bogs[a]

	Number of samples	pH	Cation sum (μeq liter^{-1})	Na$^+$	K$^+$	Ca^{2+}		Mg^{2+}
				(μeq %)				
Calcareous fen	7	7.5	2750	tr	8	0.5	79	13
Rich fen	4	6.1	547	0.2	25	1.5	59	15
Poor fen	16	5.0	343	2.9	37	3.5	26	30
Raised bog	43	3.9	283	44	29	3.2	8.8	15

[a] References 14 and 15.

Table 15.3.   Ionic Composition of Rain and Bog Waters in the North of England[a]

	Total cations	H^+	Na^{2+}	K^+	Ca^{2+}	Mg^{2+}	Cl^-	SO_4^{2-}
		(μeq liter^{-1})						
Rain (Lake District)	154	35	83	5	15	16	93	67
Bog waters (North Pennines)								
Wet weather	171	61	74	4	15	16	90	83[b]
Dry weather	462	140	187	8	58	67	166	316[b]
Ratio dry:wet	2.70	2.39	2.53	2.0	3.87	4.19	1.84	3.81

[a] References 16 and 17.

[b] These values are suspect and may include some organic anions.

nitrogen (mean pH of 7 bog waters before bubbling = 4.04, after bubbling = 4.03). In the mid-1950s Ramaut (19,20) suggested that living *Sphagnum* moss produced unknown, soluble organic acids, and Gorham (21,22) postulated the oxidation of organic sulfur compounds in the peat to sulfuric acid, to which was added sulfuric acid in rain and snow influenced by urban/industrial air pollution. In 1960 Gorham and Cragg (23) suggested that cation exchange of hydrogen ions adsorbed to peat particles for metal cations in atmospheric precipitation might be a significant source of acidity. Clymo (24) then showed that by synthesizing polygalacturonic acids in their cell walls *Sphagnum* mosses could generate metabolically the new exchange sites required by the ion-exchange theory of bog-water acidity. Recently, Kilham (25) has revived the Ramaut hypothesis of unknown, soluble organic acids produced by *Sphagnum*.

1.3.   The Present Study

The work reported here is part of a large cooperative study of the ecology and biogeochemistry of *Sphagnum* bogs. Its aim is to extend earlier work on bogs by: (1) examining the environmental factors controlling the chemistry of bog waters; (2) contrasting the chemistry of bog waters in the midcontinental and the eastern oceanic regions of North America with those of Ireland and England; and (3) investigating further the causes of their acidity.

2.   METHODS

2.1.   Choice of Sites

Bogs were selected to provide a series of ombrotrophic sites (Table 15.4; Fig. 15.3) along a broad belt transect from just east of the midcontinental forest/prairie border in Minnesota and Manitoba to extremely oceanic sites

Table 15.4. Site Locations of Bogs Sampled in North America (see Fig. 15.3)

Site	North Latitude	West Longitude
Minnesota:		
1. Red Lake	48°15′	92°37′
	48°17′	92°45′
2. Ely Lake	49°26′	92°26′
3. Arlberg	46°55′	92°47′
Manitoba:		
4. Caribou Cluster	49°24′	95°22′
5. Bullhead Bay	51°41′	96°50′
6. Beaver Point	51°41′	96°50′
Ontario:		
7. Experimental Lakes Area (ELA)	49°40′	93°43′
8. Mead	49°26′	83°56′
9. Norembego	48°59′	80°44′
10. Diamond Lake	48°52′	80°38′
Quebec:		
11. Lac Parent	48°47′	77°10′
12. Lac des Miserables	49°49′	75°01′
13. Lac St. Jean	48°54′	71°54′
	48°55′	71°47′
14. Mont Albert	48°55′	66°11′
15. Sept Iles	50°18′	66°00′
Maine:		
16. Great Sidney Heath	44°23′	69°48′
17. Bar Harbor	44°15′	68°15′
18. Carrying Place Cove	44°50′	67°00′
New Brunswick:		
19. Bull Pasture Plain	46°03′	64°20′
20. Point Sapin	46°59′	64°51′
21. Miscou Island	47°56′	64°30′
	47°58′	64°31′
22. Point Escuminac	47°04′	64°49′
Prince Edward Island:		
23. Foxley Moor	46°43′	64°04′
Nova Scotia:		
24. Cape Sable	43°28′	65°36′
25. Fourchu	45°12′	60°15′
	45°42′	60°15′
Newfoundland:		
26. Conne River Pond	48°10′	55°30′
27. Gander Bay	49°21′	54°22′
	49°22′	54°19′
28. Ladle Cove	49°28′	54°32′

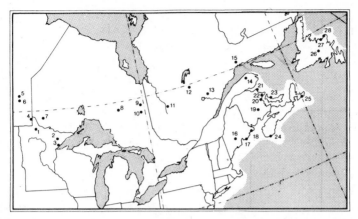

Figure 15.3.   A map of the locations of bogs sampled in North America (see Table 15.4).

in Newfoundland. For comparison, a few water samples were collected in a very extensive peatland (locally termed a "low pocosin") in the Croatan Reserve on the coastal plain of North Carolina, and from maritime bogs in Ireland (two in County Offaly, one in northwest County Mayo) and in northern England (three in the southern Pennine Mountains, where urban/industrial air pollution is severe).

2.2.   Sample Collection, Treatment, and Storage

Surface waters were collected along the transect from June through August, 1982. The "pocosin" waters were collected in November 1982, the Irish waters in September 1982, and the English waters in midwinter of 1982/1983. In wetter sites along the transect, samples were taken from large pools and small puddles, the latter usually at the bases of black spruce trees. In drier sites the waterlogged moss surface was depressed and the sample bottle was filled by water expressed from the moss. In still dryer sites a pit was dug to beneath the water table, emptied, and allowed to refill two or three times until clear. The ratio *pools and puddles: depressions and pits* was 7:3 in Ontario, 2:7 in Minnesota and Maine, and 1:3 in Manitoba. In other areas the two groups were nearly equal.

Although concentrations of different ions varied considerably within an individual bog (see Figs. 15.4 and 15.5) there were no consistent chemical differences between pools and puddles, or between depressions and pits. When waters from pools and puddles were compared to those from depressions and pits, the only notable phenomenon was a marked variability with regard to sulfate concentrations. In eight sites sulfate was higher in pools and puddles than in depressions and pits; the difference was striking in oceanic sites. In five sites the depressions and pits were higher in sulfate;

this was most striking in Minnesota and Manitoba. As we shall show (Section 3.3), sulfate reduction is very significant in these bog waters, some of which show extremely variable sulfate concentrations.

Bulk samples were taken in new, 500-mL Nalgene polyethylene bottles that had been rinsed four times with distilled deionized water, filled with the same, and left to leach. These bottles were rinsed twice with the sample before collection. Samples were stored in a cooler during automobile transport back to Minnesota. They were then refrigerated at about 4°C. Filtration was performed in the field for transect samples and in the laboratory for others, using a Millipore filtration unit attached to a hand-pumped vacuum-evacuation unit. Reeve-Angel # 934AH glass-fiber filters were prepared in the laboratory and packaged for field use. Filters for cation samples were soaked for 2 hr in 10% HCl, washed with 250 mL 10% HCl, and rinsed with 1 liter of distilled deionized water. Filters for anion samples were prepared in the same way except that soaking and rinsing were with distilled deionized water. All filters and filtration systems were rinsed twice with a small amount of sample before the remainder was filtered. All transect samples were filtered within 3 days after collection; most were filtered within a few hours. Samples from North Carolina, Ireland, and England were filtered when the samples reached the laboratory.

The filtrate was separated in the field into three fractions as follows:

One acid-washed 125-mL bottle for cation analysis.

One water-washed 125-mL bottle for anion analysis.

One water-washed 200-mL bottle for photo-oxidation studies.

Cation bottles were soaked in 10% HNO_3, rinsed five times with distilled deionized water, filled with the same, and left to leach until the time of sample collection. Bottles for anion analysis and photo-oxidation were treated in the same way as the 500-mL sampling bottles.

2.3.   Analysis

Field pH measurements were carried out on settled unfiltered samples with a Radiometer M29b field pH meter. The unit was standardized to Radiometer pH 4.01 and 7.00 buffers before each set of two or three readings. All other measurements were made in the laboratory.

Absorbance was measured at 320 nm on a Beckman DU model 2400 spectrophotometer with a slit width of 0.1 mm and a 1-cm path length. Laboratory pH was measured by a Radiometer TTTlc Autotitrator with Radiometer glass and reference electrodes. This unit was standardized to Fisher pH 4 and pH 7 buffers before each set of readings. The calibration was also checked against a low-alkalinity standard (pH 4.3) obtained from the central laboratory facilities of the National Atmospheric Deposition Program in Urbana,

Illinois. Agreement with this standard was within 0.1 pH unit. Field and laboratory pH values were in close agreement. For 31 samples tested the mean deviation (+ and −) was 0.06 units, with a range from +0.18 to −0.13 units. The mean for the "after" samples was 0.018 units higher than the "before" mean.

Laboratory analyses were done during the period January–July, 1983. Samples were stored immediately upon returning from the field in a dark cold room maintained at 4°C. Samples for cation analysis were acidified with Ultrex ultrapure nitric acid (1% by volume) at least 1 week prior to refiltration in the laboratory with Gelman A/E glass-fiber filters. Anion samples were also refiltered in the lab with Gelman filters before analysis. Cation filters were presoaked for 1 hr in, and rinsed with, 1 liter of 10% HCl. Anion filters were presoaked for 1 hr in, and rinsed with, 1 liter of distilled deionized water. Each filter was used for only one sample. All series of filtered samples included one blank of distilled deionized water; the analytical values obtained for these blanks were subtracted from values obtained for associated samples. All samples were analyzed at room temperature.

Metal cations were analyzed on a Varian Techtron 175 Atomic Absorption unit with an air–acetylene flame. Highly colored water samples are difficult to analyze owing to interference by the organic matrix; to minimize matrix effects all cation samples were run by the method of standard additions (Varian Techtron user manual). Lanthanum was added to reduce interference in samples analyzed for calcium and magnesium.

Ammonium was not analyzed in the 1982 samples. A subset of the sites sampled in 1982 was sampled by the same methods in 1983. Ammonium concentrations were measured in these waters on the fraction collected for cation analysis prior to sample acidification. The phenolhypochlorite method was employed (26), absorbance being measured in 1-cm quartz cuvettes on a Beckman Model 26 spectrophotometer.

Anion samples were analyzed by a Dionex Model 10 Ion Chromatograph against mixed standards (fluoride, chloride, phosphate, nitrate, and sulfate ions) at four or more appropriate concentrations.

"Dissolved" (filter-passing) organic carbon was measured on a subset of samples by a Beckman UV oxidation apparatus. Some samples were also photo-oxidized by UV radiation and a small amount of hydrogen peroxide. These were then analyzed for major cations and anions to see whether pH rose and bicarbonate appeared as organic acids were destroyed.

3. ENVIRONMENTAL CONTROLS ON THE CHEMISTRY OF BOG WATERS

There are currently three important sources of the major ions that reach bog surfaces via atmospheric deposition: sea spray, soil dustfall, and air pollution.

3.1. Sea Spray

Salt from the sea is responsible for very high levels of chloride and sodium in bog waters from coastal areas. Close to the ocean, chloride [Fig. 15.4(a)] sometimes exceeded 400 μeq liter^{-1} and sodium [Fig. 15.5(a)] exceeded 300 μeq liter^{-1}, in contrast to concentrations in midcontinental areas of 1–15 μeq liter^{-1} and undetectable to 42 μeq liter^{-1}, respectively. Mean concentrations of both sodium and chloride in precipitation in these continental regions are between 10 and 20 μeq liter^{-1} (27). Magnesium [Fig. 15.5(b)] was also relatively high (up to 160 μeq liter^{-1}) in coastal areas. The lowest levels of magnesium along the west–east transect (1–17 μeq liter^{-1}) were observed in northeastern Ontario and northwestern Quebec, where mean concentrations in precipitation are in the vicinity of 10 μeq liter^{-1} (27).

3.2. Dustfall from Blown Soil

Soil dustfall is characteristically high in the cultivated prairies of the midcontinent. Its influence is most evident in the bogs of Minnesota, where calcium concentrations in their waters [Fig. 15.5(c)] ranged from 36 to 119 μeq liter^{-1}. (These high values are not caused by evaporative concentration under the relatively dry midcontinental climate, because chloride and sodium do not exhibit a similar westward increase). Calcium declined eastward, and in eastern Canada the range (except for 2 of 25 cases) was between 6 and 20 μeq liter^{-1}. Occasional higher values may have been due to contamination, for instance, by shell material dropped by seabirds—a common occurrence on coastal bogs. The concentration of calcium in precipitation across the transect is primarily between 20 and 40 μeq liter^{-1} (27), with high values up to 100+ μeq liter^{-1} much farther to the west. Presumably it is

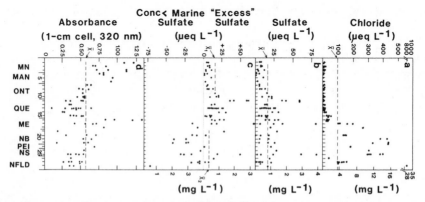

Figure 15.4. Concentrations of (a) chloride, (b) sulfate, (c) "excess" sulfate, and (d) absorbance in bog waters along a west–east transect in midtemperate North America. In (c), $\bar{x}_1$ represents only positive values, whereas $\bar{x}_2$ represents all values.

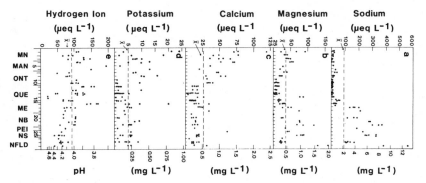

Figure 15.5. Concentrations of (a) sodium, (b) magnesium, (c) calcium, (d) potassium, and (e) hydrogen ions along a west–east transect in midtemperate North America.

dry fallout of blown-dust particulates that chiefly enriches the bog surfaces in Minnesota. In eastern Canada calcium concentrations in precipitation are much the same as those in the bog waters (27).

Magnesium was enriched not only at the coastal end of the west–east transect but also—like calcium—toward the midcontinental end close to the cultivated prairies [Fig. 15.5(b)] where dry deposition of blown-soil dust is important. Magnesium was also high (78 and 88 μeq liter^{-1}) at site no. 14, Mont Albert in Quebec, where marine influence is slight (chloride, 12 and 18 μeq liter^{-1}; sodium, 10 and 17 μeq liter^{-1}) and where calcium (18 and 27 μeq liter^{-1}) was not particularly high. This is a montane area of serpentine rocks rich in magnesium, and presumably dustfall from open soils at high altitudes is responsible for the elevated magnesium levels in the bog waters. Samples from this site were the lowest of all sites in acidity (pH 4.68) and it is possible that the peat surfaces have not yet become wholly isolated from minerotrophic ground-water influence.

In midcontinental bogs, where the influence of sea spray is minimal, there appears to be some contribution from dustfall to the very low levels of sodium in the bog waters. Whereas the ratios of sodium to chloride in maritime areas varied about the *sodium:chloride* ratio of 0.86 in sea water, they frequently rose much higher (up to 7) in the western part of the transect. Presumably this reflects the presence of sodium sulfate in many arid prairie soils. Lake waters dominated by these two ions are not uncommon in North and South Dakota (28) and Saskatchewan (29). Moreover, Verry (30) reported a correlation of sodium with sulfate ($r^2 = 0.41$) that was just as strong as its correlation with chloride ($r^2 = 0.40$) in precipitation from northern Minnesota.

Potassium was low in the bog waters along the transect [Fig. 15.5(d)], with a mean of 5 μeq liter^{-1} and an upper limit of 24 μeq liter^{-1}. Mean concentrations in precipitation are also low (about 3–6 μeq liter^{-1}) along the transect (27). However, of eight bog-water values above 10 μeq liter^{-1}, seven were from samples taken west of Quebec. Three of the four highest were collected in Minnesota, and the other was from a dry bog (no. 16) in the interior of Maine. Dustfall may well have caused these relatively high values.

3.3. Air Pollution

The influence of air pollution by sulfate and nitrate derived from urban/industrial emissions of sulfur and nitrogen oxides is not readily detectable in our transect data for two reasons. First, nitrogen—an important plant nutrient—is commonly in short supply for bog vegetation, so that inputs of nitrate in precipitation are likely to be reduced by plant uptake. Second, microbial reduction close to the water table in bogs may lower the concentrations of sulfate and nitrate substantially below those present in atmospheric precipitation. Both sulfur and nitrogen may be emitted in volatile reduced forms from the bog surface.

Both of these constraints were clearly evident in the data for bog-water chemistry. Nitrate was usually undetectable, presumably owing both to plant uptake and to microbial reduction and denitrification. It seldom amounted to more than a few μeq liter^{-1} despite concentrations in precipitation along the transect of 10–40 μeq liter^{-1} (27). Likewise, sulfate, which averaged 17.7 μeq liter^{-1} in the transect bog waters [Fig. 15.4(b)], was far below the concentrations of 50–100 μeq liter^{-1} in precipitation in the same regions (27), which would otherwise be appreciably concentrated by summer evapotranspiration from the bog surfaces. Presumably sulfate reduction was responsible for the much lower values in the bog waters.

Ammonium concentrations in bog water and precipitation followed the same pattern. Bog-water concentrations ranged from 1 to 5 μeq liter^{-1}, much less than the concentrations of 10–30 μeq liter^{-1} in precipitation. Here again, microbial transformations and plant uptake were probably responsible for the difference, as well as for the disappearance of whatever ammonium was produced by nitrate reduction. Adsorption on the cation-exchange complex of *Sphagnum* moss is an additional factor that probably contributed to low ammonium levels. There were no discernable differences in the ammonium concentrations of bog waters in maritime and continental regions.

The contrast for sulfate is even greater if the amount that is due to sea spray is subtracted from the bog-water concentration to yield nonmarine or "excess" sulfate [Fig. 15.4(c)]. "Excess" sulfate is calculated by assuming that all chloride in the bog water is marine, and subtracting *chloride* multiplied by *0.105*—the ratio of sulfate to chloride in seawater—from the total sulfate concentration. When this was done, "excess" sulfate averaged 14.9 μeq liter^{-1} in continental sites, only 0.5 μeq liter^{-1} less than total sulfate (Table 15.5). In oceanic sites total sulfate averaged 21 μeq liter^{-1}, whereas "excess" sulfate showed a slight negative value of -1.8 μeq liter^{-1} (Table

Table 15.5. The Chemical Composition of Surface Waters from North American, Irish, and English Bogs

	North American Transect[a]		North Carolina	Ireland	England
	Continental (14 bogs, 38–40 samples)	Maritime (10 bogs, 24–27 samples)	(1 bog, 5 samples)	(3 bogs, 3 samples)	(2 bogs, 6 samples)
H^+	107 ± 30	74 ± 20	217 ± 12.8	59 ± 21	160 ± 26
Na^+	11 ± 8.7	209 ± 81	126 ± 7.7	590 ± 325	144 ± 16
K^+	5.8 ± 5.3	4.5 ± 2.7	1.5 ± 1.5	24 ± 11	5.2 ± 1.8
Ca^{2+}	35 ± 26	18 ± 10	6.7 ± 2.1	68 ± 16	62 ± 15
Mg^{2+}	23 ± 20	56 ± 36	41 ± 9.5	158 ± 62	54 ± 6.5
Cl^-	5.5 ± 3.4	229 ± 113	140 ± 14	677 ± 308	180 ± 9.1
SO_4^{2-}	15.4 ± 13.9	21 ± 22	55 ± 10	83 ± 57	273 ± 30
"Excess" SO_4^{2-}	14.9 ± 14.1	-1.8 ± 23	40	12	254
Anion deficit	157 ± 63	106 ± 42	197	139	-28
DOC^b (mg liter^{-1})	$32^c \pm 13$	—	$45^d \pm 2$	26 ± 11	—
$Abs^{320\,nm}_{1\,cm}$	0.61 ± 0.26	0.51 ± 0.15	1.02 ± 0.18	0.46 ± 0.25	0.13 ± 0.05

[a] Excluding (as intermediate) sites nos. 14–16 and 26.
[b] "Dissolved" organic carbon.
[c] Only 20 samples.
[d] Only three samples.

353

15.5). In many oceanic sites, especially those most affected by sea spray, "excess" sulfate was a distinctly negative quantity suggesting that where sulfate supply is high, sulfate-reducing bacteria may be unusually active and lower the concentration of sulfate below that expected from sea spray. In this connection, Schindler et al. (31) have shown that enrichment of sulfate in a very dilute lake by addition of sulfuric acid stimulated the growth of sulfate-reducing bacteria. Alternatively, it may be that because the inland bogs undergo greater fluctuations in the water table, they experience more reoxidation of reduced sulfides during summer drawdown.

Continental bogs can be further subdivided with regard to sulfate [Fig. 15.4(c)]. In general, western bogs (sites 1–8) have lower concentrations of "excess" sulfate than eastern bogs (sites 9–13, 19). This may reflect increased exposure of the latter sites to urban/industrial sulfur pollution. In particular the unusually high sulfate concentrations in the bog at Lac Parent in Quebec (site no. 11) are probably due to its being downwind of Noranda, one of the very largest metal-smelting facilities in Canada. The oceanic sites at Bar Harbor in Maine (no. 17) and Fourchu in Nova Scotia (no. 25) exhibit very wide ranges of sulfate concentration, with the high values rivaling those at Lac Parent. The origin of these high values is not known, but may be local; the Fourchu site at least is not far from relatively large sources of air pollution in the vicinity of Sydney, Nova Scotia. More extreme effects of pollution will be discussed in Section 5.

3.4. Climate

Color and acidity of bog waters along the west–east transect are most likely related to the dryness of the site as controlled by regional climate interacting with local hydrology.

3.4.1. Water Color.

The yellow-to-brown color of bog water as measured by absorbance at 320 nm correlates well with "dissolved" (filter-passing) organic carbon (32). Along the transect, absorbance was high [Fig. 15.4(d)] in the relatively dry bogs of Minnesota and Manitoba (site nos. 1–6), where annual precipitation only slightly exceeds evaporation, and in the dry and southernmost inland Maine bog at Great Sidney Heath (site no. 16). Such dry sites exhibit a low ratio of precipitation to evaporation, and consequently a slower flow of water through the surface peat. Slower water flow in turn allows decomposition to enrich surface waters in highly colored organic solutes, which become further concentrated by evaporation.

3.4.2. Acidity.

With the exception of the Mont Albert serpentine site (no. 14), pH along the transect tended to be highest in the most oceanic bog waters [Fig.

15.5(*e*)], with pH values between 4.08 and 4.38 in Nova Scotia and New-foundland. Although high pH values were observed occasionally all along the transect, values below 3.9 were observed only in continental bogs west of the Maritime Provinces, where drier conditions probably favor the buildup of colored organic acids (see Section 5). The mean concentration of hydrogen ions along the transect (93 μeq liter^{-1}) distinctly exceeds mean concentrations in atmospheric precipitation in the same region. The westward increase of biogenically derived hydrogen ions in the bog waters also contrasts with a distinct decline in that direction of precipitation acidity caused by urban/industrial air pollution.

4. COMPARATIVE CHEMISTRY OF BOG WATERS FROM DIFFERENT REGIONS

Table 15.6 presents mean values for major ions and other chemical properties of the bog waters from all of the 13 continental and 15 more or less maritime bogs along the W–E transect in midtemperate North America. Total cations averaged 258 μeq liter^{-1}, very similar to the 283 μeq liter^{-1} for Swedish bogs (Table 15.2). The ionic proportions in the two areas are also similar, with the Swedish bogs being slightly more acid and slightly less affected by sea spray.

Table 15.5 contrasts means for the continental and maritime bog waters along the North American transect with those of surface waters collected from a "low pocosin" peatland on the coastal plain of North Carolina, from

Table 15.6   The Average Chemical Composition of Bog Waters Along a W–E Transect in North America[a]

	Number of Samples	Mean[b] Concentration	
		μeq L^{-1}	μeq %
H$^+$	80	93 ± 34	36
Na$^+$	77	96 ± 123	37
K$^+$	80	5.2 ± 4.6	2.0
Ca^{2+}	80	27 ± 21	10
Mg^{2+}	80	37 ± 33	14
Cl$^-$	78	100 ± 56	
SO$_4^{2-}$	78	18 ± 16	7.1
"Excess" SO$_4^{2-}$	75	7.3 ± 22	2.9
Anion deficit	74	136 ± 59	53
"Dissolved" org C (mg liter^{-1})	28	29 ± 13	—
Abs$_{1 cm}^{320 nm}$	80	0.57 ± 0.25	—

[a] PO$_4^{3-}$ undetectable, NO$_3^-$ and NO$_2^-$ usually undetectable, F$^-$ always very low.
[b] ± standard deviation of the mean.

three maritime Irish bogs, and from two maritime bogs in the industrialized north of England that are subject to severe air pollution (the third is omitted because of incomplete analyses).

The English bog waters (from Featherbed Moss, Axe Edge Moss, and Ringinglow Bog) were very rich in total sulfate and "excess" sulfate, owing to severe air pollution. They were also relatively high in calcium; local limestones and urban/industrial fly ash may both have been involved. The Irish bog waters (from the Redwood, Mongan, and Glenamoy Bogs) were the least acid on average, and the most subject to sea spray. They were also the richest in calcium, perhaps because of the abundance of exposed limestone in Ireland. They were low in "excess" sulfate. The waters of the "low pocosin" in North Carolina are obviously ombrotrophic, despite conventional wisdom that ombrotrophic bogs are not found south of New England in eastern North America (33). The highly colored "pocosin" waters were even more acid than those of the polluted English bogs, perhaps because of their marked lack of calcium. Potassium was also extremely low in these waters. Total and "excess" sulfate, on the other hand, were surprisingly high, and appear to match approximately the mean concentration of sulfate in local precipitation—about 40 μeq liter^{-1} (27). All of the North American transect waters were distinctly low in total sulfate. "Excess" sulfate was low in the continental American waters and actually negative on average in the maritime samples. The continental transect waters were extremely low, as expected, in sodium and chloride.

Absorbance and anion deficit, both related to "dissolved" organic carbon (see Section 5), were highest in the "low pocosin" waters. Absorbance was lowest in the English bog waters, presumably because gathered during the wet winter season. These English waters also exhibited a negative anion deficit; probably because unmeasured metal cations (e.g., Zn, Cu, Pb, etc.) from air pollution outweighed the unmeasured organic anions (see Section 5) that account for the anion deficits elsewhere.

5. THE ACIDITY OF BOG WATERS

As noted earlier, there has been a longstanding controversy as to whether the free hydrogen ions in bog waters are ascribable to organic acids (19,20,25), to sulfuric acid (21,22), or to cation exchange between metal cations in precipitation and hydrogen ions adsorbed to the polygalacturonic acids that make up the cell walls in *Sphagnum* moss and peat (23,24).

In order to shed further light upon this question we have investigated the relationships of anion deficit (DEF) and hydrogen ion concentration (H^+) in North American bog waters to both "dissolved" organic carbon (DOC) and the much more easily measured property of absorbance (ABS). Following Cronan (34) and Hemond (35), we also photo-oxidized water samples in order to examine their acidity both before and after destruction of organic acids.

Table 15.7 and Figure 15.6(a) show a strong correlation between ABS and DOC. This allows employment of ABS as an easily measured surrogate for DOC. Although ABS may be taken as the dependent variable, the regression of DOC upon it is included (Table 15.7) so that DOC may be predicted from ABS in other studies.

Anion deficit (DEF)—the difference between the sums of measured cations (H^+, Na^+, K^+, Ca^{2+}, Mg^{2+}) and anions (Cl^-, SO_4^{2-}, NO_3^-)—amounts on average to 53% of total cations in the transect waters (Table 15.6), and may be supposed to represent organic anions. (Three ions were not included in the DEF calculations. Ammonium was not measured in the 1982 samples; in 1983 samples it amounted at most to 5 μeq liter^{-1}, less than 4% of the mean calculated DEF. Total soluble phosphorus did not exceed the limit of detection, 5 μeq liter^{-1}. The concentration of fluoride was generally 2 to 3 μeq liter^{-1}, with three high values of 9.8, 17, and 29 μeq liter^{-1} at maritime sites 20, 22, and 24, respectively). If organic anions are responsible for DEF, there should be a highly significant correlation between it and both DOC and ABS. Table 15.7 and Figures 15.6(b) and 15.7(a) demonstrate that this is the case. There should also be a significant relationship between H^+ and both DOC and ABS, although less strong because the organic acids are partially neutralized by soil dustfall, absorption of ammonia from the atmosphere, and so on. Table 15.7 and Figures 15.6(c) and 15.7(b) show this to be true. The English bog waters, heavily polluted by sulfuric acid derived from urban/industrial emissions of sulfur dioxide, lie well above the regressions of H^+ upon both DOC and ABS. Although

Table 15.7. Correlations, Regressions, and Reduced Major Axes[a] for Relationships Among Anion Deficit (DEF), Hydrogen Ions (H^+), "Dissolved" Organic Carbon (DOC), and Absorbance (ABS) in Bog Waters[b]

Relationship	Number of Samples	Correlation r^2	Regression	Reduced Major Axis
ABS and DOC	28	0.76	ABS = 0.0154DOC + 0.106	—
	—	—	DOC = 49.5ABS + 1.69	—
DEF and DOC	26	0.81	DEF = 4.30DOC + 20.1	DEF = 4.78DOC + 5.73
DEF and ABS	74	0.64	DEF = 192ABS + 28.0	DEF = 241ABS + 0.49
H^+ and DOC	27	0.47	H^+ = 1.93DOC + 32.1	H^+ = 2.83DOC + 5.26
H^+ and ABS	80	0.35	H^+ = 81.4ABS + 46.7	H^+ = 137ABS + 15.1
H^+ and DEF	74	0.32	H^+ = 0.334DEF + 47.9	H^+ = 0.590DEF + 13.1

[a] The reduced major axis assumes that neither variable is dependent.
[b] DEF and H^+ as ueq liter^{-1}, DOC as mg liter^{-1}, and ABS at 320 nm in a 1-cm cell.

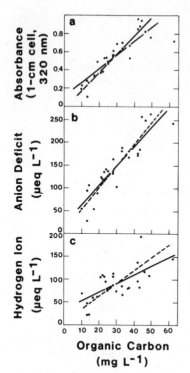

Figure 15.6. The relationships of (*a*) absorbance, (*b*) anion deficit, and (*c*) hydrogen ion concentration to the concentration of "dissolved" organic carbon in bog waters along a west–east transect in midtemperate North America. Solid lines are regressions; dashed lines are reduced major axes—assuming no dependent variable.

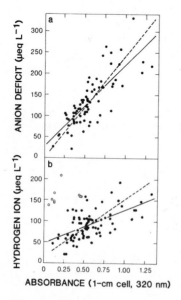

Figure 15.7. The relationships of (*a*) anion deficit and (*b*) hydrogen ion concentration to the absorbance in bog waters along a west–east transect in midtemperate North America. The open circles in (*b*) represent three bogs in northern England subject to severe air pollution. Solid lines are regressions; dashed lines are reduced major axes—assuming no dependent variable.

they are not shown, the unpolluted Irish bog waters conform well to the North American regressions, as do the waters from the ombrotrophic "low pocosin" in North Carolina.

If organic acids are important in bog waters, one might also expect a significant relationship between H^+ and DEF. Figure 15.8 illustrates this relationship ($p < 0.01$).

These regressions assume that DEF and H^+ are dependent variables. If they are instead considered as independent variables like DOC and ABS, then the reduced major axes—shown as dashed lines in Figures 15.6 through 15.8 and given as equations in Table 15.7—provide a better picture of the relationship and indeed come much closer to a zero origin in each case.

As already noted, the relationship of H^+ to DOC and ABS is less clear than the corresponding relationships of DEF to DOC and ABS, because variable amounts of H^+ are neutralized by metal cations in different sites and samples. On average, H^+ amounts to 68% of DEF (Table 15.6), with little difference between the continental and maritime bogs in this respect. H^+ may also be added to some samples as pollutant sulfuric acid. To illustrate these points, in Figure 15.8 the two samples from Mont Albert—where the influence of serpentine soils is pronounced—are the very lowest in H^+ and lie well below both the regression line and the reduced major axis. In contrast the four waters from Lac Parent—east of the smelter complex at Noranda and presumably enriched in sulfates from air pollution—have the highest H^+ concentrations, between DEF 92 and 145 μeq liter^{-1}, and lie above the equivalence line ($H^+ = DEF$). (Some other points above the equivalence line may merely represent the summation of seven separate errors in the calculation of DEF). The same situation exists in Figure 15.7 with respect to the samples from Mont Albert and Lac Parent. In Figure 15.7 the Lac Parent data are close to those for three polluted waters from Ringinglow Bog, just outside the northern English city of Sheffield.

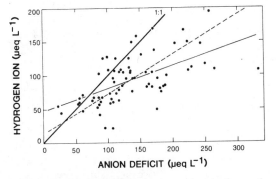

Figure 15.8. The relationship between hydrogen ion concentration and anion deficit in bog waters along a west–east transect in midtemperate North America. The heavy line represents equivalence; the thin line is the regression, and the dashed line is the reduced major axis—assuming no dependent variable.

Finally, it may be noted that in most of the unpolluted bog waters along the transect that were subjected to photo-oxidation, pH rose from primarily well below 4.7 to above 6.0, as the organic anions were oxidized to carbon dioxide and water and the metal cations associated with a part of the organic anions as metal "humates" (cf. Fig. 15.8) were converted to metal bicarbonates. In only 5 of 30 transect samples (sites 12, 13, 16, 19, and 27 from Quebec eastward) did no alkalinity result from photo-oxidation, so that pH remained very low (4.12–4.50). In these five cases large amounts of additional sulfate (84–134 μeq liter^{-1}) were produced by photo-oxidation. In the other 25 samples only small amounts of sulfate (7–29 μeq liter^{-1}) were produced, along with widely variable amounts of bicarbonate alkalinity (7–224 μeq liter^{-1}) related to the concentrations of metal cations (e.g., Ca^{2+}, Mg^{2+}) present. Evidently, the five samples yielding no alkalinity upon photo-oxidation contained substantial amounts of nonionic sulfur, capable of producing enough sulfuric acid to counterbalance the alkalinity resulting from oxidation of the metal "humates" in the sample.

After photo-oxidation the three samples from polluted Ringinglow Bog in northern England rose in pH from 3.80–3.81 to only 4.13–4.15. No alkalinity was produced, owing to the abundance of strong mineral (sulfuric) acid in the original samples. Only 39–44 μeq liter^{-1} of additional sulfate resulted from photo-oxidation. The Irish and the "low pocosin" waters behaved similarly to the normal transect waters upon photo-oxidation.

The foregoing analysis indicates clearly that the acids responsible for low pH values in unpolluted bog waters are organic in nature. Because the dissolved organic matter in bog waters is strongly yellow/brown in color, it is highly likely that the organic acids in question are products of decomposition. Recent studies by McKnight et al. (36) have shown that fulvic acids account for 67% of the DOC in the waters of Thoreau's Bog at Concord, Massachusetts. These humic substances are acid as a result of the incomplete nature of the decomposition process that leads to peat accumulation, stopping well short of complete breakdown to carbon dioxide and water. For instance, just as acetic acid is a product of the incomplete anaerobic metabolism of simple carbohydrates, so also are fulvic acids—with the empirical formula $C_{16}H_{17}O_{10}$—the breakdown products of the more complex constituents of the bog moss Sphagnum (36).

6. CONCLUDING REMARKS

Many lakes and streams on noncalcareous substrata, such as the igneous and metamorphic rocks of the Precambrian Shield areas in Canada, the United States, and Norway, are extremely dilute, with only about 50 μeq liter^{-1} of calcium in their waters (32,37). Some of these waters receive considerable acid input from bog drainage and become quite "tea colored" (DOC $\geq$ 10 mg liter^{-1}). In Wisconsin, according to Lillie and Mason (38),

38% of a random set of 553 lake waters were observed to be "brown" in color; most such waters came from lakes on Precambrian substrates in the northern half of the state. Even in cases where their bicarbonate buffering capacity can accommodate a natural acid input from bog drainage, such lake waters may be especially susceptible to acid deposition from urban/industrial sources of air pollution, because further acid input can readily drive the pH down to levels toxic to the biota.

7. ACKNOWLEDGMENTS

We thank the following for their assistance and advice: N. Urban, P. H. Glaser, A. W. H. Damman, L. A. Baker, K. B. Anderson, J. N. Stefansen, P. Epale, P. J. Newbould, F. Oldfield, N. Richardson, N. L. Christensen, F. Morel, and W. Stumm. This project was generously supported by Grant No. DEB 7922142 from the National Science Foundation.

REFERENCES

1. P. H. Glaser, G. A. Wheeler, E. Gorham, and H. E. Wright, Jr., "The Patterned Mires of the Red Lake Peatland, Northern Minnesota: Vegetation, Water Chemistry and Landforms," *J. Ecol.* **69,** 575 (1981).

2. E. Gorham, S. E. Bayley, and D. W. Schindler, "Ecological Effects of Acid Deposition upon Peatlands: A Neglected Field in Acid-Rain Research," *Can. J. Fish. Aquat. Sci.* (1984) (in press).

3. E. Gorham and W. H. Pearsall, "Acidity, Specific Conductivity and Calcium Content of Some Bog and Fen Waters in Northern Britain," *J. Ecol.* **44,** 129 (1956).

4. E. Gorham, "The Ionic Composition of Some Bog and Fen Waters in the English Lake District," *J. Ecol.* **44,** 142 (1956).

5. E. Gorham, "Chemical Studies on the Soils and Vegetation of Waterlogged Habitats in the English Lake District," *J. Ecol.* **41,** 345 (1953).

6. E. Gorham, "Some Chemical Aspects of Wetland Ecology," *Tech. Mem. Nat. Res. Counc. Canada, Assoc. Comm. Geotech. Res.*, (90), 20 (1967).

7. H. Sjörs, "Some Chemical Properties of the Humus Layer in Swedish Natural Soils," *Kungliga Skogshögskolans Skrifter* (37) 51 (1961).

8. F. H. Braekke, "Hydrochemistry in Low-pH Soils of South Norway 1. Peat and Soil Water Quality," *Medd. Nor. Inst. Skogforsk.* **36**(11), 32 (1980).

9. J. Naismith, "An Essay on Peat, Its Properties and Uses," *Trans. Highl. Soc. Scotl.* **3,** 17 (1807).

10. R. Griffith, in *Report of the Commissioners Appointed to Enquire into the Nature and Extent of Several Bogs in Ireland: and the Practicability of Draining and Cultivating Them*, (2), App. 4, 31 (1811).

11. W. Aiton, *Treatise on the Origin, Qualities, and Cultivation of Moss-Earth, with Directions for Converting It into Manure*, Wilson and Paul, Air, 1811.

12. R. Rennie, *Essays on the Natural History of Peat Moss*, III–IX. Constable, Edinburgh, 1810.

13. E. Kivinen, "Uber Elektrolytgehalt und Reaktion der Moorwässer," *Maatalouden Tutkimuskeskus Maantutkimuslaitos Agrogeol. Julk.* (38), 71 (1935).

14. M. Witting, "Katjonbestämningar i Myrvatten," *Bot. Not.* **100**, 287 (1947).

15. M. Witting, "Preliminärt Meddelande om Fortsatta Katjonbestämningar i Myrvatten Sommaren 1947," *Sven. Bot. Tidskr.* **42**, 116 (1948).

16. E. Gorham, "Factors Influencing Supply of Major Ions to Inland Waters, with Special Reference to the Atmosphere," *Geol. Soc. Am. Bull.* **72**, 795 (1961).

17. E. Gorham, "On the Acidity and Salinity of Rain," *Geochim. Cosmochim. Acta* **7**, 231 (1955).

18. S. Villeret, "Recherches sur le Role du CO_2 dans L'acidité des Eaux des Tourbières a Sphaignes," *C. R. Acad. Sci.* **232**, 1583 (1951).

19. J. L. Ramaut, "Etude de L'origine de L'acidité Naturelle des Tourbières Acids de la Baraque-Michel," *Bull. Acad. Roy. Belg.* (Classe Sci.), Ser. 5, **41**, 1037 (1955).

20. J. L. Ramaut, "Extraction et Purification de L'un des Produits de L'acidité des Eaux des Hautes Tourbières et Sécrété par *Sphagnum*," *Bull. Acad. Roy. Belg. (Classe Sci.), Ser. 5*, **41**, 1168 (1955).

21. E. Gorham, "On the Chemical Composition of Some Waters from the Moor House Nature Reserve," *J. Ecol.* **44**, 375 (1956).

22. E. Gorham, "The Influence and Importance of Daily Weather Conditions in the Supply of Chloride, Sulphate and Other Ions to Fresh Waters from Atmospheric Precipitation," *Philos. Trans. Roy. Soc. London Ser. B* **241**, 147 (1958).

23. E. Gorham and J. B. Cragg, "The Chemical Composition of Some Bog Waters from the Falkland Islands," *J. Ecol.* **48**, 175 (1960).

24. R. S. Clymo, "The Origin of Acidity in *Sphagnum* Bogs," *Bryologist* **67**, 427 (1964).

25. P. Kilham, "The Biogeochemistry of Bog Ecosystems and the Chemical Ecology of *Sphagnum*," *Mich. Bot.* **21**, 159 (1982).

26. L. Solorzano, "Determination of Ammonia in Natural Waters by the Phenolhypochlorite Method," *Limnol. Oceanogr.* **14**, 799 (1967).

27. J. W. Munger and S. J. Eisenreich, "Continental-Scale Variations in Precipitation Chemistry," *Environ. Sci. Technol.* **17**, 32A (1983).

28. E. Gorham, W. E. Dean, and J. E. Sanger, "The Chemical Composition of Lakes in the North-Central United States," *Limnol. Oceanogr.* **28**, 287 (1983).

29. D. S. Rawson and J. E. Moore, "The Saline Lakes of Saskatchewan," *Can. J. Res Sect. D* **22**, 141 (1944).

30. E. S. Verry, "Precipitation Chemistry at the Marcell Experimental Forest in North Central Minnesota,' *Water Res. Res.* **19**, 454 (1983).

31. D. W. Schindler, R. Wagemann, R. B. Cook, T. Ruszczynski, and J. Prokopowich, "Experimental Acidification of Lake 223, Experimental Lakes Area: Background Data and the First Three Years of Acidification," *Can. J. Fish. Aquat. Sci.* **37**, 342 (1980).

32. E. Gorham, "The Chemical Composition of Lake Waters in Halifax County, Nova Scotia," *Limnol. Oceanogr.* **2,** 12 (1957).

33. A. W. H. Damman, "Ecological and Floristic Trends in Ombrotrophic Peat Bogs of Eastern North America." In J. M. Gehu, (Ed.), *Colloques Phytosociologiques VII, Sols Tourbeux*, Lille, 1980, pp. 61–77.

34. C. S. Cronan, W. A. Reiners, R. C. Reynolds, and G. E. Lang, "Forest Floor Leaching: Contributions from Mineral, Organic and Carbonic Acids in New Hampshire Subalpine Forests," *Science* **200,** 309 (1978).

35. H. F. Hemond, "Biogeochemistry of Thoreau's Bog, Concord, Massachusetts," *Ecol. Monogr.* **50,** 507 (1980).

36. D. M. McKnight, E. M. Thurman, R. L. Wershaw, and H. F. Hemond, "Biogeochemistry of Aquatic Humus Substances in Thoreau's Bog, Concord, Massachusetts" (1984) (manuscript submitted for publication).

37. E. Gorham, "The Chemical Composition of Lake Waters in Halifax County, Nova Scotia," *Limnol. Oceanogr.* **2,** 12 (1957).

38. R. F. Wright, T. Dale, A. Henriksen, G. R. Hendrey, E. T. Gjessing, M. Johannessen, C. Lysholm, and E. Storen, "Regional Surveys of Norwegian Lakes and Snowpack, October 1974, March 1975 and March 1976." In A. Tollan (Ed.), *Report IR 33/77*, SNSF Project, Oslo, 1977, pp. 4.1–4.67.

39. R. A. Lillie and J. W. Mason, "Limnological Characteristics of Wisconsin Lakes," *Wisconsin Dept. Nat. Resources Tech. Bull.* (138), 116 (1983).

16

LAKE RESTORATION

René Gächter and Dieter M. Imboden

Swiss Federal Institute of Technology (ETH), Institute for Water
Resources and Water Pollution Control (EAWAG), Zurich,
Switzerland

Abstract

*A one-box model for the mass balance of total phosphorus in lakes serves
to assess the effect of various techniques to restore eutrophic lakes. It is
shown that net removal rate of phosphorus to the sediments is highly dependent on the phosphorus content of the lake. Due to this nonlinear behavior, the dynamics of the eutrophication and recovery process may differ
markedly. In those cases where the external phosphorus loading cannot be
reduced as quickly as desired or as efficiently as required, within certain
limitations lake internal measures may be useful. Hypolimnic discharge of
water may increase the tolerable mean inflow concentration by 40 mg m^{-3}
at best, and oxygen input into the hypolimnion may augment the phosphorus
retention of (mainly mesotrophic) lakes up to 80%.*

1. INTRODUCTION: LAKE EUTROPHICATION, AN UNSOLVED PROBLEM

Increasing primary productivity causes high biomass concentration in the
epilimnion of lakes, large settling rates of algae to the deeper water layers,
large oxygen consumption, and even anaerobic conditions in the hypolimnion. These changes affect the quality of the lake as an ecosystem and as a
resource for various human needs, such as drinking-water supply and rec-
reation.

Though in several countries control of lake eutrophication has been one
of the most important environmental issues since many years, success has
often remained disappointing. For instance, Switzerland spends over 10^9

365

Table 16.1.   Quality Criteria for Lakes (Swiss Regulation)

1. *Trophic state*: Oligotrophic or mesotrophic
   In particular:
   Annual primary production $\leq$ 150 g C m^{-2} yr^{-1}
   Mean phosphorus concentration $\leq$ 30 μg phosphorus per liter
2. *Oxygen concentration*: Always and everywhere $\geq$ 4 mg liter^{-1}

dollars per year for water pollution control measures ($1\frac{1}{2}$% of the gross national product). A major portion of this sum is aimed to decrease the input of phosphorus—the nutrient responsible for eutrophication in most lakes—into the surface waters. Nevertheless, in many lakes phosphorus concentrations are clearly above the limit of 30 μg liter^{-1} which is used in Switzerland as quality criterion for lakes (Table 16.1). Besides the input of sewage, intensive agriculture may significantly contribute to the phosphorus loading of lakes (1,2).

The example of 13 lakes from Switzerland (Fig. 16.1) demonstrates that the efforts to decrease the phosphorus concentration were only successful in so far as to stop and—in some cases—even to reverse eutrophication partially. However, in no case has a once-deteriorated lake completely recovered. A similar conclusion is drawn from Figure 16.2: In the hypolimnion

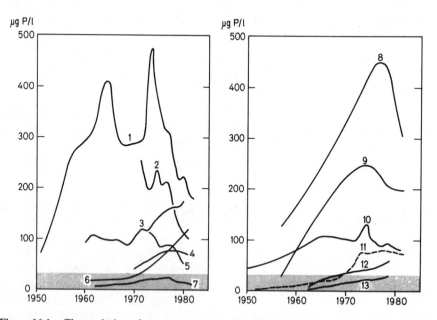

Figure 16.1.   The evolution of average concentration of soluble reactive phosphorus at spring overturn in various lakes. [Redrawn and completed from Ambühl (3).] 1, Greifensee; 2, Pfäffikersee; 3, Zugersee; 4, Lac Léman; 5, Bielersee; 6, Sempachersee; 7, Walensee; 8, Baldeggersee; 9, Hallwilersee; 10, Zürichsee; 11, Bodensee; 12, Lac de Neuchâtel; 13, Vierwaldstättersee. Shaded area indicates the water-quality goal.

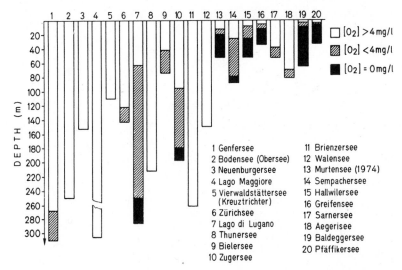

Figure 16.2.   Vertical distribution of oxygen concentration in 20 lakes at the end of the stagnation period. (From H. P. Fahrni, Swiss Federal Board of Environmental Protection).

of most lakes oxygen concentration still drops below 4 mg liter^{-1} or even to zero. This is particularly true for the lakes situated on the Swiss Central Plateau which are less than 80 m deep. Thus, it is obvious that additional efforts are necessary to reach the water-quality criteria in these lakes.

The aim of this chapter is (1) to discuss a simple model to assess the possible success of various measures against eutrophication; (2) to explain the apparent failure of certain efforts; and (3) to find solutions to complement those control measures which turned out to be only partially successful. More complex phosphorus models [e.g., Imboden and Gächter (4)] will not be discussed here, since very often the characteristics of such models are dominated by individual properties of the respective lakes thus hiding the general principles. Also, the necessary data base to employ complex models is often not available.

Some considerations will be exemplified by case studies. Complete reviews of lake restoration cases are given by Dunst et al. (5), Landner (6) and Hamm and Kucklentz (7).

2.   A SIMPLE MODEL FOR LAKE RESTORATION

The phosphorus balance in a lake is determined by the phosphorus input (M_{in}), phosphorus export through the outlet (M_{out}), and sedimentation (M_s) (see Table 16.2 for definitions of symbols):

$$V \frac{dP}{dt} = M_{in} - M_{out} - M_s \tag{1}$$

P_{in} is the average of all inlet concentrations $P_{in,i}$ (rivers, sewage pipes, etc.) weighed by the corresponding flow rates Q_i:

$$P_{in} = \frac{\sum_i P_{in,i}Q_i}{\sum_i Q_i} = \frac{M_{in}}{\sum_i Q_i} \tag{2}$$

Formally, it is always possible to write the output terms (M_{out}, M_s) as the product of the total phosphorus-content in the lake (VP) and some rate constants (see Table 16.2) and to transform Eq. 1 into

$$\frac{dP}{dt} = \rho P_{in} - (\beta\rho + \sigma)P \tag{3}$$

with the steady state solution

$$P_\infty = P_{in}\frac{\rho}{\beta\rho + \sigma} \tag{4}$$

If the parameters β, ρ, and σ are independent of P, Eq. 3 is a linear model with a simple analytical solution and one single steady state.

One goal of lake restoration is to decrease phosphorus concentration. Equation 4 serves to demonstrate the three different possibilities to achieve this goal:

1. Reduction of phosphorus loading (i.e., P_{in}) by sewage treatment or diversion, and by decrease of phosphorus loss from agricultural areas.
2. Increase of phosphorus export through the outlet by increasing β. This is achieved by diversion of hypolimnic water which, during stagnation, has higher phosphorus concentrations than the surface water.
3. Increase of phosphorus retention (increase of σ) by increasing sedimentation or decreasing phosphorus redissolution at the sediments.

The first method is the traditional approach of water pollution control by measures in the drainage area of the lake (external measure). In contrast, methods 2 and 3 involve changes within the lake (internal measures).

Minimum oxygen concentration in the hypolimnion of a lake can be an alternative criterion of lake restoration (Table 16.1). Since the redox potential influences the exchange of phosphorus at the sediment–water interface, oxygenation can indirectly affect the phosphorus balance through the size of the sedimentation parameter σ. Hypolimnic oxygen concentration can be improved by:

4. Increase of vertical mixing during the winter.
5. Input of oxygen into the hypolimnion during the stratification period.

Alternative internal restoration measures, such as mechanical removal of

Table 16.2. Definition of Symbols

Symbol	Definition	Unit of Measurement
A_0	Lake surface area.	(m^2)
B	Oxygen input.	$(g\ day^{-1})$
J_0, J_0^H, and J_0^K	Oxygen consumption rate in completely mixed lake, hypolimnion, and compartment V_K, respectively.	$(g\ m^{-3}day^{-1})$
$M_{in} = QP_{in} = \rho VP_{in}$	Phosphorus input.	$(g\ yr^{-1})$
$M_{out} = QP_0 = \rho\beta VP$	Phosphorus output.	$(g\ yr^{-1})$
$M_s = \sigma VP$	Net sedimentation of phosphorus.	$(g\ yr^{-1})$
O_2, O_2^H, and O_2^K	Oxygen concentration in the completely mixed lake, in the hypolimnion, and in a hypolimnic compartment V_K, respectively.	$(g\ m^{-3})$
O_2^{∞} and $O_2^{K,\infty}$	Steady-state oxygen concentration in a completely mixed lake and in a compartment V_K, respectively.	$(g\ m^{-3})$
O_2^s	Oxygen concentration at saturation.	$(g\ m^{-3})$
$O_2^{K,0}$	Oxygen concentration in V_K at beginning of stagnation period.	$(g\ m^{-3})$
O_2^{in}	Oxygen concentration of water flowing into compartment V_K.	$(g\ m^{-3})$
P, P_{in}, P_0, and P_b^{HD}	Mean total phosphorus concentration in lake, input, surface outflow, and in the drainage pipe.	$(mg\ m^{-3})$
P_∞ and P_∞^{HD}	Mean steady-state phosphorus concentration in a lake with natural outflow and hypolimnic drainage, respectively.	$(mg\ m^{-3})$
Q	Water input or output.	$(m^3\ yr^{-1})$
Q_{HD}	Water output via hypolimnic drainage pipe.	$(m^3\ yr^{-1})$
V, V_H, and V_K	Volume of total lake, hypolimnion, and a compartment in the deep hypolimnion, respectively.	(m^3)
Z_{max} and $\bar{Z} = V/A_0$	Maximum and average depth of a lake.	(m)
α	Transfer coefficient of oxygen at the air–water interface.	$(m\ day^{-1})$
$\beta = P_0/P$	Vertical form factor.	$(-)$
$\beta^{HD} = P_b^{HD}/P^{HD}$	Vertical form factor.	$(-)$
$\rho = Q/V$	Water renewal rate.	(yr^{-1})
ρ^K	Water renewal rate of V_K.	(yr^{-1})
σ	Net sedimentation rate.	(yr^{-1})

macrophytes, the application of herbicides, chemical precipitation of phosphorus, chemical treatment, sealing or removal of the sediments, will not be discussed in this book. Though such measures proved to be successful in small lakes, they are hardly applicable in larger lakes due to financial and logistical reasons.

3. NONLINEAR AND NONSTATIONARY BEHAVIOR OF PHOSPHORUS CONCENTRATION IN LAKES

In the traditional approach by Vollenweider (8) it is assumed that: (1) the rate constants of phosphorus-removal by export and sedimentation (β, ρ, and σ) are independent of the phosphorus-content of a lake (linear assumption); and (2) a lake is always at steady state with its actual phosphorus-loading. As a consequence of the two assumptions, phosphorus concentration within the lake (P) is always linearly related to the mean input concentration (P_{in}), at least if the seasonal variations of the involved processes are disregarded.

There is, however, theoretical and experimental evidence that the non-linearity of the model may become essential for the balance. Net sedimentation (M_s) is the difference between total phosphorus settling (S), and phosphorus release at the sediment–water interface (R). As shown schematically in Figure 16.3, the (net) sedimentation rate (σ) should decrease with increasing phosphorus concentrations for two reasons:

1. At low phosphorus concentration, primary productivity is limited by phosphorus, thus the phosphorus settling S is linearly related to phosphorus. Due to light limitation, productivity per area and thus S_{tot} reach a maximum (S_{max}) at high phosphorus values.
2. Phosphorus released by mineralization of biomass is partially fixed at the solid sediment phase and partially recycled to the lake water. The

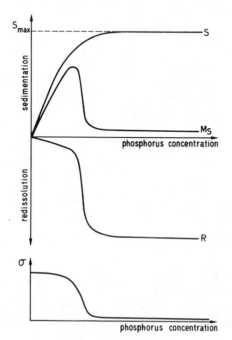

Figure 16.3. Schematic picture of the relationship between phosphorus settling S, phosphorus redissolution R, net sedimentation M_s, net sedimentation coefficient σ, and average phosphorus concentration of a lake.

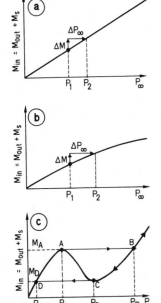

Figure 16.4. Relationship between phosphorus input and lake phosphorus concentration at steady state: (*a*) linear model, (*b*) nonlinear but monotonous model, and (*c*) model with local maxima and minima leading to a hysteresis effect (see text for further explanations).

recycled portion (R) increases with decreasing redox potential at the interface (9–11). The redox potential itself is linked to primary production by the hypolimnic oxygen demand of settling plankton.

We have not yet discussed the relationship between the trophic state and β, a vertical form factor which, considering the vertical phosphorus gradient in a lake, relates the annual mean phosphorus concentration of the surface effluent to the average phosphorus concentration of the entire lake. However, from the discussion above it becomes clear that with increasing trophic state more phosphorus becomes accumulated in the hypolimnion leading to a decrease of β.

Since at steady state total phosphorus input has to be equal to the two removal fluxes ($M_{out} + M_s$), the nonlinear relationship between phosphorus, M_s, and M_{out} may be responsible for a rather complicated behavior of the steady-state concentration P_∞ as a function of phosphorus loading. In the linear case [Fig. 16.4(*a*)], a given increment in phosphorus loading (ΔM) would eventually lead to a fixed increment in phosphorus concentration, $\Delta P_\infty = P_2 - P_1$, whereas ΔP_∞ is independent of the actual phosphorus level already reached in the lake. In the nonlinear but monotonously increasing case [Fig. 16.4(*b*)], the increment $\Delta P_\infty = P_1 - P_2$, produced by the same change of input ΔM_{in}, becomes larger with higher phosphorus levels.

The most complicated situation is met for a M_{in}/P_∞ curve with local minima and maxima [Fig. 16.4(*c*)], a situation which may evolve due to the nonmonotonous behavior of net sedimentation M_s. Assume that phosphorus loading of a lake slowly reaches the critical value M_A from below. Any further increase would produce a discontinuous jump of P_∞ to the much higher concentration P_B. Now imagine that the loading drops back to its original value (slightly below M_A); then the steady-state phosphorus concentration would not perform the reverse jump, that is, from P_B to P_A. The concentration of phosphorus would rather move along the curve from B to C and only jump to the original branch of the curve once the loading has dropped below M_D (jump from C to D). This behavior is called hysteresis. It means that given a certain phosphorus loading, the corresponding steady-state concentration P_∞ is not always unique but depends on whether the lake is in a growing or diminishing trophic development. In the later case, reduction of phosphorus input to very low values may become necessary in order to decrease P_∞ below the critical value P_A.

Because in most cases phosphorus loading rates of lakes were measured only during one year and it was rather assumed than investigated that these lakes were in steady state with their actual loading, data, which could document the schematically discussed relation between M_{in} and P_∞, are missing. However, the available information should at least demonstrate the expected decrease of sedimentation rate σ with increasing eutrophication. Two kinds of illustrations will be given: the first from a set of lakes and the second for an individual case. Since σ is affected by other factors, such as mean lake depth or hydraulic loading, we cannot expect to find a one-to-one relationship between σ and trophic state. Distributions of σ plotted in Figure 16.5 separately for "oligotrophic" (phosphorus concentration < 30 µg liter^{-1}) and "eutrophic" (phosphorus concentration > 30 µg liter^{-1}) lakes confirm the generally lower sedimentation rates of eutrophic lakes.

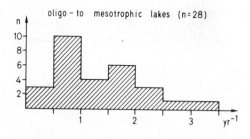

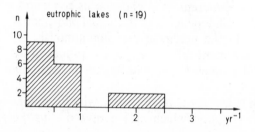

Figure 16.5. Frequency distribution of σ for oligotrophic/mesotrophic (phosphorus ≤ 30 µg liter^{-1}) and eutrophic (phosphorus > 30 µg liter^{-1}) lakes (12–17).

Table 16.3.   Characteristic Data of Lake Baldegg, Hallwil, and Sempach, Respectively

Parameter	Lake Baldegg	Lake Hallwil	Lake Sempach
A_0 (km^2)	5.2	10.0	14.4
V (10^6 m^3)	173	280	662
$\bar{Z}$ (m)	34	28	46
Z_{max} (m)	66	48	87
ρ (yr^{-1})	0.125–0.250	0.26	0.059

As a second illustration, net sedimentation for Lake Sempach, Switzerland (see Table 16.3 for characteristics) was calculated from Eq. 1 for the period 1954 to 1978 and plotted in Figure 16.6 as a function of phosphorus concentration. During the last 15 years, σ dropped from 0.35 yr^{-1} (when phosphorus concentration was below 30 μg liter^{-1}) to 0.1 yr^{-1} (phosphorus = 100 μg liter^{-1}) and is presently still decreasing. Vertical mixing in Lake Sempach during the winter is not always intensive and lasting enough to renew the hypolimnic oxygen reservoir completely. This may explain the considerable scattering in M_s during the more recent time when the annually changing hypolimnic redox potential became important for the phosphorus balance of the lake.

In addition to the varying σ, Lake Sempach also is an excellent example for a lake which is not at steady state with its actual phosphorus loading. For instance, in 1976 phosphorus export was 2.2 tons, net sedimentation 6.4 tons, and total phosphorus input 14.6 tons, leaving a net phosphorus-accumulation in the lake of 6.0 tons. Thus, if phosphorus input were reduced from 14.6 to 8.6 tons yr^{-1}, the lake concentration would not have been

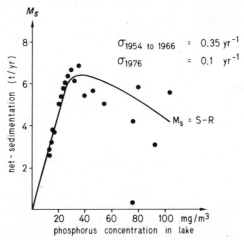

Figure 16.6.   Net sedimentation M_s as a function of phosphorus concentration in Lake Sempach.

proportionally reduced from 78 μg liter^{-1} (the value measured in 1976) to 46 μg liter^{-1}—as predicted by the linear steady-state model (Eq. 4)—but would just have remained constant.

As long as net sedimentation is proportional to phosphorus concentration ($M_s = \sigma VP$), the response time t_r (time to reach 63% of the steady-state concentration change) is

$$t_r = (\sigma + \beta\rho)^{-1} \qquad (5)$$

This time is increased to

$$t_r^* = (\rho\beta)^{-1} \qquad (6)$$

for the domain where M_s is independent of phosphorus concentration. The corresponding steady-state concentration is

$$P_\infty^* = \frac{M_{in} - M_s}{Q\beta} = \frac{1}{\beta}\left(P_{in} - \frac{M_s}{Q}\right) \qquad (7)$$

The following conclusions can be drawn from these equations:

1. At low phosphorus concentrations, where M_s is proportional to phosphorus concentration, the response time of phosphorus t_r is usually smaller than the mean residence time of the water in the lake, $t_w = \rho^{-1} = V/Q$ (see Eq. 5).

2. t_r increases with increasing trophic state since σ becomes smaller. In the ultimate case, when M_s becomes independent of phosphorus concentration the response time reaches a value t_r^* independent of M_s and larger than t_w (Eq. 6). This explains the observation on Lake Norviken by Ahlgren (18) who showed that after a reduction of P_{in} the phosphorus concentration in the lake followed a modified dilution function with $\beta = 0.77$.

3. Therefore, the restoration of a lake will take more time the higher the eutrophication process has gone already. This is not only true because of the longer way to reach moderate phosphorus concentrations, but also because response time during the initial stage of the restoration may be very large.

If the temporal change of the phosphorus input is approximated by an exponential curve of the form $M_{in} \exp(t/t_p)$, then, the relative size of t_p and t_r (or t_r^*) serves to determine whether a lake is at quasi steady state with its actual phosphorus loading. In fact, steady state occurs only as long as $t_p \gg t_r$. This condition has not been valid for Sempachersee during the more recent time. Therefore, the actual phosphorus concentration in this lake lags behind the phosphorus input; phosphorus would still increase after P_{in} became constant.

To summarize this section it has to be stressed that Eq. 1 can be misleading if uncritically used to predict the size and speed of a lake's reaction to changing phosphorus input.

4. LAKE RESTORATION BY OXYGEN INPUT

4.1. Artificial Mixing

Many lakes, especially if they are small but deep and protected from winds by hills or mountains, do not mix intensively enough in order to completely restore oxygen saturation in the deep layers during the winter. The tendency towards incomplete mixing may be enhanced by the process of eutrophication, since the biogenic accumulation of solutes in the hypolimnion may increase the stability of the lake with respect to vertical mixing (19,20).

In such lakes input of pressurized air into the deep layers during winter may serve to destroy stratification and to induce vertical mixing throughout the lake (Fig. 16.7). In Lake Baldegg, Switzerland (see Table 16.3 for characteristic data), total input of about 200 m^3 air per hour from three locations at the bottom of the lake was sufficient to destroy the chemical stratification within less than three weeks. The input rate of air corresponds to about 1.2×10^{-6} m^3 air per hour and m^3 lake volume, or 38×10^{-6} m^3 air per hour and m^2 lake surface. According to a recent summary on artificial aeration of various lakes throughout the world (21), typical input rates are between 10 and 1000×10^{-6} m^3 hr^{-1} per m^3 volume, or between 30 and $10,000 \times 10^{-6}$ m^3 hr^{-1} per m^2 surface. These lakes are generally much smaller than Lake Baldegg and have maximum depths between 2 and 20 m. Also, most

HYPOLIMNETIC OXYGENATION IN SUMMER

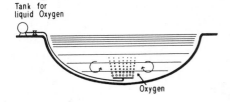

INTENSIFICATION OF MIXING IN WINTER

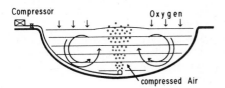

Figure 16.7. Principle of hypolimnic oxygenation in summer and artificial mixing in winter.

of the installations are designed to keep the lakes destratified also during the summer.

The process of lake reaeration can be described by a simple one-box model:

$$\frac{dO_2}{dt} = \frac{\alpha}{\overline{Z}} (O_2^s - O_2) - J_0 \tag{8}$$

O_2 and O_2^s are molecular oxygen concentration in the completely mixed lake and saturation concentration, respectively; $\overline{Z}$ is lake mean depth (volume per surface area); J_0 is oxygen consumption in the lake per volume and time; and α is the gas transfer coefficient which depends on wind velocity and on the state of the water surface. Field data on gas transfer coefficients are summerized by Emerson et al. (22) and Jähne et al. (23).

For constant coefficients, the solution of Eq. 8 is

$$O_2(t) = O_2^\infty - [O_2^\infty - O_2^0]e^{-(\alpha/\overline{Z})t} \tag{9}$$

O_2^0 is initial concentration, O_2^∞ the steady-state value

$$O_2^\infty = O_2^s - \frac{J_0\overline{Z}}{\alpha} \tag{10}$$

The temporal variation of O_2 for different initial conditions and different values for J_0 are shown in Figure 16.8 using the characteristics of Lake Baldegg. Total oxygen consumption J_0 during the winter is estimated to be about 5 tons per day in this lake. Figure and equations demonstrate that the reaeration rate of a lake increases with larger α and smaller $\overline{Z}$. Also, the steady-state oxygen concentration may significantly deviate from the saturation value O_2^s provided that the *in situ* oxygen consumption J_0 is large.

Obviously, the one-box model provides an upper limit of the reaeration process. If mixing were not fast enough to guarantee a homogeneous distribution of oxygen within the water column, a higher oxygen concentration,

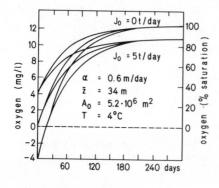

Figure 16.8. Theoretical evolution of oxygen concentration during artificial mixing for two oxygen consumption rates and three initial oxygen concentrations.

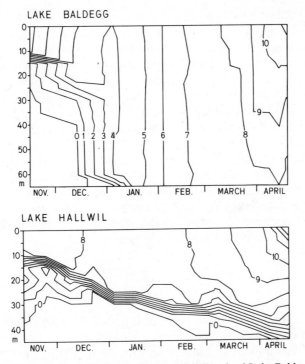

Figure 16.9. Pattern of oxygen distribution in the artificially mixed Lake Baldegg and in Lake Hallwil (control) during winter 1982/1983. Values in mg liter^{-1}.

that is, a smaller deficit, $O_2^s - O_2$, would be found at the surface. Artificial mixing mainly serves to bring the water with the lowest oxygen content to the surface and thus to maximize the natural gas transfer.

In spring 1983, after several months of aeration, oxygen concentration in Lake Baldegg reached the highest value since more than 25 years (24). However, significant differences in the oxygen content of the lake at the end of the winter have also been found in earlier years. In order to demonstrate that the situation in 1983 does not just represent an extreme meteorological year, the oxygen distribution in Lake Baldegg during the winter 1982/1983 is compared with oxygen in Lake Hallwil, a lake in the same area with similar characteristics (Table 16.3) though slightly shallower. As shown in Figure 16.9, the effect of the artificial mixing is clearly visible in the oxygen distribution of the treated lake. (Operation costs for artificial mixing were about $10,000 per season or $7 per ton of oxygen.)

4.2. Artificial Oxygen Input

From an ecological point of view, artificial mixing is only reasonable during winter. Also, much larger installations would be needed in order to mix a lake against the stabilizing force of the high solar input during summer. Most lakes have a stagnation period of at least nine months. If the hypolimnic oxygen consumption exceeds a certain value, the deeper-water layers turn anaerobic even if they reached oxygen saturation during the preceding winter. For instance, in the water volume below the 10-m depth of Lake Baldegg, the average oxygen consumption is about 1.6 mg O_2 per liters per month. Thus, even with an initial O_2 concentration of 12 mg liter^{-1}, aerobic conditions could not be sustained during nine months. Therefore, pure oyxgen is brought into the hypolimnion of Lake Baldegg at six locations at a total rate of 4.5 tons per day. The oxygen is taken from a tank at the shore filled with liquid oxygen. The small bubbles nearly completely dissolve before reaching the thermocline and therefore do not destroy the natural density stratification of the lake (Fig. 16.7). Since nearly all oxygen brought into the hypolimnion becomes dissolved in this part of the lake the change of hypolimnic oxygen concentration, O_2^H, is simply

$$\frac{dO_2^H}{dt} = \frac{B}{V_H} - J_0^H \tag{11}$$

where B is total oxygen input per unit time and V_H and J_0^H are hypolimnic volume and oxygen consumption rate, respectively.

For the hypolimnion of Lake Baldegg ($z > 10$ m), $V_H = 125 \times 10^6$ m^3, $B = 5 \times 10^6$ g day^{-1}, and $J_0^H = 0.05$ (mg liter^{-1}) day^{-1}; the average net hypolimnic oxygen decrease is 0.016 (mg liter^{-1}) day^{-1}, that is, 4.3 mg liter^{-1} for the stagnation period of about 270 days. Thus, mean oxygen concentrations in the hypolimnion would not drop below 4 mg liter^{-1} at the end of the stagnation period provided that during full overturn oxygen concentrations of 10 mg liter^{-1} or more can be achieved.

However, this conclusion may be too optimistic since spatial variation of both oxygen demand J_0 and oxygen input B may lead to an inhomogeneous O_2 distribution in the hypolimnion.

Indeed, the hypolimnic oxygen distribution during the summer 1983, the first year of full oxygen input, shows that for the thermocline and the deepest zone (10 m above bottom) oxygen input is still smaller than required (Fig. 16.10). Using the measured decrease of oxygen during the summer 1983 in combination with the depth-dependent oxygen demand calculated from data of the previous year we calculated the artificial O_2 input as a function of lake depth (Fig. 16.11). The result indicates that further improvements in the positioning and design of the diffusers will be necessary in order to increase the O_2 transfer into the deepest layer of the lake where O_2 demand is the largest.

Increased oxygen concentrations likely created an oxidized microlayer at the sediment–water interface which diminished the redissolution of phosphorus at this boundary layer. From mid-April to mid-July 1982, 6 tons of orthophosphate became accumulated in the untreated hypolimnion ($z > 15$

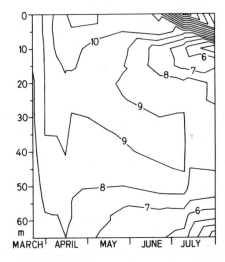

Figure 16.10. Oxygen distribution in Lake Baldegg in summer 1983, the first year of oxygenation. Values in mg liter^{-1}.

m) compared to less than 2 tons in 1983 during oxygenation. This suggests that oxygenation diminished the net release of phosphorus about three times.

Figure 16.12 compares the phosphorus retention in oligotrophic to mesotrophic and in eutrophic lakes. In the first group with average phosphorus concentrations smaller than 30 mg m^{-3}, in most cases the ratio P/P_{in} is about 0.2, whereas in eutrophic lakes the average value is about 0.4–0.5.

From this it can be deduced that in eutrophic but artificially aerated lakes in the best case this ratio may decrease to about 0.2, the value observed in oligotrophic lakes. Therefore it seems to be very unlikely that even in aerated lakes the water-quality criterion for phosphorus can be reached if P_{in} exceeds 150 mg m^{-3}.

Operation costs of oxygen input during the summer are much higher than during the winter. They are approximatively given by the costs for liquid

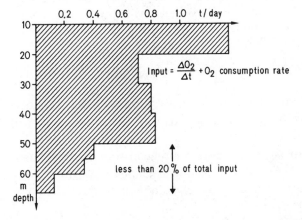

Figure 16.11.   Input of oxygen into various layers of Lake Baldegg.

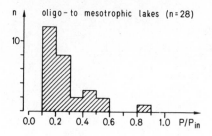

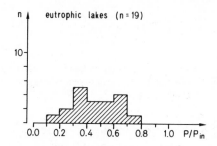

Figure 16.12. Frequency distribution of phosphorus retention for oligotrophic/mesotrophic (phosphorus ≤ 30 μg liter^{-1}) and eutrophic (phosphorus > 30 μg liter^{-1}) lakes (12–17).

oxygen (\$150 per ton O_2). Thus, for both ecological and economical reasons artificial mixing during winter is more advantageous than oxygen input during the summer; the latter method should only be employed if the former is not sufficient to guarantee the required O_2 level during the summer.

5.   DIVERSION OF HYPOLIMNIC WATER

The surface outlet of a lake can be replaced partially or completely by diversion of hypolimnic water in order to increase the export of nutrients from the lake, that is, to increase β (see Eq. 1 and Table 16.2). If the downstream end of the diversion pipe lies sufficiently below the level of the lake surface, hypolimnic diversion (HD) can be operated without energy costs.

HD has the following effects on the lake:

1.   Reduction of phosphorus concentration by increase of phosphorus export.
2.   Increase of epilimnion volume by lowering the thermocline.
3.   Removal of water with low oxygen concentration or with high oxygen demand, respectively.
4.   Reduction of mean residence time of water in hypolimnion.
5.   Removal of water enriched with solutes originating from decomposition processes in the deep hypolimnion, that is, reduction of stability of lake with respect to vertical mixing.
6.   Increase of hypolimnic water temperatures.

As a consequence of several alterations mentioned (effects 3 to 5), oxygen concentrations would increase in the hypolimnion. Due to the increased redox potential this could lead to a larger sedimentation parameter σ. Yet, the last effect (#6) could influence the hypolimnic O_2 concentration also negatively since the rate of biomass degradation increases with increasing water temperature.

5.1. Change of the Permissible Phosphorus Loading Due to HD

In a formal way, the steady-state solution of the one-box model, Eq. (4), can be employed in order to assess the quantitative effect of HD on the phosphorus concentration P_∞. As mentioned, the only parameters affected by HD may be the sedimentation rate σ and β, the ratio between mean phosphorus concentration in the outlet and in the entire lake, respectively. From the lake manager's viewpoint it is important to know to what extent hypolimnic drainage would increase the permissible P_{in}. If in Eq. 4, P_∞ is replaced by the tolerable phosphorus concentration of 30 mg m^{-3}; the permissible input concentration is

$$P_{in}^{perm} = 30 \text{ mg m}^{-3} \left(\beta + \frac{\sigma}{\rho} \right) \tag{12}$$

Hypolimnic drainage increases β, but most likely not σ because phosphorus is assumed to be always at the value of 30 mg m^{-3}. Other parameters, like average depth or flushing rate which may affect σ (8), are not altered by this kind of internal measure.

Therefore the increase of the permissible P_{in} depends only on the difference between β^{HD} and β, where β^{HD} is defined in analogy to β as the ratio between the average phosphorus concentration in the altered outlet and the average phosphorus concentration in the entire lake. In the most extreme case, that is, if the outlet is exclusively drawn from the lake bottom (with phosphorus concentration P_b^{HD}), then

$$\beta^{HD} = \frac{P_b^{HD}}{30 \text{ mg m}^{-3}}$$

compared to

$$\beta = \frac{P_0}{30 \text{ mg m}^{-3}} \tag{13}$$

$$\frac{\beta^{HD}}{\beta} = \frac{P_b^{HD}}{P_0} \tag{14}$$

According to available field data (see Table 16.4) in lakes with an aerobic

Table 16.4. Mean Total Phosphorus Concentrations At the Lake Surface P_0 and the Lake Bottom P_b in Some Swiss Lakes

Lake	Period of Observation	P_0	P_b	P_b/P_0	Z_{max}	Remarks
Horwerbucht	1970–1973	22	26	1.6	62	Permanently aerobic
Kreuztrichter	1964–1969	16	46	2.9	110	Permanently aerobic
Urnersee	1971–1973	16	23	1.4	195	Permanently aerobic
Walensee	1982	11	18	1.6	144	Permanently aerobic
Alpnachersee	1975	35	69	2.0	38	Partly anaerobic
Sempachersee	1967–1968	27	65	2.4	85	Partly anaerobic
	1975–1976	45	158	3.5	85	Partly anaerobic
Alpnachersee	1975	35	69	2.0	38	Partly anaerobic
Zürichsee	1980–1981	37	178	4.8	135	Partly anaerobic
Rotsee	1979–1980	57	566	9.9	14	Partly anaerobic

hypolimnion and with phosphorus concentrations of about 30 mg m^{-3}, β^{HD} exceeds β by a factor of 3 at the maximum. Therefore, from Eq. 12

$$P_{in}^{perm,HD} - P_{in}^{perm} = 30 \text{ mg m}^{-3}(\beta^{HD} - \beta) \le 30 \text{ mg m}^{-3}(2\beta) \quad (15)$$

According to field data collected by Fricker (16), for mesotrophic to eutrophic lakes β is about 0.7. Therefore hypolimnic siphoning increases the permissible input concentration P_{in}^{perm} by 40 mg m^{-3} at the maximum.

Curve 1 of Figure 16.13 showing the permissible P_{in} for lakes with a natural

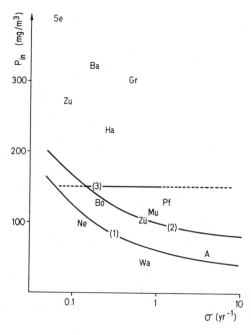

Figure 16.13. Permissible P_{in} for lakes with (1) a natural surface outflow, with (2) a hypolimnic drainage, and for (3) artificially aerated lakes.

surface outflow is derived from Vollenweider (8). Curve 2 takes into account that HD increases the permissible P_{in} at the most by 40 mg m^{-3}. The horizontal line at a level of 150 mg m^{-3} is based on the assumption that in artificially aerated lakes phosphorus retention may reach a maximum value of 80%. Although it is realized that all these estimates are only crude approximations to reality, Figure 16.13 indicates nevertheless that in many cases the water-quality goal (P $\leq$ 30 mg m^{-3}) cannot be reached with the help of lake internal measures alone if P_{in} cannot be simultaneously decreased. Yet, we have still to discuss an additional positive effect of HD—its influence on hypolimnic oxygen concentrations.

5.2. Influence of Hypolimnic Diversion (HD) on the Oxygen Balance

A simple box model serves to assess the influence of HD on the hypolimnic oxygen concentration O_2^k in a given volume section V^k:

$$V^k \frac{dO_2^k}{dt} = Q^{HD} O_2^{in} - Q^{HD} O_2^k - V^k J_0^k \qquad (16)$$

where O_2^{in} is O_2 concentration in the inflowing water (from the upper hypolimnion, thermocline, or epilimnion), Q^{HD} is the amount of water per unit time leaving the lake by HD, and J_0^k is the volumetric O_2 consumption in V^k. Division by V^k yields

$$\frac{dO_2^k}{dt} = \rho^k(O_2^{in} - O_2^k) - J_0^k, \qquad \rho^k = Q^{HD}/V^k \qquad (17)$$

where ρ^k is the water exchange rate in V^k by HD. This equation is identical to the reaeration equation 8. For constant coefficients the solution is of the same form as Eqs. 9 and 10. Again, the steady-state concentration in V^k, i.e. $O_2^{k,\infty}$, can be significantly smaller than O_2^{in} if the *in situ* consumption rate J_0^k is large (see Eq. 10). In fact, for a given J_0^k, the steady-state value $O_2^{k,\infty}$ is approaching O_2^{in} for the increasing water exchange rate ρ^k. However, since a large ρ^k may also lead to high temperatures in the hypolimnion and thus to larger oxygen consumption rates J_0^k, the maximum rate ρ^k does not necessarily yield optimum O_2 concentrations.

If $O_2^{k,\infty} > 4$ mg liter^{-1}, then O_2 concentrations in V^k always remain larger than 4 mg liter^{-1}. However, if $O_2^{k,\infty} < 4$ mg liter^{-1}, it is important to know for how long O_2^k actually would exceed 4 mg liter^{-1}. The critical time t^k, calculated from Eq. 9 with the corresponding substitution of variables, is

$$t^k = \frac{1}{\rho^k} \ln \frac{O_2^{k,0} - O_2^{k,\infty}}{4 \text{ mg liter}^{-1} - O_2^{k,\infty}} \qquad (18)$$

where $O_2^{k,0}$ is O_2 concentration at the beginning of the stagnation period.

In most large and deep lakes duration of the stagnation period is about 9 months. Under ideal conditions, $O_2^{k,0}$ and O_2^{in} correspond to the saturation concentration at about 4°C (about 12 mg liter^{-1}). As shown in Figure 16.14, for consumption rates of 2 to 5 mg liter per month typically found in eutrophic lakes, water exchange rates ρ^k have to range between 3 and 7 yr^{-1} to keep O_2^k above 4 mg liter^{-1}. Thus, HD is only efficient if a significant portion (or all) of the outlet can be diverted from the hypolimnion and if the water renewal rate of the lake is high.

Since for given ρ^k the absolute amount of through-flowing water, Q_{in}, grows in proportion to lake volume V^k, in large lakes HD would require immense diversion pipes. Ideal objects for HD are therefore smaller lakes with high flushing rates.

5.3. Examples of Hypolimnic Diversion

HD has first been proposed by Thomas (25) as a measure of lake restoration. The first installation was built in Kortowo Lake (Poland) (Olszewski, 1961, 1963). Its volume is 5.3×10^6 m^3, separated into two basins (depth 17 and 16 m, respectively) by a sill of 6 m. The HD pipe takes water from the southern basin (at a depth of 13 m) at a rate of 5.9×10^6 m^3 yr^{-1}. The small flushing rate ($\rho = 1.1$ yr^{-1}), the moderate depth of the pipe, and the division of the lake into two separate basins explain why the installations did not significantly effect the trophic state of the lake. Still, as Mientki (26) showed, the anaerobic period in the southern basin is shorter than in the northern basin by 30 days. Also, PO_4^{3-} and NH_4^+ concentrations are lower in the northern basin. As Sikorowa (27) concludes, HD has at least resulted in a gradual improvement of the lake (though not in a complete restoration), in spite of the steady increase of nutrient input during the last years. Anaerobic

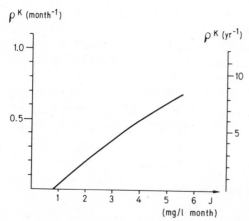

Figure 16.14. Minimum flushing rate ρ^k required in order to prevent oxygen concentration to drop below 4 mg liter^{-1}.

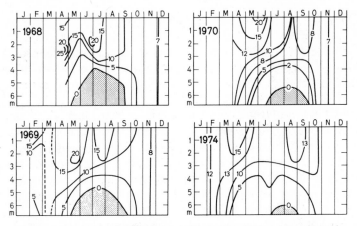

Figure 16.15. Oxygen distribution in Mauensee. Values in mg liter^{-1}.

phases became shorter, and hydrogen sulfide has completey disappeared from the lake.

In Mauensee, a small lake in Central Switzerland ($V = 2 \times 10^6$ m^3, Q_{in} = 3.3 $\times$ 10^6 m^3 yr^{-1}) HD has been installed in spring 1968 through which 2.1 $\times$ 10^6 m^3 yr^{-1} water is discharged from a depth of 7 m the maximum depth of the lake. The water volume below 4 m is replaced about twice during stagnation. Oxygen consumption rate is about 5 mg liter^{-1} per month. In order to prevent O$_2$ concentration from dropping below 4 mg liter^{-1}, at this oxygen consumption rate the hypolimnion should be flushed at least six times during stagnation period (see Figure 16.14), explaining why the water-quality criterion for oxygen could not be reached in the first year of HD operation. Yet, observations show, that the volume of the anaerobic zone and duration of anaerobic conditions at the lake bottom decreased from year to year (Fig. 16.15). During the years 1968 and 1969, all hypolimnic nitrate had been consumed by denitrification. In contrast, between 1971 and 1974 there was always nitrate present even at the deepest point of the lake.

As estimated from mass balance calculations, during the period 1968 to 1974 phosphorus export exceeded input by about 3.7 t. This quantity corresponds to phosphorus input of about 10 years. Since the total phosphorus content in the lake water never exceeded 200 kg (with the exception of short periods during redissolution events), this excess export of 3.7 t must have caused a corresponding loss of phosphorus from the sediments. As a consequence, maximum observed net phosphorus redissolution rates decreased from 370 kg phosphorus per day in 1968 to about 5 kg phosphorus per day in 1974.

Positive results from HD were also reported from Burgäschisee, Switzerland (Ambühl, personal communication), Piburgersee and Hechtsee, Austria (29,30). However, in no case did HD succeed to completely reach the quality criteria listed in Table 16.1.

6. SUMMARY AND CONCLUSIONS

The increasing input of phosphorus to lakes represents a perfect example of a chemical perturbation of aquatic systems. As an important lesson on the response behavior of lakes we learned that a balanced pattern of mass fluxes is often altered in a nonlinear manner, that is, that the relative size of, for instance, different removal pathways (flushing, sedimentation) is changed with increasing external mass input. As an example, net sedimentation rate of phosphorus was found to be highly dependent on the phosphorus-concentration in the lake. A second point concerned the importance of a dynamic concept to evaluate and understand the behavior of lakes under changing external conditions. It was concluded that an uncritical application of linear steady-state mass balance models to predict the response of a lake to the change of chemical mass inputs can be highly misleading. An important reason for this shortcoming is our lack of knowledge about the factual chemical and biological mechanisms which lie behind such empirical quantities as the rate of sedimentation.

Yet, the simple one-box model can still have its proper value, especially if critically used in combination with appropriate information on the range of behavior of different lakes as, for instance, compiled in the OECD reports on eutrophication (12,16,17). Based on such knowledge, the one-box concept was employed to derive the usefulness and limitations of the lake internal measures against eutrophication like artificial mixing, hypolimnic oxygen input, and discharge of water from the hypolimnion. It was shown that hypolimnic discharge can increase the permissible mean phosphorus input concentration P_{in} by 40 mg m^{-3} at best; oxygen input to the hypolimnion may decrease the concentration ratio between lake and input, P/P_{in}, to the minimum value of about 0.2. Thus, due to these limitations internal measures cannot fully replace the necessity to decrease the external phosphorus input.

Internal measures are only effective as long as they are in operation. If they are abandoned without reduction of the external input, in most cases the lake will, sooner or later, move to the same unsatisfactory conditions which had triggered the measures at the beginning. For the rare cases where a lake is in the unstable situation between two or more stationary states [Fig. 16.4(c)], a temporal application of internal measures could lead to a permanent improvement.

Hypolimnic diversion and artificial mixing do not cause large operation costs; they may thus become permanent installations. However, input of oxygen is a very expensive operation and cannot be recommended as a permanent solution. It is only meaningful to accelerate the recovery of a lake after the phosphorus input has been reduced. The high costs of oxygenation may be advantageous in so far as to represent an additional incitement to lower the external nutrient loading.

As a consequence of the points discussed above, it becomes clear that we should first fight the causes and only then may try to relieve the undesired symptoms. Thus, if a lake does not reach the water-quality criteria, the external nutrient loading should be lowered first. Only in those cases where this cannot be realized as quickly as desired or as efficiently as required, lake internal measures should be considered.

REFERENCES

1, R. Gächter and O. Furrer. "Der Beitrag der Landwirtschaft zur Eutrophierung der Gewässer in der Schweiz," *Schweiz. Z. Hydrol.* **34**, 41 (1972).

2. R. Gächter, D. Imboden, H. Bührer, and P. Stadelmann, "Möglichkeiten zur Sanierung des Sempachersees," *Schweiz. Z. Hydrol.* **45**, 246 (1983).

3. H. Ambühl, "Eutrophierungskontrollmassnahmen an Schweizer Mittelland-seen," *Z. f. Wasser- und Abwasserforschung* **15**, 113 (1982).

4. D. M. Imboden and R. Gächter, "A Dynamic Lake Model for Trophic State Prediction," *Ecol. Modelling* **4**, 77 (1978).

5. R. C. Dunst, D. M. Born, P. T. Uttormark, S. A. Smith, S. A. Nichols, I. O. Peterson, D. R. Krauer, S. L. Dernns, D. R. Winter, and T. L. Wirth, "Survey of Lake Rehabilitation. Techniques and Experiences," Technical Bulletin 75, Department of Natural Resources, Madison, Wisconsin (1974).

6. L. Landner, "Eutrophication of Lakes. Causes, Effects and Means for Control with Emphasis on Lake Rehabilitation," World Health Organization ICP/CEP 210 (1976).

7. A. Hamm and V. Kucklentz, *Möglichkeiten und Erfolgsaussichten der Seen-restaurierung. Materialien 15.* Bayerisches Staatsministerium für Landesent-wicklung und Umweltfragen (1981).

8. R. A. Vollenweider, "Advances in Defining Critical Loading Levels for Phosphorus in Lake Eutrophication," *Mem. Inst. Ital. Idrobiol.* **33**, 53 (1976).

9. W. Einsele, "Ueber die Beziehung des Eisenkreislaufs zum Phosphorkreislauf im eutrophen See," *Arch. Hydrobiol.* **29**, 664 (1936).

10. C. H. Mortimer, "The Exchange of Dissolved Substances between Mud and Water in Lakes," *J. Ecol.* **29**, 280 (1941).

11. C. H. Mortimer, "The Exchange of Dissolved Substances between Mud and Water in Lakes," *J. Ecol.* **30**, 147 (1942).

12. R. A. Vollenweider, "Scientific Fundamentals of the Eutrophication of Lakes and Flowing Waters with Particular Reference to Nitrogen and Phosphorus as Factors in Eutrophication," Technical Report DAS/CSI/68.27, Organization for Economic Cooperation and Development (OECD) Paris (1968).

13. P. J. Dillon and R. H. Rigler, "A Test of a Simple Nutrient Budget Model Predicting the Phosphorus Concentration in Lake Water," *J. Fish. Res. Board Can.* **31**, 1771 (1974).

14. R. A. Vollenweider and P.-J. Dillon, "The Application of the Phosphorus Loading Concept to Eutrophication Research," NRRC Report No. 13690 (1974).

15. D. P. Larsen and H. T. Mercier, "Phosphorus Retention Capacity of Lakes," *J. Fish. Res. Board Can.* **33**, 1742 (1976).

16. H. J. Fricker, "OECD Eutrophication Programme. Regional Project Alpine Lakes," Swiss Federal Board of Environmental Protection & OECD (1980).

17. S.-O. Ryding, "Monitoring of Inland Waters. OECD Eutrophication Programme, The Nordic Project," Nordforsk, Secretariat of Environmental Sciences Publ. 1980:2, (1980).

18. I. Ahlgren, "Role of Sediments in the Process of Recovery of A Eutrophicated Lake." In H. L. Golterman (Ed.), *Interactions between Sediments and Fresh Water,* Proceedings of an International Symposium held at Amsterdam, the Netherlands. September 6–10, 1976, p. 372.

19. J.-P. Pelletier, "Un lac méromictique, le Pavin (Auvergne)". *Ann. Station Biol. Basseen-Chandesse* **3,** 147 (1968).

20. Th. Joller, "Untersuchungen Vertikaler Mischungsprozesse mit Chemisch-Physikalischen Tracern im Hypolimnion des Eutrophen Baldeggersees," Ph.D. thesis, ETH Zürich, 1984, in preparation.

21. R. A. Pastorok, Th. C. Ginn, and M. W. Lorenzen, "Evaluation of Aeration/Circulation as a Lake Restoration Technique," EPA-600/3-81-014.

22. S. Emerson, W. S. Broecker, and W. Schindler, "Gas Exchange Rates in a Small Lake as Determined by the Radon Method," *J. Fish. Res. Board Can.* **30,** 1475 (1973).

23. B. Jähne, K. O. Munnich, and U. Siegenthaler, "Measurements of Gas Exchange and Momentum Transfer in a Circular Wind–Water Tunnel," *Tellus* **31,** 321 (1979).

24. P. Stadelmann, T. Joller and D. M. Imboden, "Die Auswirkungen von Internen Massnahmen im Baldeggersee; Zwangszirkulation und Sauerstoffbegasung des Hypolimnions," *Verh. Internat. Verein. Limnol.* **22,** (1984) (in press).

25. E. A. Thomas, "Ueber Massnahmen Gegen die Eutrophierung Unserer Seen und zur Förderung ihrer Biologischen Produktionskraft, *Schweiz. Fisch. Ztg.* **52,** 161, 198 (1944).

26. C. Mientki, "Chemical Properties of Kortowskie Lake Waters after An 18 Year's Experiment on its Restoration," *Pol. Arch. Hydrobiol.* **24,** 1, 13, 25, 37, 49 (1977).

27. A. Sikorowa, "Possibility of Protection Against Excessive Eutrophication of Lakes Using the Method of the Removal of Hypolimnion Waters," *Pol. Arch. Hydrobiol.* **24,** 123 (1977).

28. R. Gächter, "Die Tiefenwasserableitung, ein Weg zur Sanierung von Seen," *Schweiz. Z. Hydrol.* **38,** 1 (1976).

29. R. Pechlaner, "Erfahrungen mit Restaurierungsmassnahmen an Eutrophierten Badeseen Tirols," *Oesterr. Wasserwirtschaft* **30,** 112 (1978).

30. R. Pechlaner, "Responses of the Eutrophied Piburger See to Reduced External Loading and Removal of Monimolimnic Water," *Arch. Hydrobiol. Beitr. Ergebn. Limnol.* **13,** 293 (1979).

17

KINETICS OF CHEMICAL PROCESSES OF IMPORTANCE IN LACUSTRINE ENVIRON- MENTS

James J. Morgan

Environmental Engineering Science, California Institute of Technology, Pasadena, California 91125, USA

Alan T. Stone

Geography and Environmental Engineering, The Johns Hopkins University, Baltimore, Maryland 21218, USA

"Old aquatic chemists may never reach equilibrium, but they will surely produce a great deal of entropy trying."

ANONYMOUS, τ

"Every bond you break, every step you take, I'll be watching you."

THE POLICE, 1983

Abstract

The characteristic reaction times of chemical processes, $AB \rightarrow A + B$, treated by first-order rate laws range from nanoseconds (e.g., $MnSO_4$ aq. $\rightarrow Mn^{2+} + SO_4^{2-}$) to teraseconds (e.g., racemization of aminoacids). For chemical processes described by second-order rate laws, $A + B \rightarrow AB$, and reactant concentrations at the millimolar level, characteristic reaction times *are known to range from about 10 ns to far in excess of decades, under chemical conditions encountered in freshwater environments. Such a wide range in characteristic reaction times suggests that* kinetic *models of lake water chemical processes are required under conditions where the characteristic time of reaction,* τ_{ch}, *is comparable to the residence time* τ_R, *of the epilimnetic or hypolimnetic environment of a lake. For* $\tau_{ch} < \tau_R$, *an* equilibrium *model of the chemical process can yield useful information on the speciation in the lake water. For very slow processes, with* $\tau_{ch} > \tau_R$, *these reactions can be neglected. In general, there are chemical reactions of interest in hypolimnetic and epilimnetic waters (e.g., oxidation–reduction, hydrolysis, precipitation or dissolution) that are characterized by a wide range of reaction times. These reaction times often depend strongly on the chemical speciation of the lake water. Simultaneous application of kinetic and equilibrium descriptions for those processes with* $\tau_{ch} \simeq \tau_R$ *and those for which* $\tau_{ch} < \tau_R$, *respectively, yields "constrained equilibrium" or "pseudo-equilibrium" models. This can result in considerable simplification for the prediction of physical–chemical speciation in natural waters. The required information comprises the initial concentration conditions for all components of interest, the rate laws or slow chemical processes (or corresponding transport expressions or diffusionally-limited processes), and thermodynamic data for species involved in* fast *reactions.*

A goal of this chapter is to examine the relationship between kinetic and equilibrium models of chemical processes important in lakes. A second objective is to discuss some of the available experimental rate laws for aqueous chemical reactions within the framework of a transition state theory of reaction rates. This discussion yields a better understanding of the role that chemical speciation plays in the dynamics of reactions in lakes and points to the importance of more data on the temperature dependence of reaction rates.

1. INTRODUCTION

Lakes are open systems. Chemical reactions in lake waters and sediments are driven by fluxes of matter and energy from the surroundings (Fig. 17.1). Entropy production within lacustrine environments is a result. Chemical composition of the lake is regulated by processes and reactions which add and remove dissolved and particulate *species*. (Our analytical chemical information at present often extends only to elemental concentrations in operationally defined dissolved and particulate fractions, to oxidized versus reduced forms of elements, to free versus bound forms of metal ions, and to some molecular-weight fractions of polymers.) Many chemical species are transformed at appreciable rates through homogeneous or heterogeneous reactions which are almost always complex and which are in some cases catalyzed by organisms, other aqueous species, light, or surfaces. For *biogenic* elements such as C, N, O, P, S, and Si and the biochemically essential

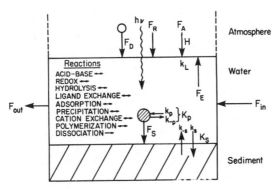

Figure 17.1. Generalized dynamic description of reactions and fluxes regulating composition in the lacustrine environment. F_{in} and F_{out} represent hydraulic inputs and outputs; F_R, precipitation flux; F_D, dry deposition; F_A, gas solution (k_L an exchange coefficient and H a Henry's law constant); F_E, evaporative fluxes; F_S, sedimentation flux; and k_p, k_{-p}, k_s, and k_{-s} represent rate coefficients for heterogeneous transfer reaction processes between solid phases and water.

metals, for example, iron, manganese, copper, and calcium, kinetics of removal from and return to the water and consequent vertical distributions in lake waters and sediments are strongly influenced by rates of biochemical processes such as photosynthesis, respiration, nitrate reduction, sulfate reduction, and fermentation. Kinetics of abiotic reactions, for example, SiO_2 dissolution, $CaCO_3$ precipitation, reduced metal oxygenations, photocatalyzed metal reductions, and ion-exchange reactions, can also exercise major influences on geochemical cycles in lakes.

The time invariant conditions of the lake as a chemical system is the steady state, not equilibrium. Although complete equilibrium is not found in lakes, near-equilibrium or quasi-equilibrium conditions are frequently realized, a result of the differences in the rates of reactions, or time constants, among coupled chemical reactions, and differences between the time scales for reactions and those of physical transport processes. The existence of local equilibria (in space or time) or quasi-equilibria can lead to greatly simplified models for lake chemistry.

2. PHYSICAL AND CHEMICAL RATES

Chemical composition of the water and particulate matter in lakes varies in space and time because of chemical reactions and transport processes which jointly determine the rates of change in total concentration of chemical elements and concentration of individual species. The resulting accumulation of a chemical species can be expressed as:

accumulation = inputs from surroundings − outputs to surroundings

$$\pm \text{ production and loss via chemical reactions} \qquad (1)$$

For a chemical species in solution, application of the material balance principle at a particular point in lake water or sediment yields the rate of change of aqueous concentration, C_i, of species i (1,2):

$$\frac{\partial C_i}{\partial t} = -\overline{U} \cdot \nabla C + \sum_1^3 \frac{\partial}{\partial x_k} \left(\epsilon_k \frac{\partial C_i}{\partial x_k} \right) + \sum_j^n \nu_{ij} R_j \qquad (2)$$

where $\overline{U}$ is the three-dimensional advection velocity vector (length per time), k stands for the three spatial coordinates, ∇C is the gradient of concentration, ϵ_k is a Fickian diffusive mixing parameter for each spatial coordinate (length^2 per time), R_j is the *rate* (moles per volume per time) of the jth reaction in parallel producing or consuming species i, and ν_{ij} is the stoichiometric coefficient (dimensionless, positive for production and negative for consumption) of species i in the reaction j. For the jth of the parallel overall reactions, the reaction can be represented as

$$\nu_{1j} A_1 + \nu_{2j} A_2 + \cdots + \nu_{ij} A_i + \cdots + \nu_{mj} A_m = 0 \qquad (3a)$$

or, generally, as

$$\sum_i^m \nu_{ij} A_i = 0 \qquad (3b)$$

If concentration gradients are significant only in the vertical direction z,

$$\frac{\partial C_i}{\partial t} = -U_z \frac{\partial C}{\partial z} + \frac{\partial}{\partial z} \left(\epsilon_z \frac{\partial C_i}{\partial z} \right) + \sum_j^n \nu_{ij} R_j \qquad (4)$$

The total rate of change in concentration is the algebraic sum of advective, diffusive, and reaction terms, that is,

$$\left(\frac{dC_i}{\partial t} \right)_{tot} = \left(\frac{dC_i}{\partial t} \right)_{adv} + \left(\frac{dC_i}{dt} \right)_{diff} + \left(\frac{dC_i}{dt} \right)_{chem} \qquad (5)$$

Dividing the chemical reaction rate by the local concentration,

$$\frac{1}{C_i} \left(\frac{dC_i}{dt} \right)_{chem} = \frac{1}{C_i} \sum_j^n \nu_{ij} R_j \equiv \frac{1}{\tau_{chem}} \qquad (6)$$

gives a time scale for the *chemical* processes bringing about the production or disappearance of species i. For example, if there is a single overall reaction (complex or elementary) resulting in the disappearance of species A, with a differential equation $R = -[A]/dt = k[A]$ describing the rate, then $\tau_{chem} = 1/k$. Comparison of chemical reaction time scales is helpful in understanding the relative importance of transport and reaction rates in determining distributions of chemical properties in lakes.

For an assumed well-mixed lake volume, that is, for $\nabla C \simeq 0$—for example, a lake's epilimnion with relative rapid vertical and horizontal mixing—a simplified model can be used to compare input and output fluxes of a species with rates of production or loss in the lake through chemical reaction. If F_{ik} represents the total molar flux (moles per time) of species i added to the well-mixed volume V of the lake by process k and Q represents a steady hydraulic flow rate in and out, the rate of change in the concentration in the well-mixed volume is

$$\frac{dC_i}{dt} = \frac{1}{V} \sum_k^p F_{ik} - \frac{Q}{V} C_i + \sum_j^n v_{ij}R_j \qquad (7)$$

where v_{ij} and R_j have the same meanings as in Eq. 2. We note that R_j, the rate of the jth homogeneous or heterogeneous overall reaction, is quite general, and may represent zero-order, first-order, or second-order differential equations for irreversible reactions, complex irreversible reactions, or complex reversible reactions. The goal of experimental kinetics is to obtain

$$R_j = \frac{1}{v_{ij}} \frac{dC_i}{dt} = f(C_i, C_m, \ldots, T, I, h\nu, A_s, \ldots).$$

Some specific forms of R_j are discussed later in the chapter. The flux terms are, in general, *functions* and depend on the concentration of species i in inflowing streams and in other volumes of the lake, on sediment reactions, on chemistry of atmospheric deposition, and so on. Because aqueous chemical reactions can be coupled through the simultaneous rates of change of common species, $1, 2, \ldots, i, \ldots, l$, there will, in general, be a number of equations with the form of Eq. 7 needed to describe temporal variation or time-invariant conditions for related species.

If we identify V/Q as the mean water residence time τ_R, $(\sum_k^p F_{ik})/Q$ as the equivalent mean input concentration of i, $\overline{C}_i$, and $(\sum_j^n v_{ij}R_j)/C_i$ as $1/\tau_{\text{chem}}$, then Eq. 7 can be rewritten simply as

$$\frac{1}{C_i} \frac{dC_i}{dt} = \frac{1}{\tau_R} \frac{\overline{C}_i}{C_i} - \frac{1}{\tau_R} + \frac{1}{\tau_{\text{chem}}} \qquad (8)$$

For long characteristic times of reaction (various slow homogeneous and heterogeneous reactions) in comparison to water residence times, changes in concentrations of chemical species are governed by hydraulic factors; for shorter τ_{chem}, reaction rates may govern changes in concentration.

The objective of this chapter is to consider chemical reactions in lakes from a kinetic point of view. A kinetic perspective on lake chemistry is necessary if we are to arrive at a quantitative description of the wide range of homogeneous and heterogeneous reactions involving components of the natural biogeochemical cycles as well as flows of synthetic pollutants in lakes. An earlier review by Brezonik (3) provides an excellent introduction to kinetic concepts in relation to natural water chemistry. In emphasizing *rates* of chemical reactions in aquatic systems we aim to encourage a general kinetic approach for modeling the distributions of chemical properties in lakes, for example, distribution of elements, distribution of dissolved and particulate forms of elements, and distribution of chemical *species* in solution and in particulate phases.

A key to understanding overall chemical composition distributions in lake waters and sediments is *chemical speciation.* In equilibrium models of natural water systems, the activities of aqueous-, particulate-, surface-, and gaseous-phase species are related through equilibrium constants, which in turn depend upon standard *chemical potentials* of species. Kinetic models depend equally on a knowledge of chemical species, with the concept of species *reactivity* occupying a fundamental place in the analysis of rate processes.

For reversible chemical reactions (those capable of proceeding at appreciable rates in forward and reverse directions), thermodynamic and kinetic descriptions provide complementary views of the equilibrium state. Because lakes are *open* systems, thermodynamic descriptions, strictly speaking, are excluded. Total equilibrium is not attained. However, quasi-equilibrium (near equilibrium) conditions are of great importance in describing overall rate processes in lakes. An additional goal of this paper is to identify the connections between kinetic and equilibrium concepts for applications to rate processes in lakes. For *irreversible* processes, of course, only kinetic descriptions are possible.

3. CHARACTERISTIC TIME SCALES

The scope for application of kinetic descriptions in lake chemistry is indicated by comparing the *time scales* of several transport processes in lakes with the *characteristic times* for chemical reactions in lakes. Imboden and Lerman (4) and Schwarzenbach and Imboden (5) have discussed information on estimated times of mixing by various mechanisms in lakes. Water renewal times for whole lakes are estimated to range from 10^{-1} to 10^2 yr, times of horizontal mixing in epiliminia to range from 10^{-4} to 10^{-2} yr, mixing times for epilimnetic and hypolimnetic waters to range from 10^{-2} to 10^2 yr, sedimentation times to range from 10^{-3} to 10^{-1} yr, and so forth for various mechanisms (5). Broadly speaking, physical transport time scales for lake systems span a range of approximately 10^{-4} to 10^2 yr. For lakes which are stratified annually, an hypolimnion isolation time is about half a year.

Chemical reaction rates and reaction half-lives for a wide variety of reactions of potential importance in freshwater environments have been summarized recently by Hoffmann (6) and by Pankow and Morgan (7). For aqueous reactions with first-order kinetics, that is, $-d[A]/dt = k_1[A]$, experimentally determined values of k_1 range from 10^9 s^{-1} for dissociation of MnSO$_4$(aq) (near the diffusion-controlled limit) to 10^{-12} s^{-1} for amino acid racemization. The range in characteristic times (half-lives) is thus from nanoseconds to over 10,000 yr. For first-order hydrolysis of alkyl halides (an important class of organic pollutant compounds), characteristic times of reaction at neutral pH and 25°C range from approximately 10^2 s (t-butyl chloride) to 10^{11} s, or about 3000 yr (trichloromethane) (8).

For a variety of second-order aqueous reaction rates, for example, for $-d[A]/dt = 2k_2[A]^2$, or $-d[A]/dt = k_2[A][B]$, Pankow and Morgan (7) summarized experimental values of second-order rate constants, k_2 at 25°C ranging from 10^{11} M^{-1} s^{-1} (diffusion-controlled limit for "fast" biomolecular processes involving univalent ions of opposite charge, e.g., H$^+$ + OH$^-$ → H$_2$O, H$^+$ + HS$^-$ → H$_2$S) to 10^{-2} M^{-1} s^{-1} [for complex, pseudo-second-order oxygenation of reduced manganese, Mn(II) + O$_2$ → Mn(III), IV) + ⋯ (pH ~ 9)]. For 10^{-5} M reactant concentrations, this range of second-order rate constants leads to characteristic times for reaction ranging from a microsecond to 10 Ms, or about 0.3 yr (characteristic times for second-order processes depend on the concentration of reactant). The review of kinetic information of potential applicability to aquatic systems by Hoffmann (6) revealed first-order metal-ion hydrolysis characteristic times ranging from approximately 10^{-10} to 10^3 s [Mn(CN)$_6^{3-}$]. Ligand substitution reactions for a number of transition metals with a variety of ligands have characteristic times (second-order kinetics, 10^{-5} M concentration) in a range from milliseconds to 10^8 seconds. Second-order electron transfer reactions with Fe(II) as the reductant at 10^{-5} M concentrations revealed characteristic times ranging from approximately 1 to 10^8 s.

The foregoing summary of mixing time scales in lakes and characteristic times for reactions of interest indicates the following overall ranges for comparison:

Mixing time scales ~10^3–10^9 s.
First-order reaction times ~10^{-9}–10^{12} s.
Second-order reaction times (10^{-5} M) ~10^{-6}–10^8 s.

Schematically,

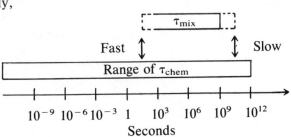

It is clear that some chemical reactions with short characteristic times, $\tau_{chem} \ll \tau_{mix}$, will proceed completely to products or to equilibrium in lakes, whereas other reactions with long characteristic times, $\tau_{chem} \gg \tau_{trans}$, progress only to a slight extent. Reactions for which $\tau_{chem} \sim \tau_{mix}$, for example, with τ_{chem} in the range from 10^3 to 10^9 s, require a *kinetic* description in order to account for element and chemical species distributions. For species involved in coupled chemical reactions with both short ($\tau_{chem} \ll \tau_{mix}$) and longer ($\tau_{chem} \sim \tau_{mix}$) time scales, *quasi-equilibrium* descriptions of *fast* reactions may be combined with kinetic descriptions of *slow* reactions to yield "constrained equilibrium" or "pseudoequilibrium" models (9,10).

4. AQUEOUS CHEMICAL KINETICS

The aim of this discussion is to outline a framework for treating chemical reactions in lakes kinetically. The subject of kinetics requires an approach different from that taken in energetic and equilibrium thermodynamic treatments of aquatic systems. *Time* is now a variable. Recognizing the *paths* or *mechanisms* whereby reactants become products is essential to understanding rates of reactions. There is a fundamental distinction between *overall reactions,* for example, Eq. 3a, which describe the overall stoichiometry of reactants being converted to products (i.e., the conventional chemical equation) and *elementary reactions,* which describe single chemical steps in *molecular* terms, usually unimolecular or biomolecular processes. For more detailed considerations of the kinetic topics summarized here, we suggest the discussion of rate laws and transition state theory by Lasaga (11), and the wider treatment of kinetics by Gardiner (12), Moore and Pearson (13), Hammes (14), Adamson (15), Weston and Schwarz (16), and Benson (17). Chemical kinetics has three aspects: *phenomenology, reaction mechanism,* and the *theory of elementary reactions.*

4.1. Phenomenological Description of Reaction Rates

The goal is establishing the *rate law*, a mathematical function, specifically a differential equation, which describes the reaction rate as a function of system composition (i.e., concentrations of aqueous, solid, and surface *species*), temperature, ionic strength, surface area of solids, light intensity, and possibly mixing energy (if diffusion is a potential limiting process). Determining the correct stoichiometry of the reaction is closely related to establishing the rate law. The phenomenological aspect of kinetics deals with macroscopic data. Detecting reaction intermediates is a further experimental aspect of kinetics, which may help to identify the reaction mechanism.

An example of an overall chemical reaction of interest in lakes is the oxygenation of ferrous iron, which in mildly acidic to mildly alkaline solutions can be described by the stoichiometry

$$4Fe^{2+} + O_2 + 10H_2O = 4Fe(OH)_3(s) + 8H^+ \qquad (9)$$

The experimental rate law (under low reactant iron concentration conditions when heterogeneous catalysis by product is negligible) has been found to be

$$R = -\frac{1}{4}\frac{d[Fe^{2+}]}{dt} = k[O_2][Fe^{2+}][H^+]^{-2} \qquad (10)$$

in which k is a rate constant depending on temperature.

4.2. Reaction Mechanisms

The *chemical mechanism* of a reaction is a proposed set of *elementary* (molecular) reactions which provide a sequential path or a number of parallel paths which account for both the stoichiometry and the observed rate law of the overall reaction. If the reaction mechanism is *simple*, it consists of a single elementary step (apart from molecular diffusion of reactants and products, which is always a step in aqueous reactions) capable of accounting for the overall reaction. Experimental rate laws often point to *complex* mechanisms, that is, a sequence of elementary steps, or two or more such sequences in parallel. Complex mechanisms frequently introduce *intermediate* species, that is, neither reactants nor products. An energetic or equilibrium description of an overall reaction deals only with reactant and product species, whereas a *mechanistic* description of reaction kinetics must recognize, in addition, ground-state catalyst species, intermediate species in the ground state, and excited (high-energy) electronic states created by absorption of photons.

An example of a proposed mechanism which, in part, accounts for the rate law for aqueous ferrous iron oxygenation (Eq. 10) comprises the following elementary steps (18,19):

$$Fe^{2+} + H_2O \rightleftharpoons FeOH^+ + H^+ \qquad (i)$$

$$FeOH^+ + O_2 \rightleftharpoons Fe(OH)O_2{}^+ \qquad (ii)$$

$$Fe(OH)O_2{}^+ \rightarrow FeOH^{2+} + O_2^{\cdot-} \qquad (iii)$$

$$O_2^{\cdot-} + H^+ \rightleftharpoons HO_2^{\cdot} \qquad (iv)$$

$$HO_2 + Fe^{2+} + H_2O \rightarrow FeOH^{2+} + H_2O_2 \qquad (v)$$

$$H_2O_2 + Fe^{2+} \rightarrow FeOH^{2+} + OH^{\cdot} \qquad (vi)$$

$$OH^{\cdot} + Fe^{2+} \rightarrow FeOH^{2+} \qquad (vii)$$

Further hydrolysis of $FeOH^{2+}$ yields (with nucleation) $Fe(OH)_3(s)$. We note that the radical species O_2^-, HO_2^-, and OH^-, along with hydrogen peroxide, appear as reaction intermediates in the proposed mechanism. The importance of speciation in kinetics and the mechanism thus goes beyond consideration of reactants and products. If step (ii) or (iii) is the slowest, hence rate-determining step, some of the important features of the experimental rate law can be accounted for. Alternative mechanisms are not, however, excluded.

4.3.   Theories of Elementary Reactions

This aspect of kinetics is concerned with the intimate mechanism of chemical-bond making and breaking, the energetic and geometric requirements for reaction, and the ways in which reaction species approach one another and products separate. An elementary bimolecular solution reaction

$$A + B \rightarrow products \tag{11}$$

in which reactants must encounter one another prior to transformation into products may be viewed in simplified fashion as a two-step sequence:

$$A + B \xrightarrow{k_E} [A,B]^* \xrightarrow{k_R} products \tag{12}$$

where k_E is a rate constant for the *encounter* (collision) step which brings the solute reactants into first contact, say in a "cage" of solvent molecules or as a weak "pair," denoted by $[A,B]^*$, and k_R is a rate constant for the *chemical* step as such. The rate of encounter is then $k_E[A][B]$ and the rate of the chemical reaction is $k_R[A,B]^*$. If the intrinsic rate of the chemical step is very *fast*, then the encounter step is rate determining. An upper limit to the "fast" elementary reaction rate is set by molecular diffusion and can be described by the Smoluchowski–Debye theories of solute molecule or ion collision frequency in water (16,20). The equation for the diffusion-controlled rate constant is

$$k_E = \frac{4\pi N}{1000} (D_A + D_B)(r_A + r_B)f \tag{13}$$

where N is Avogadro's number, D is the diffusion coefficient, r the solute species radius, and f a factor which accounts for long-range forces, for example, electrostatic, between the approaching reactants. At 25°C, for aqueous neutral species with radii of approximately 0.5 nm, k_E is about 10^{10} $M^{-1}s^{-1}$; for singly charged ions of opposite charge, k_E is about $10^{11}\,M^{-1}s^{-1}$. Thus, the characteristic time scale for "fast" biomolecular solution reactions, at $10^{-5}\,M$ concentrations, is measured in μs.

If the specific rate of the chemical step is *slow* compared to the encounter rate, that is, $k_R \ll k_E \sim 10^{10} \, M^{-1} s^{-1}$, then the rate of the elementary reaction, A + B → products, is governed by chemical reaction. [Similar considerations enter in the case of heterogeneous reactions, for example, dissolution or adsorption, where, for spherical particles, the quantities of interest are k_s (cm s^{-1}) and D/a (cm s^{-1}), where k_s is a rate constant and a the radius (20,24)].

Theoretical understanding of slow elementary chemical reactions is a central problem in chemical dynamics. A fundamental quantitative molecular picture of elementary reaction rates in aqueous solutions and at interfaces has proved extremely difficult to construct because of the role of the solvent, that is, the variety of physical and chemical effects resulting from interactions between reactants and solvent water molecules. The presence of ions at significant concentrations, extensive hydration of most species, and participation of water molecules in reactions as catalyst, intermediate, reactant, or product all serve to complicate mechanistic interpretation of slow aqueous reactions (13,15). Nonetheless, useful qualitative and quantitative interpretation is possible through considering known effects of hydration on thermodynamic properties.

5. ACTIVATED COMPLEX THEORY

The most widely used theoretical framework for elementary solution and interfacial reactions is *activated complex theory* (ACT), also referred to as absolute reaction rate theory, or transition-state theory (TST). The term *activated complex* refers to a high-energy ground-state species formed from reactants, for example,

$$ A + B \underset{}{\overset{K^{\ddagger}}{\rightleftharpoons}} AB^{\ddagger} \tag{14} $$

and at local equilibrium ($K^{\ddagger}$ is a kind of equilibrium constant) with them. Figure 17.2 illustrates the essential features of a potential energy profile for the elementary reaction. The profile is meant to represent a lowest-energy path on a three-dimensional surface. The potential energy state of the activated complex is the *transition state* (in transition between reactants and products). The activated complex then falls apart to yield products

$$ AB^{\ddagger} \rightarrow \text{products} \tag{15} $$

The TST was developed originally by Eyring and others on the basis of statistical mechanics [see, e.g., Lasaga (11) or Moore and Pearson (13)]. The fundamental result is a bimolecular rate constant for an elementary process expressed in terms of (1) the total molecular partition functions per unit volume (q_i) for reactant species and for the activated complex species ($q^{\ddagger}$),

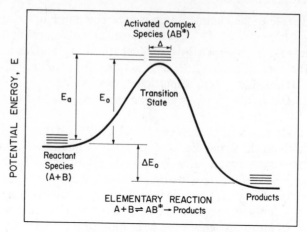

Figure 17.2. Potential energy along a one-dimensional reaction coordinate for an elementary reaction. E_0 is the difference in zero-point energies between reactants and activated complex, $AB^{\ddagger}$; ΔE_0 is the energy difference between reactants and products. $E_a \simeq E_0$, the activation energy of the elementary process.

and (2) the difference in zero-point potential energies between the activated complex and reactants (E_0):

$$k_2 = \frac{kT}{h} \frac{q^{\ddagger}}{q_A q_B} \exp(-E_0/kT) \tag{16}$$

where k is Boltzmann's constant, T is absolute temperature, and h is Planck's constant. If $q^{\ddagger}$ could be evaluated from molecular properties and E_0 calculated from potential energy surfaces, an elementary rate constant could then be obtained for any bimolecular reaction. Evaluation of total partition functions by calculation on the basis of fundamental translational, vibrational, and rotational properties proves to be too difficult for the complicated solvated species in solution. Activated complex theory, ACT, must then be applied in a different way, using a *thermodynamic* formulation instead of that in Eq. 16. Briefly, the term kT/h is regarded as the frequency of decomposition of the activated complex species, so that the rate of reaction becomes

$$R = \frac{kT}{h} [AB^{\ddagger}] = \frac{kT}{h} K^{\ddagger}[A][B] \frac{\gamma_A \gamma_B}{\gamma_{\ddagger}} \tag{17}$$

with $K^{\ddagger}$ viewed as the *thermodynamic* formation constant for $AB^{\ddagger}$ and the γ's are activity coefficients relating activities and concentrations (functions of ionic strength). The elementary rate constant k_2 is then

$$k_2 = \frac{kT}{h} K^{\ddagger} \frac{\gamma_A \gamma_B}{\gamma_{\ddagger}} = \frac{kT}{h} \frac{\gamma_A \gamma_B}{\gamma_{\ddagger}} \exp(-\Delta G^{\ddagger 0}/RT) \qquad (18)$$

where $\Delta G^{\ddagger 0}$ is a standard free energy of activation.
    Because $\Delta G^{\ddagger 0} = \Delta H^{\ddagger 0} - T\Delta S^{\ddagger 0}$,

$$k_2 = \frac{kT}{h} \frac{\gamma_A \gamma_B}{\gamma_{\ddagger}} \exp(\Delta S^{\ddagger 0}/R) \exp(-\Delta H^{\ddagger 0}/RT) \qquad (19)$$

Figure 17.3 illustrates schematically the essential features of the thermodynamic formulation of ACT. If it were possible to evaluate $\Delta S^{\ddagger 0}$ and $\Delta H^{\ddagger 0}$ from a knowledge of the properties of aqueous and surface species, the elementary bimolecular rate constant could be calculated. At present, this possibility has been realized for only a limited group of reactions, for example, certain (outer-sphere) electron transfers between ions in solution.

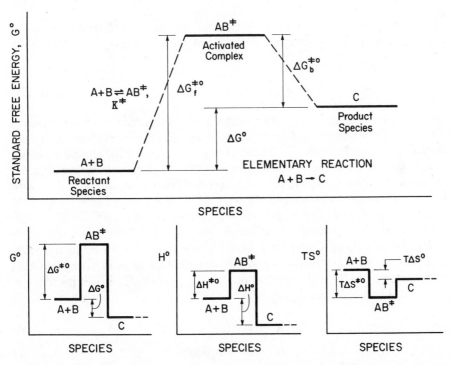

Figure 17.3. Schematic illustration of the essential features of thermodynamics of activated complex theory for an elementary reaction. Free energies, entropies, and enthalpies of activation are shown: $\Delta G^{\ddagger 0} = \Delta H^{\ddagger 0} - T\Delta S^{\ddagger 0}$, and $\Delta G^0 = \Delta G_f^{\ddagger 0} - \Delta G_b^{\ddagger 0}$, where f and b denote the forward and back reactions. $K^{\ddagger}$ is the equilibrium constant for activation.

The ACT framework finds wide use in interpreting experimental bimolecular rate constants for elementary solution reactions and for *correlating*, and sometimes interpolating, rate constants within families of related reactions. It is noted that a parallel development for *unimolecular* elementary reactions yields an expression for k_1 analogous to Eq. 19, with appropriate $\Delta S^{\ddagger 0}$.

It is essential to limit quantitative interpretation of the ACT relationship, for example, Eqs. 18 and 19, to slow *elementary reactions*. Complex reactions, for example, a sequence of elementary steps, need to be first resolved through examination of the rate law and consideration of the mechanism in order to identify possible rate-determining steps and obtain entropies and enthalpies for activated complex formation.

6. ARRHENIUS EQUATION AND PARAMETERS

The Arrhenius equation (1889) is

$$k = A \exp(-E_a/RT) \qquad (20)$$

in which k is a rate constant, $A = A(T)$, in general, and E_a is the *activation energy*. A and E_a are referred to as the Arrhenius parameters. The logarithmic form of Eq. 20,

$$\ln k = \ln A - E_a/RT \qquad (21)$$

suggests plotting logarithms of experimental rate constants versus reciprocal absolute temperatures $(1/T)$ in order to estimate the preexponential factors A and activation energies E_a. The Arrhenius parameters for complex overall reactions will require analysis in terms of the reaction mechanism, that is, in terms of elementary reactions. For an *elementary* solution reaction, comparison of the Arrhenius equation with ACT equation 19 (with $\gamma_A \gamma_B/\gamma_{\ddagger}$ of unity) shows (20) that $A \equiv e(kT/h) \exp(\Delta S^{\ddagger 0}/R)$, with $\Delta H^{\ddagger 0} = E_a - RT$. Theoretical interpretation of Arrhenius parameters is limited strictly to elementary reactions. Complex reaction sequences need to be analyzed in terms of possible mechanisms and rate-determining steps in order to obtain and interpret ACT activation parameters. For example, if the mechanism of a certain overall reaction is

$$A + B \overset{K}{\rightleftharpoons} C \qquad (i)$$

$$C + D \overset{k}{\rightarrow} \text{products} \qquad (ii)$$

with (i) at local equilibrium, that is, *fast* compared to (ii), the rate-determining step, then the *apparent* (observed) rate constant will be kK, and the *apparent* activation energy, E_{app}, will be $E_{a,ii} + \Delta H_i^0$. Knowledge of the overall temperature dependence *and* the temperature dependence of K, that is, ΔH_i^0, allows determination of E_a for elementary step (ii), which in turn yields $\Delta H^{\ddagger 0}$ and $\Delta S^{\ddagger 0}$ (or A). Resolution of the complex reaction is a prerequisite for theoretical interpretation of observed rates. From a practical viewpoint, use of E_{app} for slow, overall reactions is nonetheless of considerable value for interpolation of rate constants.

It is to be noted that rate constants for fast (diffusion-controlled) steps are also temperature dependent, since the diffusion coefficient depends on temperature. The usual experimental procedure, suggested by the Arrhenius equation, of plotting $\ln k$ vs $1/T$ will indicate *apparent* activation energies for diffusion control of approximately 12–15 kJ mole^{-1}. For fast heterogeneous chemical reactions in which intrinsic chemical and mass transfer rates are of comparable magnitude, care will need to be taken in interpretation of apparent activation energies of the overall process.

7. SLOW ELEMENTARY REACTIONS

In the framework of activated complex theory, aqueous solution reactions and heterogeneous reactions are "slow" because they have large $\Delta H^{\ddagger 0}$ or small $\Delta S^{\ddagger 0}$, or both. At 25°C, the quantity ekT/h is 1.7×10^{13} s^{-1}. An *activation energy* of 30 kJ mole^{-1} reduces the rate constant by a factor of 6.1×10^{-6}; an *activation entropy* of -80 JK^{-1} mole^{-1} reduces the rate constant by a factor of 6.5×10^{-5}. A negative entropy of activation is associated with an activated complex which is more solvated and highly ordered than the reactants. For aqueous reactions, experimentally determined activation entropies cover a range from about -160 to $+160$ JK^{-1} mole^{-1}. (The numerical values of $\Delta S^{\ddagger 0}$ reflect units of the bimolecular or unimolecular rate constant, and depend on the concentration scale.) Aqueous reaction activation energies for low-temperature reactions ($E_a = \Delta H^{\ddagger 0} + RT \simeq \Delta H^{\ddagger 0} + 2.5$ kJ mole^{-1}) range from about 10–15 kJ mole^{-1} at the lower end up to about 150 kJ mole for very slow elementary reactions. For an uncatalyzed unimolecular reaction at 25°C with $\Delta S^{\ddagger 0} \sim -50$ JK^{-1} mole^{-1} and $\Delta H^{\ddagger 0} \sim 100$ kJ mole^{-1}, the time constant is $\sim$900 yr, very long for the lake time scales of interest. Biological catalysis can speed reactions by factors of from 10^7 to 10^{14} (e.g., the urease enzyme, $\sim 10^{13}$), with an associated lowering of $\Delta H^{\ddagger 0}$ and/or increasing of $\Delta S^{\ddagger}$. Less specific homogeneous or heterogeneous catalysts (e.g., transition metals in solution, acids or bases in solution, metal or ligand sites on solid surfaces) can accelerate slow aqueous reactions significantly (e.g., by 10^3–10^7) by providing alternate lowered energy or increased entropy reaction pathways. [In addition to *thermal* reactions, with species in their ground electronic states, *photochemical* reactions may provide the needed activation energy for bond breaking in a reaction by absorption of a photon, either by a reactant itself, for example, an organic molecule or a metal–ligand complex, thereby generating high-energy electronic states, or by other species capable of transferring energy to reactant species (indirect photolysis). The energy per mole of photons ($h\nu$) *absorbed* in the visible range is from 170 kJ to 300 kJ, corresponding to wavelengths from 700 nm to 400 nm.] Morel (9) has provided a concise discussion of the elements of photochemical processes in aquatic systems.

An illustrative diagram of the general pattern of ACT relationships among the kinetic parameters, $\Delta H^{\ddagger 0}$ and $\Delta S^{\ddagger 0}$, temperature ($\sim$5–30°C), and the rate constants for bimolecular reactions is presented in Figure 17.4. As shown, fast (diffusion-controlled) reactions are little influenced by temperature (E_a $\sim$ 15 kJ mole^{-1}), and have characteristic times of 10–100 μs at μM concentrations. A reaction with an activation enthalpy of 40 kJ mole^{-1} and zero activation entropy would have a reaction half-life of $\sim$1 s at 25°C for μM concentrations. At 5°C, the corresponding reaction half-life is $\sim$3 s. For

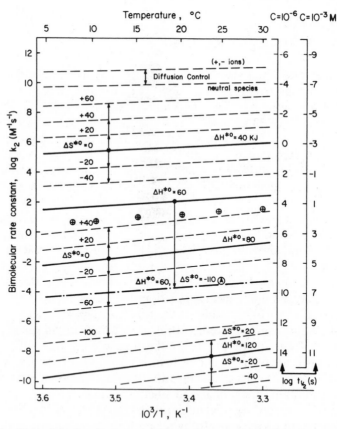

Figure 17.4. Illustration of predictions of activated complex theory for elementary bimolecular reactions as a function of temperature for typical ranges of $\Delta H^{\ddagger 0}$ and $\Delta S^{\ddagger 0}$. For each $\Delta H^{\ddagger 0}$, several values of $\Delta S^{\ddagger 0}$ are indicated. Half-life times, $t_{1/2}$, are shown for $C = 10^{-6}$ M and $C = 10^{-3}$ M. (Point A is for $CO_2 + H_2O \rightarrow H_2CO_3$ (reaction 5, Table 17.1). Data indicated by $\oplus$ are for the experimental rate constants of $H_2CO_3 \rightarrow CO_2 + H_2O$ at various temperatures.)

reactions of specified activation enthalpies, for example, 40 kJ mole^{-1}, the bimolecular reaction rate depends on the activation entropy. Thus, a *negative* entropy of 60 JK^{-1} mole^{-1} *slows* the reaction a 1000-fold, that is, lowers the preexponential Arrhenius factor A from $\sim 10^{13}$ to 10^{10} $M^{-1}s^{-1}$.

Reactions with larger $\Delta H^{\ddagger 0}$ values are affected more strongly by temperature variation. For example, for $\Delta H^{\ddagger 0}$ = 120 kJ mole^{-1}, lowering temperature from 30° to 5°C slows the rate by a factor of 70, whereas for $\Delta H^{\ddagger 0}$ = 60 kJ mole^{-1}, the same temperature change slows the rate by a factor of about 9.

A reaction of importance in lake waters, the hydration of CO_2,

$$CO_2 + H_2O \rightarrow H_2CO_3$$

is characterized kinetically by the parameters E_a = 63.0, $\Delta H^{\ddagger 0}$ = 60.5 (kJ mole^{-1}), $\Delta S^{\ddagger 0}$ = -107 JK^{-1} mole^{-1}, and logk_2 = -3.3 $(M^{-1}s^{-1})$ at 25°C (point A, Fig. 17.4). The slowness of this reaction ($t_{1/2} \sim 3$ min at 10°C) is associated with a large energy barrier as well as a highly negative activation entropy. The free energy of activation, $\Delta G^{\ddagger 0}$, is 92 kJ mole^{-1}, giving $K^{\ddagger} \simeq 10^{-16}$ M^{-1}, and indicating that the quasi-equilibrium concentration of the activated complex, $[H_2O,CO_2^{\ddagger}]$, for a 10^{-5} M CO_2 solution would be $\sim 10^{-19}$ M! For the corresponding dehydration reaction, $H_2CO_3 \rightarrow CO_2 + H_2O$, experimental values of the *unimolecular* rate constants (21) are also graphed on Figure 17.4. The dehydration constant varies from about 3 to 30 s^{-1} over the temperature range 5–30°C. Interpretation of these data in terms of ACT yields $\Delta H^{\ddagger 0}$ = 61.5 kJ mole^{-1} and $\Delta S^{\ddagger 0}$ = -15 JK^{-1} mole^{-1} for dehydration of H_2CO_3, consistent with ΔH^0 = -1 and ΔS^0 = -92 for the reversible *overall* reaction, and with an equilibrium constant, $K = k_f/k_b$, equal to the ratio of the hydration and dehydration rate constants, a consequence of the principle of detailed balancing (microscopic reversibility) (13).

The range of rate constants found for a large number of elementary aqueous reactions at low temperatures (5–30°C) is outlined in Figure 17.4, that is, from approximately 10^{-10} $M^{-1}s^{-1}$ to 10^{10} $M^{-1}s^{-1}$, with corresponding millimolar concentration time scales from approximately nanoseconds to teraseconds. A similar overall range applies to unimolecular reaction rates (for $\Delta S^{\ddagger -}$ = 0, ACT gives an identical relationship between $\Delta H^{\ddagger 0}$ and the unimolecular rate constant k_1).

8. EXAMPLES OF ELEMENTARY AQUEOUS REACTIONS

To illustrate some typical values of rate constants for elementary processes in water we have collected pertinent information on kinetic parameters for a few reactions investigated experimentally. Table 17.1 presents information on rate constants, activation energies, and activation entropies for some aqueous solution reactions. While most of these are thought elementary,

Table 17.1.   Kinetic Parameters for Elementary Aqueous Reactions[a]

Reactants	$\log k$[b] $(M^{-1}s^{-1})$	E_a (kJ mole^{-1})	$\Delta S^{\ddagger o}$ (JK^{-1} mole^{-1})	Refer- ences
Solvent exchange[c]:				
1 $Cu^{2+}(aq) + H_2O \rightarrow$	8.9	26	7	(22)
2 $Al^{3+}(aq) + H_2O \rightarrow$	−1.8	116	99	(22)
3 $Cr^{3+}(aq) + H_2O \rightarrow$	−7.3	112	−18	(22)
Neutralization:				
4 $CO_2(aq) + OH^- \rightarrow$	4.0	56	12	(13)
Hydration:				
5 $CO_2(aq) + H_2O \rightarrow$	−3.3	63	−107	(21)
Substitution:				
6 $Co(NH_3)_5Cl^{2+} +$ $OH^- \rightarrow$	0.2	112	132	(13)
Redox:				
7 $S_2O_3^{2-} + SO_3^{2-} \rightarrow$	−4.3	61	−115	(13)
Substitution:				
8 $CH_3I(aq) + OH^- \rightarrow$	−4.1	93	−22	(13)
Redox:				
9 $Fe^{2+}(aq) + H_2O_2 \rightarrow$	1.8	31	−116	(23)
10 $Fe^{2+} + Fe^{3+} \rightarrow$	0.5 (21.6°C)	41	−105	(22)
11 $Fe^{2+} + FeOH^{2+} \rightarrow$	3.4 (21.6°C)	31	−75	(22)
12 $Fe^{3+} + Cr^{2+} \rightarrow$	3.4	24	−117	(22)
13 $FeOH^{2+} + Cr^{2+} \rightarrow$	6.5	22	−54	(22)
14 $CoOH^{2+} + Fe^{2+} \rightarrow$	2.6	69	34	(22)
15 $Co(NH_3)_5Cl^{2+} +$ $Fe^{2+} \rightarrow$	−2.9	55	−126	(22)
Dissociation:				
16 $(CH_3)_3CCl(aq) \rightarrow$ $(CH_3)_3C^+ + Cl^-$	−1.5 s^{-1}	95	37	(24)
17 $H_2CO_3 \rightarrow CO_2 + H_2O$	1.2 s^{-1}	64	−15	(21)

[a] 25°C, except as noted.

[b] M^{-1} s^{-1}, except for reactants 16 and 17, unimolecular, s^{-1}.

[c] Solvent exchange formulated as second order; for first order, $k = k_2(55/6)$, that is, [H$_2$O] $\simeq$ 55 M.

some are actually complex, e.g., reactions 6 and 8. The reader should consult original sources for detailed interpretations and for measures of experimental uncertainty. By inspecting Table 17.1 it will be readily seen that both entropic and enthalpic factors strongly influence the rate constants of elementary reactions. The examples of solvent exchange rates for three cations, Cu^{2+}, Al^{3+}, and Cr^{3+}, indicate a small energy barrier for solvent exchange with Cu^{2+} (aq), but large and comparable barriers for both Al^{3+} (aq) and Cr^{3+} (aq). The Al^{3+} water exchange is almost a million times faster than for Cr^{3+}, however, because of a highly favorable entropy for the former, indicative of a dissociative activation step ($\Delta S^{\ddagger 0} \gg 0$). Reduction of $FeOH^{2+}$ by Cr^{2+} is considerably faster than for Fe^{3+}; the hydrolyzed oxidant species is associated with a smaller negative activation entropy. Bridging by OH^- groups in the activated complex is probably an important factor, as is also the case in $FeOH^{2+}$ and Fe^{2+} isotopic exchange. Reactions between ions of like charge are generally accompanied by large negative entropies, for example, reactions 7, 10, and 12, which can be interpreted qualitatively as loss of freedom of motion of water molecules in the neighborhood of the activated complex (25). Entropies of activation are often correlated with volumes of activation, positive values of each viewed as diagnostic of the *dissociative* mechanism.

A summary of the foregoing information on representative elementary reactions is presented in Figure 17.5, in which the experimental bimolecular rate constants (k_2, 25°C) are related to experimental activation energies (E_a) in the framework of ACT. Half-lives of reactions are shown for $C = 10^{-6}$

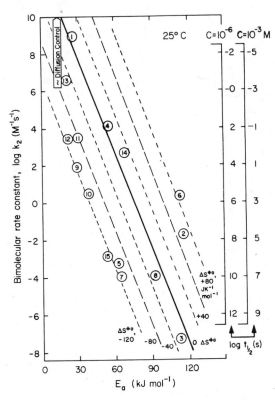

Figure 17.5. Bimolecular rate constants for representative aqueous reactions as a function of activation energy and activation entropy. The numbered points correspond to reactions in Table 17.1.

M and $C = 10^{-3} M$ (except for reactions 1, 2, 3, and 5 which involve water at ~55 M as a reactant, and for which the half-lives are 0.1 ns, 6 s, 2 × 10⁶ s, and 35 s, respectively). Figure 17.5 (if appropriately expanded to include other reactions) can be helpful in comparing examples of related reactions, for example, ligand substitution, redox reactions, dissociation, and so on, and for sorting "slow" and "fast" elementary reactions.

The implications of the specific rates of such elementary reactions for observed chemical time scales in natural waters will depend on the chemical mechanisms in which the elementary reactions play a part, that is, on their contribution to complex reactions. Of particular importance are the consequences of elementary unimolecular versus bimolecular rate-determining steps for rate laws. Solvent exchange, metal–ligand complex formation, and substitutions at carbon atoms are all examples of nucleophilic substitution reactions (attack by an electron-rich nucleophilic group on an electron-poor center). When the substitution involves a slow prior dissociation, for example, reaction 16, Table 17.1,

$$(CH_3)_3CCl \rightarrow (CH_3)_3 C^+ + Cl^- \quad \text{(slow)} \tag{i}$$

$$(CH_3)_3C^+ + OH^- \rightarrow (CH_3)_3COH \quad \text{(rapid)} \tag{ii}$$

the mechanism is said to be "S_N1" (substitution, nucleophilic, unimolecular) and the rate is first order in the dissociating reactant. Similarly, a dissociative mechanism is considered likely for complex formation with aqueous Al^{3+}. For a pure dissociation ("D") mechanism,

$$Al(H_2O)_6^{3+} \rightarrow Al(H_2O)_5^{3+} + H_2O \quad \text{(slow)} \tag{iii}$$

$$Al(H_2O)_5^{3+} + SO_4^{2-} \rightarrow Al(H_2O)_5SO_4^+ \quad \text{(rapid)} \tag{iv}$$

For a dissociative interchange ("I_d") mechanism, the steps are visualized as

$$Al(H_2O)_6^{3+} + SO_4^{2-} \rightleftharpoons Al(H_2O)_6^{3+}, SO_4^{2-} \quad \text{(rapid)} \tag{v}$$

$$Al(H_2O)_6^{3+}, SO_4^{2-} \rightarrow Al(H_2O)_5SO_4^+ + H_2O \quad \text{(slow)} \tag{vi}$$

with *dissociation* of the outer-sphere complex the slow step. In contrast, an "S_N2" (substitution, nucleophilic, bimolecular) mechanism [or a purely associative ("A") mechanism] is limited by the entering nucleophile. For example

$$OH^- + CH_3Cl \rightarrow [HOCH_3Cl]^- \quad \text{(slow)} \tag{vii}$$

$$[HOCH_3Cl] \rightarrow CH_3OH + Cl^- \quad \text{(fast)} \tag{viii}$$

The rate is then first order in OH^- *and* in the alkyl halide, and second order overall. Strong nucleophiles promote associative mechanisms.

For dissociative mechanisms ("D," "I_d") of complex formation between a metal ion and various ligands, a direct correlation of complex-formation rates with those found for solvent exchange is anticipated, and has been confirmed for most metals (6,9,22). It is found, for example, that the constant of the interchange rate of $AlSO_4^+$ formation (reaction vi above) is ~ 1 s^{-1}, compared to 0.13 s^{-1} for water exchange. Solvent-exchange rates and ligand substitution rates are usually accelerated by the presence of other ligands, for example, hydroxide ion, in the metal-ion's inner coordination sphere. The reader is referred to Burgess (22) for a comprehensive review of data on elementary reactions of aqueous metals.

9. RAPID AQUEOUS REACTIONS

The bimolecular and unimolecular rate constants for many aqueous reactions approach the diffusion-controlled limits, that is, approximately 10^{10} to 10^{11} M^{-1} s^{-1} and approximately 10^{10} s^{-1}, respectively (eq. 13, Figs. 17.4 and 17.5). Important examples are neutralization reactions and reactions of some free-radical intermediates. As illustrated by water exchange on Cu^{2+}(aq) (reaction 1, Table 17.1), certain exchange and substitution reactions are also extremely rapid. [Although perhaps not directly relevant to redox mechanisms in nature, reactions are rapid for hydrated electrons, e^-(aq), and metal ions, e.g., $Cu^{2+} + e^-$(aq) $\rightarrow Cu^+$, $\log k_2 = 10.5$ $M^{-1}s^{-1}$; $Mn^{2+} + e^-$(aq) $\rightarrow Mn^+$, $\log k_2 = 7.9$ $M^{-1}s^{-1}$ (19). Morel (9) discusses photochemical production of e^-(aq).]

9.1. Coordination

The reaction of aqueous protons and hydroxide ions is diffusion controlled. With $f \sim 10$ (eq. 13), $H^+ + OH^- \rightarrow H_2O$, $\log k_2 = 11.2$ at 25°C. Formation of the Mn^{2+}, SO_4^{2-} outer-sphere complex has a bimolecular rate constant $\log k_2 \simeq 10.6$; the $SO_4^{2-}-H_2O$ bimolecular constant for the sequential interchange, Mn^{2+}, $SO_4^{2-} \rightarrow MnSO_4 + H_2O$ has $\log k_2 \simeq 6.6$ $M^{-1}s^{-1}$. Reactions between hydroxide and many aqueous acid species are rapid, for example, $OH^- + HCO_3^- \rightarrow CO_3^{2-} + H_2O$, $\log k_2 \simeq 9.8$ $M^{-1}s^{-1}$. However, unfavorable activation energies and entropies slow some neutralizations considerably, for example, $OH^- + CO_2 \rightarrow HCO_3^-$, $\log k_2 \simeq 4.0$ $M^{-1}s^{-1}$, $E_a = 56$ (reaction 4, Table 17.1). [For examples of rate constants of a number of rapid coordination reactions (neutralization, complexation, etc.), see, e.g., Adamson (15), Weston and Schwarz (16), and Laidler (26).]

9.2. Redox Intermediates

There is increasing evidence that some reactive redox intermediates, for example, 'superoxide' (O_2^- and $HO_2^{\cdot}$), peroxide (H_2O_2 and HO_2^-), and hydroxyl radical ($OH^\cdot$ and $O^{\div}$) may play significant roles in the mechanisms and kinetics of overall redox reactions in atmospheric water and in surface waters (27–30). Bimolecular rate constants for reactions of some redox intermediates are near the diffusion-controlled limit at 25°C, for example, $OH^\cdot$ + $OH^\cdot \rightarrow H_2O_2$, log k_2 = 9.7 $M^{-1}s^{-1}$; $O_2^{\div}$ + $OH^\cdot \rightarrow OH^-$ + O_2, log k_2 = 9.9 $M^{-1}s^{-1}$. Some redox intermediate and free-radical reactions are rather slow, for example, $HO_2^{\div}$ + $H_2O_2 \rightarrow OH^\cdot$ + O_2 + H_2O, log k_2 = $-0.3\ M^{-1}s^{-1}$. Trace metals can interact with redox intermediates, thereby generating complex chain reactions. For example, Fe^{3+} + $H_2O_2 \rightarrow Fe^{2+}$ + $HO_2^{\div}$ + H^+, or Fe^{2+} + $H_2O_2 \rightarrow FeOH^{2+}$ + $OH^\cdot$ (reaction 9, Table 17.1 and Fig. 17.5, log k_2 = 1.8 $M^{-1}s^{-1}$. It is to be emphasized that the chemical time scales of overall redox processes in lake waters are unlikely to reflect directly the bimolecular rate constants of inherently rapid steps; actual rates are determined by appropriate functions of rate constants and actual reactant and intermediate concentrations, for example, $k_2[OH^\cdot]^2$, $k_2[Fe^{3+}][H_2O_2]$, $k_2[Fe^{2+}][OH^\cdot]$, and so on. For review of kinetics of known elementary aqueous redox intermediate and free-radical reactions see, for example, Graedel and Wechsler (28).

10.   IONIC STRENGTH AND ELEMENTARY REACTIONS

Activated complex theory provides a basis for estimating the effect of ionic strength on the rate constant of an elementary reaction, known as the *primary* salt effect (*secondary* salt effects are associated with changes in overall rate because of influences of ionic strength on prior, rapid elementary reactions which are at local or pseudoequilibrium with respect to the slow step). According to Eq. 18,

$$k_2 = \frac{kT}{h} K^{\ddagger} \frac{\gamma_A \gamma_B}{\gamma_{\ddagger}}$$

so that the rate constant k_2 compared to that at infinite dilution is k_2 = $k_0(\gamma_A \gamma_B / \gamma_{\ddagger})$. Models based on the Debye–Hückel theory, for example, the extended law (11) or the Davies equation (20), yield simple expressions for k_2 as a function of ionic strength I and the charge numbers Z_A and Z_B of the ionic reactants. For example, using the Davies equation

$$\log(k_2/k_0) = 2AZ_A Z_B \left(\frac{I^{1/2}}{1 + I^{1/2}} - 0.2I \right) \tag{22}$$

Increased ionic strength *increases* k_2 over the infinite dilution ($I \sim 0$) value for reactions between like-charged ions, for example, Fe^{2+} and Cu^{2+}, and *decreases* k_2 for reactions between ions of opposing charge, for example, Fe^{2+} and $C_2O_4^{2-}$.

The range of ionic strength of lake waters is from about $10^{-4}\ M$ to about

1 M (saline lakes). For a reaction in which $Z_A Z_B = 1$, simplified activity coefficient models for ion–ion elementary reactions predict a variation in k_2/k_0 from approximately 1.3 at $I = 10^{-2}$ M, to 1.8 at $I = 0.1$ M and to 2.6 at $I = 0.5$ M, with reciprocal values for $Z_A Z_B = -1$. For $Z_A Z_B = 4$, the corresponding increasing ratios of k_2/k_0 are 2.3 at $I = 10^{-2}$ M, 9.1 at $I = 0.1$ M, and ~40 at $I = 0.5$ M. These are merely rough estimates, based on the Davies equation. Reactions such as 7, 11, and 13–15 (Table 1.1) would thus be accelerated only about twofold or less at typical lake-water ionic strengths, but would experience much greater accelerations in saline environments. A reaction such as 6 in Table 1.1, for which $Z_A Z_B = -2$, would be slowed to 70% at $I = 10^{-2}$ M, 30% at $I = 0.1$ M, and 15% at $I = 0.5$ M (with respect to infinite dilution rate).

Reactions between ions and neutral molecules also experience increases or decreases in rate because of electrolyte ion effects (13). These nonelectrostatic salt effects are more difficult to predict, but are also small, for example, only about $\pm 20\%$ at $I \sim 1$ M.

11.   COMPLEX REACTIONS

Most reactions in natural systems have mechanisms that involve several elementary steps and hence are *complex reactions*. Familiar examples are *reversible* reactions, *parallel* or *concurrent* reactions, and *series* or *consecutive* reactions. Other complex mechanisms involve various combinations of series and parallel, irreversible and reversible reactions, for example, competitive, coupled, chain, and catalyzed reactions. Some of the more important complex reaction types have been treated extensively by many authors and both differential and integrated forms of rate laws obtained for the more important reaction types [see, among others, Lasaga (11), Moore and Pearson (13), Hammes (14), Weston and Schwarz (16), Benson (17), Stumm and Morgan (20), Laidler (26)]. As our interest here centers on the *rates* of important overall reactions, R_j, in lake systems, interpreted or predicted in terms of chemical mechanisms where possible, we will limit our discussion of complex reactions to a few essential concepts, illustrating them with some experimental or hypothetical examples. Of particular importance in natural waters are the effects of pH, metal ions, and ligands on the chemical *speciation* of reactants and intermediates.

11.1.   Formulation of the Rate

For an aqueous reaction, R_j (in M s^{-1}, e.g.) is given by an expression in terms of concentrations, or absolute mole numbers and solution volume,

$$R_j = \frac{1}{v_{ij}} \frac{d[A_i]}{dt} = \frac{1}{v_{ij}} \frac{1}{V} \frac{dn_i}{dt} \tag{23}$$

For a heterogeneous reaction, the reaction rate is often expressed on an area basis, for example,

$$R_j^s = \frac{1}{\nu_{ij}} \frac{1}{A_s} \frac{dn_i}{dt} = \frac{1}{A_s} \frac{d\xi}{dt} \tag{24}$$

where A_s is surface area and $d\xi/dt$ is the reaction rate expressed as moles of reaction advancement (extent) per time. The rate of heterogeneous reaction expressed in terms of rate of change in solution is just $R_j = R_j^s A/V$. (We remark that the rate of *internal entropy production* per unit volume, θ_j, associated with a particular chemical reaction can be expressed as

$$\theta_j = \frac{1}{V} \frac{dS_j}{dt} = -\frac{\Delta G_j}{T} R_j \tag{25}$$

where S_j is the internal entropy change of the reaction and ΔG_j is the actual free energy of reaction.) The use of the reaction rate variable R_j for an overall reaction is convenient but arbitrary, in that the stoichiometric coefficient ν_{ij} can be arbitrarily chosen. The experimentally meaningful chemical quantities are n_i and $[A_i]$. Another qualification: steady conditions in a reaction sequence are required in order that R_j be characteristic of the overall reaction. Thus, for $A_1 \rightarrow A_i \rightarrow \ldots \rightarrow A_p$, a uniform reaction rate requires quasi-steady conditions for the intermediates, A_i, so that $d[A_1]/dt = d[A_p]/dt$.

11.2. Reversible Reactions

An elementary reaction is reversible when E_a and $\Delta S^{\ddagger 0}$ values of the opposing reaction, and the temperature, are such as will allow an appreciable back rate. Denoting by R_f the *forward* rate and by R_b the *back* rate, the *net* rate is $R = R_f - R_b$. For example, the hydration of CO_2 by the path $CO_2 + H_2O \rightarrow H_2CO_3$ and the dehydration of CO_2 by the path $H_2CO_3 \rightarrow CO_2 + H_2O$ have rate constants $k_{f0} = 5 \times 10^{-4} M \text{ s}^{-1}$ and $k_{b0} = 16 \text{ s}^{-1}$, respectively. At *equilibrium*, $R_f = R_b$, so that $k_{f0}[CO_2][H_2O] = k_{b0}[H_2CO_3]$, and the ratio k_{f0}/k_{b0} is $[H_2CO_3]/[CO_2][H_2O]$. Hence, $k_{f0}/k_{b0} = K$, the equilibrium constant for an elementary process and its reverse. This is an example of the principle of chemical detailed balancing (13), which applies to each reversible *elementary* process within a complex reaction at equilibrium. Thus, taking as an example the reversible sequence comprising hydration and ionization,

$$CO_2 + H_2O \underset{k_{b0}}{\overset{k_{f0}}{\rightleftarrows}} H_2CO_3 \underset{k_{b1}}{\overset{k_{f1}}{\rightleftarrows}} HCO_3^- + H^+ \tag{26}$$

application of detailed balance at complete equilibrium yields $k_{f0}/k_{b0} = K_0$,

$k_{f1}/k_{b1} = K_1$, and $K = K_0 K_1 = (k_{f0}/k_{b0})(k_{f1}/k_{b1})$, where $K = ([HCO_3^-][H^+]/[CO_2][H_2O])_{equil}$.

A *kinetic* description of overall reaction 26 requires the values of k_{f0}, k_{b0}, k_{f1}, and k_{b1}, and the concentrations of reactants and products. With this information, exact equations for the rates of change of $[CO_2]$, $[H_2CO_3]$, $[HCO_3^-]$, and $[H^+]$ can be written and solved analytically or numerically. [See, e.g., Moore and Pearson (13), Pankow and Morgan (7).] A simplification to be considered would be a *quasi-steady-state* approximation (qssa) for the intermediate $[H_2CO_3]$, namely, assuming $d[H_2CO_3]/dt \approx 0$, that is, *small* with respect to the rate of reaction. An even greater simplification obtains if either the first or second step can be considered as being at *constrained equilibrium* (local, pseudoequilibrium or quasiequilibrium) while the other sets the rate. In the particular case of reaction 26, with $k_{f0} = 5 \times 10^{-4} M^{-1}s^{-1}$, $k_{b0} = 16 s^{-1}$, $k_{f1} = 8 \times 10^6 s^{-1}$, and $k_{b1} = 5 \times 10^{10} M^{-1}s^{-1}$ (diffusion controlled), the second elementary reaction can be considered at constrained or quasiequilibrium, and the first, slower reaction is the rate-determining step. The rates are then $R_f = k_{f0}[CO_2][H_2O]$ and $R_b = (k_{b0}k_{b1}/k_{f1})[HCO_3^-][H^+]$, and we can identify $k_{f0} \equiv k_f$ and $k_{b0}k_{b1}/k_{f1} \equiv k_b$; k_f and k_b are the overall reaction rate constants in the forward and reverse directions. Evidently, $k_f/k_b = K_1 K_1 = K$. It is also seen that $R_f/R_b = K/Q$, where Q is the reaction quotient. This is an appealingly simple result, and is a specific example of a more general relationship which exists between kinetics and equilibria, hence thermodynamics, for reversible complex reactions [Boudart (31)]. The general result for an overall reaction comprising a sequence of reversible elementary reactions is

$$\frac{k_f}{k_b} = K^n \tag{27}$$

and

$$\frac{R_f}{R_b} = \left(\frac{K}{Q}\right)^n \tag{28}$$

where $n > 0$. The value of n depends upon the details of the reaction mechanism; specifically n depends on the number of times each elementary reaction must occur to yield the overall reaction stoichiometry and upon the relative magnitudes of the elementary reaction rates under steady conditions. For reaction 26, we found $n = 1$, the stoichiometric number of each step being unity. This is not always the case. A simple example (13) will illustrate the idea. If the overall reversible reaction A = B, K proceeds by the *mechanism* $2A \underset{k_{-1}}{\overset{k_1}{\rightleftharpoons}} 2B$, the relationships are

$$\frac{k_f}{k_b} = \left(\frac{[B]^2}{[A]^2}\right)_{equil} = K^2 \tag{i}$$

and

$$\frac{R_f}{R_b} = \left(\frac{K}{Q}\right)^2 \tag{ii}$$

In energetic terms, the foregoing *general* relationship between R_f and R_b can be expressed in terms of ΔG, the reaction free energy:

$$R = R_f - R_b = R_f \left[1 - \exp\left(\frac{n\Delta G}{RT}\right)\right] \tag{29}$$

For conditions very close to equilibrium, that is, for (ΔG) *small,* the expression is approximated by

$$R \simeq -R_{\text{exch}} \frac{n\Delta G}{RT} \tag{30}$$

and the *net* velocity is proportional to $-\Delta G$. Far from equilibrium, where $-n\Delta G \gg RT$, $R \simeq R_f$, and R is independent of ΔG.

Sequential reversible reaction mechanisms, can, in general, be treated kinetically by one of four approaches, depending upon the characteristic times of the component steps (which depend, in general, on *both* rate constants and concentrations). These approaches include (1) *analytical* solution (limited to a few simpler mechanisms); (2) *numerical* solution (always available, in principle); (3) *quasi-steady-state approximation,* $d[A_i]/dt \simeq 0$, for all intermediates; and (4) existence of a *rate-determining step* in the sequence, with other steps at constrained equilibrium. The reaction sequence: $ML \rightleftharpoons M + L$ followed by $M + S \rightleftharpoons MS$ was treated numerically by Pankow and Morgan (7). Exact solutions are available for the sequential mechanism $A_1 \rightleftharpoons A_2 \rightleftharpoons A_3$ (13). Keck (10) has discussed relative merits of constrained equilibrium and steady-state approaches.

11.3. Reversible and Irreversible Reaction Sequences

Reactions of this type are among the most frequently encountered in natural water rate processes (and in aqueous solution kinetics generally). A typical mechanism entails a reversible isomerization, dissociation, or association step followed by one or more irreversible steps. An example is the dissociation of a ferrous iron complex, followed by irreversible oxidation:

$$Fe(II)L \underset{k_{-1}}{\overset{k_1}{\rightleftharpoons}} Fe(II) + L \tag{i}$$

$$Fe(II) \xrightarrow[O_2]{k_2} Fe(III) \tag{ii}$$

where L is a ligand and k_2 a conditional rate constant, $k_2 = k[O_2][H^+]^{-2}$, incorporating constant concentrations of O_2 and H^+. The rate expression for step (ii) may be written

$$\frac{d[\text{Fe(III)}]}{dt} = k_2[\text{Fe(II)}] \qquad \text{(iii)}$$

The set of coupled, nonlinear differential equations describing the concentration variations of Fe(II)L, Fe(II), L, and Fe(III) with time has been solved numerically in order to explore the kinetics of a model system (7). The detailed kinetic behavior is strongly dependent on values of k_1, k_{-1}, k_2, Fe_T, and L_T. Under two simplifying assumptions, quasi-steady-state or constrained equilibrium, simple rate laws can be obtained for mechanisms such as (i) and (ii). Whether either assumption yields accurate results depends upon the particular parameters. Numerical analysis results (7) for complex reaction (i) and (ii) show that equilibrium or steady-state approximations for (i) and (ii) are valid under some conditions. The *qssa* can be expressed as

$$k_1[\text{Fe(II)L}] \simeq k_{-1}[\text{Fe(II)}][\text{L}] + k_2[\text{Fe(II)}] \qquad \text{(iv)}$$

which gives the approximately steady-state [Fe(II)] as

$$[\text{Fe(II)}] \simeq \frac{k_1[\text{Fe(II)L}]}{k_{-1}[\text{L}] + k_2} \qquad \text{(v)}$$

and the *qssa* rate

$$R' \simeq \frac{d[\text{Fe(III)}]}{dt} = \frac{k_2 k_1[\text{Fe(II)L}]}{K_{-1}[\text{L}] + k_2} \qquad \text{(vi)}$$

If $k_{-1}[\text{L}] \gg k_2$, then

$$R' \simeq k_2 \frac{K_1[\text{Fe(II)L}]}{[\text{L}]} \qquad \text{(vii)}$$

These approximations can prove valuable for examining proposed mechanisms in relation to experimental rate laws, but care must be taken in justifying assumptions made in different situations, by examining rate constants and concentrations (13,16).

Among other important aquatic chemical reaction mechanisms with reversible and irreversible (or nearly so) steps are: enzyme-catalyzed reactions (the Michaelis–Menten model) (9), dissolution of solid phases in which rapidly occurring acid–base or exchange reaction steps precede a slower dissociation or coordination step [e.g., in Al_2O_3 dissolution (32)] or electron transfer [e.g., in MnOOH reductive dissolution (33)], and homogeneous solution reactions catalyzed by acids, bases, metal ions, or complexes (13,34).

The Michaelis–Menten rate law for enzyme catalysis of conversion of substrate S to product P by the mechanism

$$S + E \underset{k_{-1}}{\overset{k_1}{\rightleftharpoons}} ES \quad (\text{rapid}) \qquad (i)$$

$$ES \xrightarrow{k_2} P + E \quad (\text{slow}) \qquad (ii)$$

is

$$R = \frac{d[P]}{dt} = \frac{k_2[E]_T[S]}{K_m + [S]} \qquad (31)$$

in which $K_m = (k_{-1} + k_2)/k_1$, and $K_m \simeq k_{-1}/k_1$ for the equilibrium approximation. The *form* of the Michaelis–Menten law, based on site saturation, with $R_{\max} = k_2[E]_T$, generally expressed as

$$R = \frac{R_{\max}[A]}{K + [A]} \qquad (32)$$

is found applicable to a variety of surface and solution reactions describable by mechanisms analogous to (i) and (ii) (20).

11.3.1. Temperature and Sequential Mechanisms

The influence of temperature on complex reactions can be understood in terms of the activated complex framework, through knowledge of E_a, or $\Delta H^{\ddagger 0}$, and $\Delta S^{\ddagger 0}$ for each component step (eq. 19). In general, there is a temperature dependence for each elementary reaction. For *rapid* equilibria, that is, with rate constants large compared to those of the rate-determining step, the temperature effect on k is determined by $\Delta H_0 = \Delta H_f^{\ddagger 0} - \Delta H_b^{\ddagger 0}$ (Fig. 17.3). For a *qssa* description, the overall temperature dependence is determined by combined effect of the $\Delta H_i^{\ddagger 0}$ for the l elementary rate constants included in the rate equation. For example, in the two-step mechanism

$$A \underset{k_{-1}}{\overset{k_1}{\rightleftharpoons}} B \qquad (i)$$

$$B \xrightarrow{k_2} C \qquad (ii)$$

the *qssa* yields

$$R \simeq \frac{k_2 k_1}{k_{-1} + k_2} [A] \qquad (iii)$$

and three $\Delta H^{\ddagger 0}$ values are needed to describe the effect of temperature on rate. If $k_{-1} \gg k_2$, then

$$R \simeq K_1 k_2 [A]$$

and ΔH_1^0 and $\Delta H_2^{\ddagger 0}$ are needed.

The reduction of Fe(III) by Cr(II) in the absence of complexing ligands (reactions 12 and 13, Table 17.1) proceeds by a mechanism representative for many metal redox reactions (for which pH domain varies). The hydrolysis step

$$Fe^{3+} + H_2O \xrightarrow{K_1} FeOH^{2+} + H^+ \qquad \text{(i)}$$

is very rapid (22), with log K (at 25°C) of -2.8 and ΔH^0 of 43 kJ mole^{-1}. The electron transfer step

$$FeOH^{2+} + Cr^{2+} \xrightarrow{k_2} Fe^{2+} + Cr^{3+} + OH^- \qquad \text{(ii)}$$

has log $k_2 = 6.5$ (25°C, $M^{-1}s^{-1}$) and $E_a = 22$ kJ mole^{-1} (Table 17.1). The constrained equilibrium approximation gives

$$R = -\frac{d[Cr^{2+}]}{dt} = \frac{K_1 k_2 [Fe^{3+}][Cr^{2+}]}{[H^+]} \qquad \text{(iii)}$$

The effective rate constant is $k_1 k_2$ at a given temperature, determined by $\Delta H_1^0 + \Delta H_2^{\ddagger 0} = 43 + 22 = 65$ kJ mole^{-1}. At a given $[H^+]$, an apparent rate constant, $k_{app} = K_1 k_2/[H^+]$, can be defined for $[Fe^{3+}]$ and $[Cr^{2+}]$ in the rate law. The influence of temperature on k_2, K_1, and k_{app} at given $[H^+]$ is determined by application of the van't Hoff equation (20) and Eq. 19 (or 20), assuming ΔH_1^0 and $\Delta H_2^{\ddagger 0}$ are constant. Figure 17.6 shows the respective variations of k_2, K_1, and k_{app} (for $[H^+] = 10^{-2}$ M) for the Cr(II)/Fe(III) reaction. The point is made that in interpreting complex reactions, various influences on an *apparent* activation energy (including concentrations of species in the reaction system) may require attention.

11.4. Concurrent Reactions

Overall reactions can proceed by mechanisms comprising parallel paths, so that the total rate is described by a sum, for example, $R = R_1 + R_2 + \cdots + R_p$. A parallel reaction mechanism is the essence of catalysis, for example, hydrolysis of some alkyl halides taking place via a step involving reaction with H_2O as well as by a step involving reaction with OH^-, so that $k = k_{H_2O} + k_{OH} \cdot [OH^-]$. The observed reaction rate from parallel pathways, for example, $R = R_{H_2O} + R_{OH^-} + R_{H^+}$ is often dominated by a single large flux along one path. Thus, the "rate-determining step" (if there be one) in concurrent mechanisms is the *faster* one (compare sequential mechanisms).

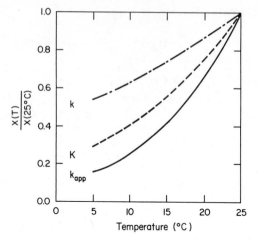

Figure 17.6. Complex influence of temperature on the rate of the reaction Fe(III) + Cr(II) → Fe(II) + Cr(III) in acid solution. The overall temperature effect reflects variation in k for electron transfer (through $\Delta H^{\ddagger 0}$) and K of prior hydrolysis (through ΔH^0).

The hydration of aqueous CO_2 in water can take place by two different elementary steps (see reactions 4 and 5, Table 17.1). The total forward rate of reaction is

$$R_f = k_{f4}[CO_2][H_2O] + k_{f5}[CO_2][OH^-] \qquad (33)$$

and the overall rate R is

$$R = R_f - R_b = -\frac{d[CO_2]}{dt} = k_{f4}[CO_2]H_2O] + k_{f5}[CO_2][OH^-]$$
$$- k'_{b4}[HCO_3^-][H^+] - k_{b5}[HCO_3^-] \qquad (34)$$

where the effective rate constants k_{f4} and k'_{b4} were derived earlier from a constrained equilibrium approximation in discussing reaction 26). We note that one path, 4, consists of a two-step sequence. The numerical values at 25°C are: $k_{f4} = 5 \times 10^{-4}\ M^{-1}s^{-1}$, $k_{f5} = 1 \times 10^4\ M^{-1}s^{-1}$, $k'_{b4} = 9 \times 10^4\ M^{-1}s^{-1}$, and $k_{b5} = 2 \times 10^{-4}\ s^{-1}$. At pH 7 and [CO_2] of $10^{-5}\ M$, the H_2O hydration rate is $3 \times 10^{-7}\ M\ s^{-1}$ and the rate on the OH^- path $\sim 1 \times 10^{-8}$ $M\ s^{-1}$. At higher pH, ~ 9, the forward OH^- path rate will dominate and become rate determining. Similar considerations apply to the relative fluxes making up R_b.

11.5. Radical and Chain Reaction Mechanisms

Our discussion has emphasized mechanisms built on sequences and parallel paths of rather simple kinds, involving thermal reactions and nonradical intermediate species. Earlier, in presenting a suggested mechanism for the aqueous Fe(II) + O_2 reaction (Eq. 9, and reactions (i)–(vii) in Section 4.2, the reactive intermediates $O_2^{\pm}$, $HO_2^{\cdot}$, H_2O_2 and $OH^{\cdot}$ were identified. Some elementary reaction rates of these species were briefly discussed. We note here that these species can participate in radical chain reactions, for example, the Fe(II) catalyzed decomposition of peroxide (19):

$$Fe^{2+} + H_2O_2 \rightarrow FeOH^{2+} + OH^{\cdot} \qquad \text{(i)}$$

$$OH^{\cdot} + H_2O_2 \rightarrow H_2O + HO_2^{\cdot} \qquad \text{(ii)}$$

$$HO_2^{\cdot} \rightleftharpoons H^+ + O_2^{\pm} \qquad \text{(iii)}$$

$$O_2^{\pm} + H_2O_2 \rightarrow O_2 + OH^- + OH^{\cdot} \qquad \text{(iv)}$$

$$Fe^{3+} + HO_2^{\cdot} \rightarrow Fe^{2+} + H^+ + O_2 \qquad \text{(v)}$$

Chain reaction mechanisms involve a combination of sequential and concurrent steps. A rate expression is obtained by applying the quasi-steady-state approximation to reactive intermediates $HO_2^{\cdot}$ and $OH^{\cdot}$.

The production of $O_2^{\pm}$ and H_2O_2 is related to (among other processes) photochemical oxidation and reduction processes such as, for example,

$$X + h\nu \rightarrow X^{\cdot} + e^-(aq) \qquad \text{(vi)}$$

$$e^-(aq) + O_2 \rightarrow O_2^{\pm} \qquad \text{(vii)}$$

$$M^n + O_2 + h\nu \rightarrow M^{n+1} + O_2^{\pm} \qquad \text{(viii)}$$

$$M^n, OH^- + h\nu \rightarrow M^{n-1} + OH^{\cdot} \qquad \text{(ix)}$$

and

$$2O_2^{\pm} + 2H^+ \rightarrow H_2O_2 + O_2 \qquad \text{(x)}$$

Reduction of metals in photochemically active complexes, for example,

$$M(II)L \quad \text{or} \quad M(III)L + h\nu \rightarrow M(I) \quad \text{or} \quad M(II) + L^{\cdot} \qquad \text{(xi)}$$

has been proposed for several metals. Photochemically initiated reactions involving iron reduction, oxidative destruction of humic compounds, and oxygen consumption in moderately acidic lakes have been described by Miles and Brezonik (35). Photoreduction of Fe(III) in low-pH lakes [slow reoxidation of Fe(II)] has been reported by Collienne (36). The reduction of $MnO_2(s)$ and $MnOOH(s)$ by *marine* humic materials under mildly acidic to mildly alkaline conditions has been documented (33,37). It is likely that similar processes occur in lakes. For a wider discussion of some general concepts and models related to photochemical redox processes see Morel (9) and Waite and Morel (38).

12. IMPORTANCE OF pH IN AQUEOUS KINETICS

Many other reactions in lake waters and sediments are strongly affected in their kinetic behavior through pH effects on reactant speciation. The example of Fe(III) reduction by Cr(II) illustrated the greater reactivity of $FeOH^{2+}$ in comparison to Fe^{3+}. Mechanisms of Fe(II) oxygenation (Eqs. 9 and 10), Mn(II) oxygenation (20), and uranium (IV) oxygenation (19) each point to a strong influence of hydrolyzed metal species on rates. In an analogous way, some oxide surface hydroxide groups, $\equiv SOH$ [presumably possessing π-donor properties (18) and capable of promoting electron transfer], catalyze Mn(II) oxygenations (39,40). A reasonable hypothesis appears to be that O and OH groups coordinated to metal ions both in solution and on surfaces can catalyze electron transfers.

Dissolution reaction rates of oxides are influenced by pH, with rate expressions of the form $R_H^s = k_s[H^+]^m$, which is viewed as reflecting the overall effect of pH on proton-related surface species, $\theta_{H,i}^n S_T$, for example, $\equiv SOH_2^+$ and SOH (32,33), over a limited pH-domain. Parameter m is not, in general, a constant, only approximately so. The key role of pH with respect to redox and coordination reaction kinetics in homogeneous solution is extensively documented [e.g., Espenson (41)].

The following simple example is representative of patterns encountered in H^+ catalysis of reactions. [More elaborate rate laws will reflect increased complexity of equilibrium or quasi-steady-state proton-related (and metal–ligand related) speciation in solution.] Two paths for dissociation of a metal–ligand complex, MX^{2+}, are assumed:

$$MXH^{3+} \xrightarrow{k_1} M^{3+} + XH \tag{i}$$

$$MX^{2+} \xrightarrow{k_2} M^{3+} + X^- \tag{ii}$$

and a prior proton transfer step

$$MXH^{3+} \overset{K_3}{\rightleftharpoons} MX^{2+} + H^+ \tag{iii}$$

is rapid. The rate is

$$R_T = R_1 + R_2 = \frac{d[M^{3+}]}{dt} = k_1[MXH^{3+}] + k_2[MX^{2+}] \tag{iv}$$

which, assuming (iii) is at equilibrium, yields

$$R_T = \frac{k_1[H^+] + k_2K_3}{K_3 + [H^+]} M_T = k_H M_T \qquad (v)$$

Figure 17.7 summarizes the overall influence of pH for $[MX]_T = 10^{-3}\ M$, for $k_1 = 10\ s^{-1}$, $k_2 = 1\ s^{-1}$, and $pK_3 = 7$, and shows the respective contributions of the two paths. Note that R_T is not of uniform order in $[H^+]$, showing zero-order behavior at low pH and high pH, with a *variable* order in between {compare reported $R_s = k_s[H^+]^m$, above, or $R = (k' + k''[H^+])[Fe^{3+}][Eu^{2+}]$, for oxidation of Fe(III) by Eu(II) via two paths in acid solution (41)}. The key to mechanistic interpretation of rate laws for aqueous reactions is the recognition of *predominant reactant species,* whether in aqueous solution or in solid phases (or other nonaqueous phases) and the identification of reactive intermediate species formed from the predominant species. For proton transfer reactions, the various species are related to one another through the aqueous species ionization fractions α_0, $\alpha_1, \ldots, \alpha_i, \ldots, \alpha_n$ of a component protolyzable component X and through analogous surface proton species fractions $\theta_{H,0}, \theta_{H,1}, \ldots, \theta_{Hi}, \ldots$. The concentration of species i is then $\alpha_i X_T$ or $\theta_i S_T$. Rate laws for a wide variety of homogeneous and heterogeneous reactions of importance in lakes need to be interpreted in terms of proton-dependent speciation of reactants, as well as ligand-dependent speciation, $\alpha_L X_T$, and metal-dependent speciation, $\alpha_M X_T$ (and $\theta_L S_T, \theta_M S_T$).

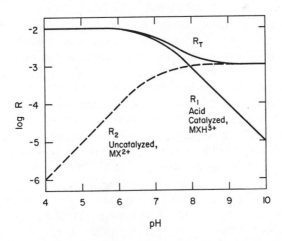

Figure 17.7. Influence of pH on dissociation of an MX complex by two paths, $MX^{2+} \xrightarrow{k_2}$ and $MXH_3^+ \xrightarrow{k_1}$. The equilibrium $MXH^{3+} \rightleftharpoons MX^{2+} + H^+$ is rapidly established. $[MX]_T = 10^{-3}\ M$.

13. A CONCLUDING KINETIC OVERVIEW

Our aim has been to consider chemical reactions in lakes from a kinetic viewpoint. A kinetic approach to modeling the distribution of chemical properties in lakes and sediments is a natural complement to equilibrium models. We argue that approaching equilibrium through kinetics has the advantage of seeing the connections between dynamic models, steady-state models, and local equilibrium models of lake chemistry. Equation 7 provides a simple but reliable starting point for preliminary study of lake geochemical cycles and the fate of pollutants in lake systems. Our emphasis is on R_j, the rate of any chemical reaction, and we have accordingly emphasized some of the general concepts of chemical kinetics as the basis for interpreting experimental observations on rates and estimating rates in lakes from proposed mechanisms. The implied context for chemical modeling in any lake system is the existence of biogeochemical cycles of carbon, nitrogen, phosphorus, sulfur, and the other biogenic elements. Available kinetic information on abiotic transformation of inorganic and organic compounds must be applied within the framework of photosynthesis, respiration, and other major rate processes in the lacustrine environment. Understanding the interplay of uncatalyzed reactions with reactions catalyzed by organisms, surfaces, and solution species will be an increasingly important goal of research directed at understanding the fate, distribution, and residence time of pollutants in lakes. The time scales of many different processes and reactions, $\tau_{chem}^{-1} = (\sum v_i R_j)/C_i$ (Eq. 8), need to be evaluated and related to the physical time scales of the lake system.

We conclude with an overview, not exhaustive, but we hope exemplary, of some characteristic times for processes and reactions of potential relevance in relation to chemical dynamics in lakes. Figure 17.8 collects some of the available rate data, expressed as a characteristic time, for physical and chemical processes. The range of times is great, from seconds to years. The specific conditions assumed for illustration are of necessity somewhat arbitrary, for example, those for ligand exchange, electron transfer, mineral dissolution, photolysis, oxygenation, and so on. (The reader will, for reactions of interest, wish to consult original sources of information.) And the diagram is incomplete. It might also identify some characteristic times for photosynthesis (10^4–10^6 s), organic degradation (10^5–?), photoreduction of ferric iron (10^2–10^5 s), sulfate reduction (10^6–10^8 s), and so on. The particular illustrations given are meant to serve a *qualitative* as well as a *quantitative* purpose. The qualitative one is that the kinetic range is quite wide, and that the chemical times should be seen in relation to the various mixing processes of the lake. The quantitative point, of course, is that the specific conditions of a lake (pH, temperature, sunlight, input chemicals, etc.) need to be examined carefully in carrying models forward.

There is both need and opportunity to sharpen and extend this overview of kinetics in relation to the lacustrine environment. The last row of Figure 17.8 is vacant!

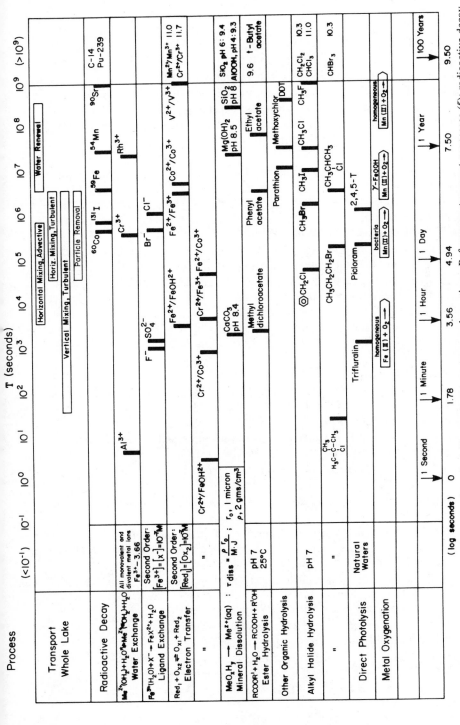

Figure 17.8. Examples of characteristic times for lake transport processes and reactions. References: transport processes (5); radioactive decay (42); water exchange of metal ions (22,43); ligand exchange on Fe^{3+} (aq) (22,43,50); electron transfer between metal ions (22); mineral dissolution (44); ester hydrolysis (8); other organic hydrolysis (8,45); alkyl halide hydrolysis (8); direct photolysis (46,51); and metal oxygenation (39,47-49).

ACKNOWLEDGMENTS

The authors appreciate the strong personal encouragement of Werner Stumm, and the catalytic influence of Charles O'Melia in facilitating a bimolecular association. We acknowledge the stimulus to further examine chemical time scales provided by the earlier work of A. Lerman, D. Imboden, and R. Schwarzenbach on lake models. We thank J. F. Pankow and M. R. Hoffmann for advice and discussions related to aqueous kinetics. We are grateful to Elaine Granger for her skill and adaptability in generating a typescript in a "characteristic time" of $\log t < 6$ s.

REFERENCES

1. A. Lerman, *Geochemical Processes: Water and Sediment Environments,* Wiley–Interscience, New York, 1979.
2. H. B. Fischer, E. J. List, R. C. Y. Koh, J. Imberger, and N. H. Brooks, *Mixing in Inland and Coastal Waters,* Academic Press, New York, 1979.
3. P. L. Brezonik, "Chemical Kinetics and Dynamics in Natural Water Systems." In *Waste and Water Pollution Handbook,* L. L. Ciaccio (Ed.), Marcel Dekker, New York, 1974.
4. D. M. Imboden and A. Lerman, "Chemical Models of Lakes." In *Lakes: Chemistry, Geology, Physics,* A. Lerman (Ed.), Springer, New York, 1978.
5. R. P. Schwarzenbach and D. M. Imboden, "Modelling Concepts for Hydrophobic Pollutants in Lakes," *Ecol. Modelling* **22,** 171 (1984).
6. M. R. Hoffmann, "Thermodynamic, Kinetic, and Extrathermodynamic Considerations in the Development of Equilibrium Models for Aquatic Systems," *Environ. Sci. Technol.* **15,** 345–353 (1981).
7. J. F. Pankow and J. J. Morgan, "Kinetics for the Aquatic Environment," *Environ. Sci. Technol.* **10,** 1155–1164 and 1306–1313 (1981).
8. I. J. Tinsley, *Chemical Concepts in Pollutant Behavior,* Wiley–Interscience, New York, 1979.
9. F. M. M. Morel, *Principles of Aquatic Chemistry,* Wiley–Interscience, New York, 1983.
10. J. C. Keck, "Rate-Controlled Constrained Equilibrium Method for Treating Reactions in Complex Systems." In *Maximum Entropy Formalism,* R. D. Levine and M. Tribus (Eds.), MIT Press, Cambridge, 1978.
11. A. C. Lasaga, "Rate Laws of Chemical Reactions," Chapter 1, and "Transition State Theory," Chapter 4. In *Reviews in Mineralogy 8,* Kinetics of Geochemical Processes, Mineralogical Society, Washington (1983).
12. W. C. Gardiner, *Rates and Mechanisms of Chemical Reactions,* Benjamin, Menlo Park, California, 1969.
13. J. W. Moore and R. G. Pearson, *Kinetics and Mechanism,* 3rd ed., Wiley–Interscience, New York, 1981.
14. G. G. Hammes, *Principles of Chemical Kinetics,* Academic Press, New York, 1978.

15. A. W. Adamson, *A Textbook of Physical Chemistry,* 2nd ed., Academic Press, New York, 1979.

16. R. E. Weston and H. A. Schwarz, *Chemical Kinetics,* Prentice-Hall, Englewood Cliffs, New Jersey, 1972.

17. S. W. Benson, *The Foundations of Chemical Kinetics,* McGraw-Hill, New York, 1960.

18. S. Fallab, "Reactions with Molecular Oxygen," *Angew. Chem. Internat. Edit.* **6,** 496 (1967).

19. D. Benson, *Mechanisms of Inorganic Reactions in Solution,* McGraw-Hill, London, 1968.

20. W. Stumm and J. J. Morgan, *Aquatic Chemistry,* 2nd ed., Wiley–Interscience, New York, 1981.

21. J. T. Edsall, "Carbon Dioxide, Carbonic Acid, and Bicarbonate Ion: Physical Properties and Kinetics of Interconversion." In *CO_2: Chemical, Biochemical and Physiological Aspects,* R. E. Forster et al. (Eds.), NASA SP-188, Washington, D.C. 1969.

22. J. A. Burgess, *Metal Ions in Solution,* Ellis Horwood Ltd., Sussex, England, 1978.

23. C. F. Wells and M. A. Salam, "Complex Formation between Fe(II) and Inorganic Anions", *Trans. Faraday Soc.* **63,** 620 (1967).

24. E. A. Moelwyn-Hughes, *The Chemical Statics and Kinetics of Solutions,* Academic Press, 1971.

25. K. J. Laidler, *Reaction Kinetics,* Vol. 2, *Reactions in Solution,* MacMillan, New York, 1963.

26. K. J. Laidler, *Chemical Kinetics,* 2nd ed., McGraw-Hill, New York, 1965.

27. T. Mill, D. G. Hendry, and H. Richardson, "Free-Radical Oxidants in Natural Waters," *Science* **207,** 886–888 (1980).

28. T. E. Graedel and C. J. Wechsler, "Chemistry Within Aqueous Atmospheric Aerosols and Raindrops," *Rev. Geophys. Space Phys.* **19,** 505–539 (1981).

29. W. J. Cooper and R. G. Zika, "Photochemical Formation of Hydrogen Peroxide in Surface and Ground Waters Exposed to Sunlight, *Science* **220,** 711–712 (1983).

30. M. R. Hoffmann and S. D. Boyce, "Catalytic Autoxidation of Aqueous Sulfur Dioxide in Relationship to Atmospheric Systems." In *Trace Atmospheric Constituents,* S. E. Schwartz (Ed.), Wiley, New York, 1983, pp. 147–189.

31. M. Boudart, "Consistency between Kinetics and Thermodynamics," *J. Phys. Chem.* **80,** 2869 (1976).

32. G. Furrer and W. Stumm, "The Role of Surface Coordination in the Dissolution of δ-Al_2O_3 in Dilute Acids," *Chimia* **37,** 338 (1983).

33. A. T. Stone and J. J. Morgan, "Reduction and Dissolution of Mn(III) and Mn(IV) Oxides by Organics," *Environ. Sci. Technol.* **18,** 450 (1984).

34. M. R. Hoffmann and B. C. H. Lim, "Kinetics and Mechanism of the Oxidation of Sulfide by Oxygen: Catalysis by Homogeneous Metal Phthalocyanine Complexes," *Environ. Sci. Technol.* **13,** 1406 (1979).

35. C. J. Miles and P. L. Brezonik, "Oxygen Consumption in Humic-Colored

Waters by a Photochemical Ferrous–Ferric Catalytic Cycle," *Environ. Sci. Technol.* **15**, 1089 (1981).

36. R. H. Collienne, "Photoreduction of Iron in the Epilimnion of Acidic Lakes," *Limnol. Oceanogr.* **28**, 83 (1983).

37. W. G. Sunda, S. A. Huntsman, and G. R. Harvey, "Photoreduction of Manganese Oxides in Seawater and its Geochemical and Biological Implications," *Nature* **301**, 234 (1983).

38. T. D. Waite and F. M. M. Morel, "Photoreductive Dissolution of Colloidal Iron Oxides in Natural Waters," *Environ. Sci. Technol.* (1984) (in press).

39. W. Sung and J. J. Morgan, "Oxidative Removal of Mn(II) from Solution Catalyzed by the γ-FeOH Surface," *Geochim. Cosmochim Acta* **45**, 2377 (1981).

40. J. J. Morgan, "Dynamics of Iron and Manganese in Marine Systems: Implications of Laboratory Investigations," *Thalassia Jugoslav.* **18**, 313 (1982).

41. J. H. Espenson, "Homogeneous Inorganic Reactions." In E. S. Lewis (Ed.), *Investigation of Rates and Mechanisms,* Part 1, 3rd ed., Wiley–Interscience, New York, 1974.

42. General Electric: Chart of Nuclides.

43. R. G. Wilkins. *The Study of Kinetics and Mechanisms of Reactions of Transition Metal Complexes,* Allyn and Bacon, Boston, 1974.

44. R. C. Bales, *Surface Chemical and Physical Behavior of Chrysotile Asbestos in Natural Waters and Water Treatment,* Ph.D. thesis, California Institute of Technology, Pasadena, California, 1984.

45. N. L. Wolfe, R. G. Zepp, D. F. Paris, G. L. Baughman, and R. C. Hollis, "Methyoxychlor and DDT Degradation in Water: Rates and Products," *Environ. Sci. Technol.* **11**, 1077 (1977).

46. R. G. Zepp and D. M. Cline, "Rates of Direct Photolysis in Aquatic Environment," *Environ. Sci. Technol.* **11**, 359 (1977).

47. W. Davison and G. Seed, "The Kinetics of the Oxidation of Ferrous Iron in Synthetic and Natural Waters," *Geochim. Cosmochim. Acta* **47**, 67 (1983).

48. S. Emerson, S. Kalhorn, L. Jacobs, B. M. Tebo, K. H. Nealson, and R. A. Rosson, "Environmental Oxidation Rate of Manganese(II): Bacterial Catalysis," *Geochim. Cosmochim. Acta* **46**, 1073 (1982).

49. W. Sung, *Catalytic Effects of the γ-FeOOH Surface on the Oxygenation Removal Kinetics of Fe(II) and Mn(II),"* Ph.D. Thesis, California Institute of Technology, Pasadena, California, 1981.

50. D. Benson, "Substitution of Labile Metal Ions," in *Comprehensive Chemical Kinetics,* **18**, C. H. Bamford and C. F. H. Tipper (Eds), Elsevier, 1976.

51. Y. Skurlator, R. G. Zepp, and G. L. Baughman, "Photolysis Rates of (2,4,5-trichlorophenoxy)acetic acid and 4-amino-3,5,6-trichloropicolinic acid in natural waters," *J. Agric. Food Chem.* **31**, 1065 (1983).

INDEX

NEW TECHNOLOGY OF PEST CONTROL
Carl B. Huffaker, Editor

THE SCIENCE OF 2,4,5-T AND ASSOCIATED PHENOXY HERBICIDES
Rodney W. Bovey and Alvin L. Young

INDUSTRIAL LOCATION AND AIR QUALITY CONTROL: A Planning Approach
Jean-Michel Guldmann and Daniel Shefer

PLANT DISEASE CONTROL. Resistance and Susceptibility
Richard C. Staples and Gary H. Toenniessen, Editors

AQUATIC POLLUTION
Edward A. Laws

MODELING WASTEWATER RENOVATION: Land Treatment
I. K. Iskandar, Editor

AIR AND WATER POLLUTION CONTROL: A Benefit Cost Assessment
A. Myrick Freeman, III

SYSTEMS ECOLOGY: An Introduction
Howard T. Odum

INDOOR AIR POLLUTION: Characterization, Prediction, and Control
Richard A. Wadden and Peter A. Scheff

INTRODUCTION TO INSECT PEST MANAGEMENT, Second Edition
Robert L. Metcalf and William H. Luckman, Editors

WASTES IN THE OCEAN—Volume 1: Industrial and Sewage Wastes in the Ocean
Iver W. Duedall, Bostwick H. Ketchum, P. Kilho Park, and Dana R. Kester, Editors

WASTES IN THE OCEAN—Volume 2: Dredged Material Disposal In the Ocean
Dana R. Kester, Bostwick H. Ketchum, Iver W. Duedall and P. Kilho Park, Editors

WASTES IN THE OCEAN—Volume 3: Radioactive Wastes and the Ocean
P. Kilho Park, Dana R. Kester, Iver W. Duedall, and Bostwick H. Ketchum, Editors

LEAD AND LEAD POISONING IN ANTIQUITY
Jerome O. Nriagu

INTEGRATED MANAGEMENT OF INSECT PESTS OF POME AND STONE FRUITS
B. A. Croft and S. C. Hoyt

PRINCIPLES OF ANIMAL EXTRAPOLATION
Edward J. Calabrese